ADVANCED MATHEMATICS

ADVANCED MATHEMATICS

[As per IV Semester (EE and EC Branch) of RTU and Other Universities]

C.B. Gupta
M.Sc., Ph.D.
Professor
Department of Mathematics
Birla Institute of
Technology and Science
Pilani, Rajasthan

A.K. Malik
M.Sc.
Lecturer in Mathematics
BK Birla Institute of
Engineeneering and
Technology
Pilani, Rajasthan

V. Kumar
M.Sc., Ph.D.
Faculty of Mathematics
BK Birla Institute of Engineering
and Technology
(BK BIET), Pilani, (Rajasthan)

PUBLISHING FOR ONE WORLD

NEW AGE INTERNATIONAL (P) LIMITED, PUBLISHERS
New Delhi • Bangalore • Chennai • Cochin • Guwahati • Hyderabad
Jalandhar • Kolkata • Lucknow • Mumbai • Ranchi
Visit us at **www.newagepublishers.com**

Published by New Age International (P) Ltd., Publishers
First Edition: 2009

Branches:
- 36, Malikarjuna Temple Street, Opp. ICWA, Basavanagudi, **Bangalore**. ✆ (080) 26677815
- 26, Damodaran Street, T. Nagar, **Chennai**. ✆ (044) 24353401
- Hemsen Complex, Mohd. Shah Road, Paltan Bazar, Near Starline Hotel, **Guwahati**. ✆ (0361) 2543669
- No. 105, 1st Floor, Madhiray Kaveri Tower, 3-2-19, Azam Jahi Road, Nimboliadda, **Hyderabad**. ✆ (040) 24652456
- RDB Chambers (Formerly Lotus Cinema) 106A,1st Floor, S.N. Banerjee Road, **Kolkata**. ✆ (033) 22275247
- 16-A, Jopling Road, **Lucknow**. ✆ (0522) 2209578
- 142C, Victor House, Ground Floor, N.M. Joshi Marg, Lower Parel, **Mumbai**. ✆ (022) 24927869
- 22, Golden House, Daryaganj, **New Delhi**. ✆ (011) 23262370, 23262368

ISBN : 978-81-224-2685-4

Rs. 195.00

C-09-02-3452

Printed in India at Ram Printograph, Delhi.
Typeset at Goswami Associates, Delhi.

PUBLISHING FOR ONE WORLD
NEW AGE INTERNATIONAL (P) LIMITED, PUBLISHERS
4835/24, Ansari Road, Daryaganj, New Delhi-110002
Visit us at **www.newagepublishers.com**

Preface

We feel happy and honoured while presenting this book **"Advanced Mathematics"** for engineering students studying in **B. Tech. IV Semester (EE and EC Branch)** of Rajasthan Technical University and all Indian Universities. In this book we have presented the subject matter in very simple and precise manner. The treatment of the subject is systematic and the exposition easily understandable. All standard examples have been included and their model solutions have also been given.

This book falls into five units:

In first and second unit we have discussed the Numerical Analysis. The unit I deals with Finite Difference—Forward, Backward and Central difference, Newton's formula for Forward and Backward differences, Interpolation, Stirling's formula and Lagrange's interpolation formula. Solution of non-linear equations in one variable by Newton-Raphson method, Simultaneous algebraic equation by Gauss and Regula-Falsi method, Solution of simultaneous equations by Gauss elimination and Gauss Seidel methods, Fitting of curves (straight line and parabola of second degree) by method of least squares are also discussed.

In unit II, we have discussed Numerical differentiation, Numerical Integration, Trapezoidal rule, Simpson's one-third and three-eighth rules. Numerical solution of ordinary differential equations of first order, Picard's method, Euler's and modified Euler's methods. Miline's method and Runga-Kutta fourth order method, Simple linear difference equations with constant coefficients are also discussed in the unit.

Unit III deals with the special functions, Bessel's functions of first and second kind, Simple recurrence relations, Orthogonal property of Bessel's transformation and generating functions. Legendre's function of first kind, Simple recurrence relations, Orthogonal property and generating function are also discussed.

In unit IV, the basic principles of probability theory is given in order to prepare the background for its application to various fields. Baye's theorem with simple applications, Expected value, Theoretical probability distributions—Binomial, Poisson and Normal distributions are discussed.

Unit V deals with Lines of regression, concept of simple Co-relation and Rank correlation. Z-transforms, its inverse, simple properties and application to difference equations are also discussed.

We are grateful to New Age International (P) Limited, Publishers and the editorial department for their commitment and encouragement in bringing out this book within a short span of period.

AUTHORS

Acknowledgement

We are thankful to Prof. L. K. Maheshwari, Vice-Chancellor, Prof. R. K. Mittal, Deputy Director (Administration), Prof. G. Raghurama, Deputy Director (Academic) of Birla Institute of Technology & Science (BITS), Pilani for their encouragement and all over support in completing the book. The authors are also highly thankful to Dr. P. S. Bhatnagar, Director, BK Birla Institute of Engineering & Technology (BKBIET), Pilani for his motivation, time to time support and keen interest in the project. Dr. S. R. Singh Pundir, Reader, D. N. College Meerut, deserves for special thanks.Thanks are also due to Mr. Anil and Mr. Rahul of BKBIET for providing necessary help during the project.

We also place our thanks on record to all those who have directly or indirectly helped us in completion of the project.

At the last but not in the least we are very much indebted to our family members without whom it was not possible for us to complete this project in time. Thanks are also due to M/s NEW AGE INTERNATIONAL (P) LTD. PUBLISHERS and their editorial department.

AUTHORS

Syllabus

ADVANCED MATHEMATICS

B. Tech. (IV Semester) of EE Branch

UNIT 1 : NUMERICAL ANALYSIS–I

- Finite differences – Forward, Backward and Central differences.
- Newton's forward and backward differences, interpolation formulae.
- Stirling's formula, Lagrange's interpolation formula.
- Solution of non-linear equations in one variable by Newton Raphson method and Simultaneous algebraic equation by Gauss and Regula Falsi method.
- Solution of simultaneous equations by Gauss elimination and Gauss Seidel methods.
- Fitting of curves (straight line and parabola of second degree) by method of least squares.

UNIT 2 : NUMERICAL ANALYSIS–II

- Numerical differentiation,
- Numerical integration trapezoidal rule, Simpson's one-third and one eighth rule.
- Numerical Integration of ordinary differential equations of first order,
- Picard's method, Euler's and modified Euler's methods.
- Miline's method and Runga Kutta fourth order method.
- Simple linear difference equations with constant coefficients.

UNIT 3 : SPECIAL FUNCTIONS

- Bessel's functions of first and second kind, simple recurrence relations,
- Orthogonal property of Bessel's, Transformation, Generating functions,
- Legendre's function of first kind, Simple recurrence relations,
- Orthogonal property, Generating function.

UNIT 4 : STATISTICS AND PROBABILITY

- Elementary theory of probability,
- Baye's theorem with simple applications, Expected value,
- Theoretical probability distributions-Binomial, Poisson and Normal distributions.
- Lines of regression, co-relation and rank correlation.

UNIT 5 : CALCULUS OF VARIATIONS AND TRANSFORMS

- Z-transforms, its inverse, simple properties and application to difference equations.

Syllabus

MATHEMATICS-IV

B. Tech. (IV Semester) of EC Branch

UNIT I : NUMERICAL ANALYSIS–I

- Finite differences – Forward, Backward and Central differences.
- Newton's forward and backward differences, interpolation formulae.
- Stirling's formula, Lagrange's interpolation formula.

UNIT II : NUMERICAL ANALYSIS–II

- Integration-Trapezoidal rule, Simpson's one-third and three-eighth rules.
- Numerical solution of ordinary differential equations of first order.
- Picard's method, Euler's and modified Euler's methods.
- Miline's method and Runga-Kutta fourth order method.
- Differentiation.

UNIT III : SPECIAL FUNCTIONS

- Bessel's functions of first and second kind, simple recurrence relations,
- Orthogonal property of Bessel's, Transformation, Generating functions,
- Legendre's function of first kind, Simple recurrence relations,
- Orthogonal property, Generating function.

UNIT IV : STATISTICS AND PROBABILITY

- Elementary theory of probability,
- Baye's theorem with simple applications, Expected value,
- Theoretical probability distributions-Binomial, Poisson and Normal distributions.
- Lines of regression, co-relation and rank correlation.

UNIT V : CALCULUS OF VARIATIONS

- Functional, strong and weak variations simple variation problems,
- The Euler's equation.

Contents

*Not for EC branch students.

UNIT II : NUMERICAL ANALYSIS-II

*Not for EC branch students.

UNIT III : SPECIAL FUNCTION

*Not for EC branch students.

*Not for EC branch students.

*Not for EC branch students.

UNIT I

NUMERICAL ANALYSIS-I

In this unit, we shall discuss finite differences, forward, backward and central differences. Newton's forward and backward differences interpolation formula, Stirling's formula, Lagrange's interpolation formula.

The unit is divided into five chapters:

The chapter first deals with the forward, backward, central differences and relation between them, fundamental theorem of the difference calculus, factorial notation and examples.

Chapter second deals with interpolation formula of Newton's forward, Newton's backward, Stirling's for equally width of arguments, and Lagrange's formula for unequally width of arguments.

Chapter third deals with solution of linear simultaneous equation by Gauss elimination and Gauss-Seidel method.

Chapter fourth deals with solution of algebraic and transcendental equation by Regula-Falsi and Newton-Raphson method.

Chapter fifth deals with fitting of curves for straight line and parabola of second degree by method of least squares.

CHAPTER 1

Calculus of Finite Differences

INTRODUCTION

Numerical analysis has great importance in the field of Engineering, Science and Technology etc.

In numerical analysis, we get the result in numerical form by computing methods of given data. The base of numerical analysis is calculus of finite difference which deals with the changes in the dependent variable due to changes in the independent variable.

1.1. FINITE DIFFERENCES

Suppose the function $y = f(x)$ has the values $y_0, y_1, y_2, \ldots\ldots y_n$ for the equally spaced values $x = x_0, x_0 + h, x_0 + 2h, \ldots\ldots x_0 + nh$. If $y = f(x)$ be any function then the value of the independent variable 'x' is called argument and corresponding value of dependent variable y is called entry. To determine the value of y and $\frac{dy}{dx}$ for some intermediate values of x, is based on the principle of finite difference. Which requires three types of differences.

1.2. FORWARD DIFFERENCES

The differences $y_1 - y_0,\ y_2 - y_1, \ldots\ldots y_n - y_{n-1}$ are called the first forward differences of the function $y = f(x)$ and we denote these difference by $\Delta y_0, \Delta y_1 \ldots\ldots \Delta y_n$, respectively, where Δ is called the descending or forward difference operator.

In general, the first forward differences is defined by

$$\Delta y_x = y_{x+1} - y_x$$

The differences of the first forward differences are called the second forward differences and denoted by $\Delta^2 y_0, \Delta^2 y$, etc.

Therefore, we have $\Delta^2 y_0 = \Delta [y_1 - y_0]$

$$= \Delta y_1 - \Delta y_0$$
$$= (y_2 - y_1) - (y_1 - y_0)$$
$$= y_2 - 2y_1 + y_0$$

Similarly, $\Delta^2 y_1 = \Delta[y_2 - y_1] = (y_3 - y_2) - (y_2 - y_1) = y_3 - 2y_2 + y_1$

In general, we have $\Delta^2 y_x = \Delta y_{x+1} - \Delta y_x$

Again, the differences of second forward differences are called third forward differences and denoted by $\Delta^3 y_0$, $\Delta^3 y_1$ etc.

Therefore, we have $\Delta^3 y_0 = \Delta^2 y_1 - \Delta^2 y_0$

$$= (y_3 - 2y_2 + y_1) - (y_2 - 2y_1 + y_0)$$
$$= y_3 - 3y_2 + 3y_1 - y_0 \quad \text{and so on}$$

In general, the nth forward difference is given by

$$\Delta^n y_x = \Delta^{n-1} y_{x+1} - \Delta^{n-1} y_x$$

Forward Difference Table

Argument x	*Entry* $y = f(x)$	*First Differences* Δy	*Second Differences* $\Delta^2 y$	*Third Differences* $\Delta^3 y$	*Fourth Differences* $\Delta^4 y$
x_0	y_0				
		$y_1 - y_0 = \Delta y_0$			
$x_0 + h$	y_1		$\Delta y_1 - \Delta y_0 = \Delta^2 y_0$		
		$y_2 - y_1 = \Delta y_1$		$\Delta^2 y_1 - \Delta^2 y_0 = \Delta^3 y_0$	
$x_0 + 2h$	y_2		$\Delta y_2 - \Delta y_1 = \Delta^2 y_1$		$\Delta^3 y_1 - \Delta^3 y_0 = \Delta^4 y_0$
		$y_3 - y_2 = \Delta y_2$		$\Delta^2 y_2 - \Delta^2 y_1 = \Delta^3 y_1$	
$x_0 + 3h$	y_3		$\Delta y_3 - \Delta y_2 = \Delta^2 y_2$		
		$y_4 - y_3 = \Delta y_3$			
$x_0 + 4h$	y_4				

1.3. BACKWARD DIFFERENCES

The differences $y_1 - y_0, y_2 - y_1, ..., y_n - y_{n-1}$ are called the first backward differences of the function $y = f(x)$ and we denote these differences by ∇y_1, ∇y_2, ..., ∇y_n, respectively, where ∇ is called the ascending or backward differences operator.

In general, the first backward difference is defined by

$$\nabla y_x = y_x - y_{x-1}$$

The differences of the first backward differences are called second backward differences and denoted by $\nabla^2 y_2$, $\nabla^2 y_3$, etc.

Therefore $\nabla^2 y_2 = \nabla(y_2 - y_1) = \nabla y_2 - \nabla y_1$

$$= (y_2 - y_1) - (y_1 - y_0) = y_2 - 2y_1 + y_0$$

In general, we have $\nabla^2 y_x = \nabla y_x - \nabla y_{x-1}$

Again the differences of second backward differences are called third backward differences and denoted by $\nabla^3 y_3$, $\nabla^3 y_4$ etc.

Therefore, we have $\nabla^3 y_x = \nabla^2 y_x - \nabla^2 y_{x-1}$

In general, the nth backward differences is given by

$$\nabla^n y_x = \nabla^{n-1} y_x - \nabla^{n-1} y_{x-1}$$

Backward Differences Table

Argument x	Entry $y = f(x)$	First Diff. ∇y	Second Diff. $\nabla^2 y$	Third Diff. $\nabla^3 y$	Fourth Diff. $\nabla^4 y$
x_0	y_0				
		$y_1 - y_0 = \nabla y_1$			
$x_0 + h$	y_1		$\nabla y_2 - \nabla y_1 = \nabla^2 y_2$		
		$y_2 - y_1 = \nabla y_2$		$\nabla^2 y_3 - \nabla^2 y_2 = \nabla^3 y_3$	
$x_0 + 2h$	y_2		$\nabla y_3 - \nabla y_2 = \nabla^2 y_3$		$\nabla^3 y_4 - \nabla^3 y_3 = \nabla^4 y_4$
		$y_3 - y_2 = \nabla y_3$		$\nabla^2 y_4 - \nabla^2 y_3 = \nabla^3 y_4$	
			$\nabla y_4 - \nabla y_3 = \nabla^2 y_4$		
$x_0 + 3h$	y_3				
		$y_4 - y_3 = \Delta y_4$			
$x_0 + 4h$	y_4				

1.4. CENTRAL DIFFERENCES

The differences $y_1 - y_0 = \delta y_{1/2}, y_2 - y_1 = \delta y_{3/2}, \ldots, y_n - y_{n-1} = \delta y_{n-1/2}$ are called central differences and δ is called central difference operator.

Similarly $\quad \delta y_{3/2} - \delta y_{1/2} = \delta^2 y_1$

$$\delta y_{5/2} - \delta y_{3/2} = \delta^2 y_2$$

and $\quad \delta^2 y_2 - \delta^2 y_1 = \delta^3 y_{3/2}$ and so on.

The Central Difference Table

Argument x	Entry $y = f(x)$	First Diff. δy	Second Diff. $\delta^2 y$	Third Diff. $\delta^3 y$	Fourth Diff. $\delta^4 y$
x_0	y_0				
		$\delta y_{1/2}$			
x_1	y_1		$\delta^2 y_1$		
		$\delta y_{3/2}$		$\delta^3 y_{3/2}$	
x_2	y_2		$\delta^2 y_2$		$\delta^4 y_2$
		$\delta y_{5/2}$		$\delta^2 y_{5/2}$	
x_3	y_3		$\delta^2 y_3$		
		$\delta y_{7/2}$			
x_4	y_4				

1.5. SHIFT OPERATOR *E*

The shift (increment) operator E is defined as

$$Ey_x = y_{x+h}$$
$$E^2 y_x = y_{x+2h}$$
$$\vdots \quad\quad \vdots$$
$$E^n y_x = y_{x+nh}$$

The inverse operator E^{-1} is defined as

$$E^{-1} y_x = y_{x+(-h)} = y_{x-h}.$$

1.6. RELATIONS BETWEEN THE OPERATORS

(*i*) We know that $\Delta y_x = y_{x+h} - y_x = Ey_x - y_x$

$$\Delta y_x = (E-1)\, y_x$$

$\therefore$ $\Delta = E - 1$

or $\boxed{E = \Delta + 1}$

(*ii*) We have $\nabla y_x = y_x - y_{x-h} = y_x - E^{-1} y_x$

$$\nabla y_x = (1 - E^{-1})\, y_x$$

$\therefore$ $\nabla = 1 - E^{-1}$

or $\boxed{E^{-1} = 1 - \nabla}$

(*iii*) We have $E\nabla y_x = E(y_x - y_{x-h}) = Ey_x - Ey_{x-h} = y_{x+h} - y_x$

$$E\nabla y_x = \Delta y_x$$

$$\boxed{E\nabla = \Delta} \quad \quad ...(i)$$

Again $\nabla E y_x = \nabla y_{x+h} = y_{x+h} - y_x$

$$\nabla E y_x = \Delta y_x$$

or $\boxed{\nabla E = \Delta}$ ...(*ii*)

By (*i*) and (*ii*) $\boxed{E\nabla = \nabla E = \Delta}$

(*iv*) Since $\delta\, y_x = y_{x+h/2} - y_{x-h/2} = E^{1/2} y_x - E^{-1/2} y_x$

$$\delta\, y_x = (E^{1/2} - E^{-1/2})\, y_x$$

Thus $\boxed{\delta = E^{1/2} - E^{-1/2}}$

(*v*) We have $\Delta y_x = y_{x+h} - y_x$

$$= Ey_x - y_x = (E-1)\, y_x$$
$$= (E^{1/2} - E^{-1/2})\, E^{1/2} y_x$$

$$\Delta y_x = \delta\, E^{1/2}\, y_x$$

$$\Rightarrow \qquad \boxed{\Delta = \delta\, E^{1/2}}$$

(*vi*) We have $\quad E f(x) = f(x + h)$

$$= f(x) + hf'(x) + \frac{h^2}{2!} f''(x) + \ldots \qquad \text{[By Taylor's Theorem]}$$

$$= f(x) + hD\, f(x) + \frac{h^2}{2!} D^2 f(x) + \ldots.$$

$$= \left(1 + hD + \frac{h^2}{2!} D^2 \ldots..\right) f(x)$$

$$E f(x) = e^{hD} f(x)$$

$$\boxed{E = e^{hD}}$$

(*vii*) We have $\quad (1 + \Delta)(1 - \nabla) f(x) = (1 + \Delta)(f(x) - \nabla f(x))$

$$= (1 + \Delta)[f(x) - (f(x) - f(x - h))]$$

$$= (1 + \Delta) f(x - h) = E f(x - h)$$

$$(1 + \Delta)(1 - \nabla) f(x) = f(x)$$

Thus $\quad (1 + \Delta)(1 - \nabla) = 1$

1.7. FUNDAMENTAL THEOREM OF THE DIFFERENCE CALCULUS

If $f(x)$ be a polynomial of nth degree in x, then the nth difference of $f(x)$ is constant and $\Delta^{n+1} f(x) = 0$.

Proof. Consider the nth degree polynomial

$$f(x) = A_0 + A_1 x + A_2 x^2 + \ldots.. + A_n x^n$$

where $A_0, A_1, A_2, \ldots.. A_n$ are constants and n is a positive integer.

By the definition, we have

$$\Delta f(x) = f(x + h) - f(x)$$

$$= [A_0 + A_1 (x + h) + A_2 (x + h)^2 + \ldots.. A_n (x + h)^n] - [A_0 + A_1 x + A_2 x^2 + \ldots.. + A_n x^n]$$

$$= A_1 h + A_2 [(x + h)^2 - x^2] + A_3 [(x + h)^3 - x^3] + \ldots... + A_n [(x + h)^n - x^n]$$

$$= A_1 h + A_2 [x^2 + {}^2C_1\, x h + h^2 - x^2] + A_3[x^3 + {}^3C_1\, x^2 h + {}^3C_2\, x h^2 + h^3 - x^3] + \ldots..$$

$$+ A_n [x^n + {}^nC_1\, x^{n-1} h + {}^nC_2\, x^{n-2} h^2 + \ldots + {}^nC_n\, h^n - x^n)$$

$$\Delta f(x) = B_1 + B_2 x + B_3 x^2 + \ldots.. + B_{n-1} x^{n-2} + n A_n h\, x^{n-1} \qquad \ldots(1.1)$$

where $B_1, B_2 \ldots.. B_{n-1}$ are constants

By (1.1), we see that the first difference of a polynomial of degree n is again a polynomial of degree $(n - 1)$.

Again $\Delta^2 f(x) = \Delta f(x+h) - \Delta f(x)$

$$= B_1 + B_2(x+h) + B_3(x+h)^2 + ... + B_{n-1}(x+h)^{n-2} + nA_n h(x+h)^{n-1}$$
$$- [B_1 + B_2 x + B_3 x^2 + ... + B_{n-1} x^{n-2} + n A_n h x^{n-1}]$$
$$= B_2 h + B_3[(x+h)^2 - x^2] + B_4[(x+h)^3 - x^3] + ...$$
$$+ B_{n-1}[(x+h)^{n-2} - x^{n-2}] + n A_n h[(x+h)^{n-1} - x^{n-1}]$$
$$= B_2 h + B_3[x^2 + {}^2C_1 xh + h^2 - x^2] + B_4[x^3 + {}^3C_1 x^2 h$$
$$+ {}^3C_2 x h^2 + h^3 - x^3] + ... + B_{n-1}[x^{n-2} + {}^{n-2}C_1 x^{n-3} h$$
$$+ {}^{n-2}C_2 x^{n-3} h^2 + ... + {}^{n-2}C_{n-2} h^{n-2} - x^{n-2}]$$
$$+ nA_n h\,[x^{n-1} + {}^{n-1}C_1 x^{n-2} h + {}^{n-1}C_2 x^{n-3} h^2 + ... + {}^{n-1}C_{n-1} h^n - x^{n-1}]$$
$$= C_2 + C_3 x + C_4 x^2 + ... + C_{n-1} x^{n-3} + n(n-1) h^2 A_n x^{n-2} \quad ...(1.2)$$

where $C_2, C_3, ..., C_{n-1}$ are constants

By (1.2) we see that the second difference of a polynomial of degree n is again a polynomial of degree $(n-2)$

Proceeding in the same way, we will get a zero degree polynomial for the nth difference

i.e., $\Delta^n f(x) = n(n-1)\ (n-2) .... 1\ h^n a_n x^{n-n} = n!\ h^n a_n$

Thus, nth difference is constant

Now $\Delta^{n+1} f(x) = \Delta[\Delta^n f(x)] = \Delta[n!\ h^n a_n] = 0$ $[\because\ \Delta C = 0]$

1.8. FACTORIAL FUNCTION

A product of the form $x(x-h)(x-2h) ...... (x-(n-1)h)$ is called factorial function and denoted by $x^{(n)}$.

Thus $x^{(n)} = x(x-h)(x-2h) ..... (x-(n-1)h)$.

If interval of differencing is unity.

Then $x^{(n)} = x(x-1)(x-2)(x-3) ..... (x-(n-1))$

1.9. TO SHOW THAT $\Delta^n x^{(n)} = n!\ h^n$ AND $\Delta^{n+1} x^{(n)} = 0$

Proof. By the definition of Δ we have

$$\Delta x^{(n)} = (x+h)^{(n)} - x^{(n)}$$
$$= (x+h)(x+h-h)(x+h-2h) ..... (x+h-(n-1)h)$$
$$- x(x-h)(x-2h) ...... (x-(n-1)h)$$
$$= (x+h)x(x-h)(x-2h) ..... (x-(n-2)h)$$
$$- x(x-h)(x-2h) ..... (x-(n-2)h)(x-(n-1)h)$$
$$= x(x-h)(x-2h) ..... (x-(n-2)h)((x+h)-(x-(n-1)h)$$
$$= x^{(n-1)} nh = nh\ x^{(n-1)}$$

Again $\Delta^2 x^{(n)} = \Delta\Delta x^{(n)} = \Delta[\, nhx^{(n-1)}] = nh\,\Delta\, x^{(n-1)} = nh\,[(x+h)^{(n-1)} - x^{(n-1)}]$

$$= nh\,[(x+h)(x+h-h)(x+h-2h) \,.....\, (x+h-(n-2)h)$$

$$- x\,(x-h)(x-2h) \,......\, (x-(n-2)\,h]$$

$$= nh\,[(x+h)\,x\,(x-h)(x-2h) \,.....\, (x-(n-3)h)$$

$$- x\,(x-h)(x-2h) \,.....\, (x-(n-3)\,h)(x-(n-2)\,h)]$$

$$= nh\,x(x-h)(x-2h) \,......\, (x-(n-3)h)\,[x+h-(x-(n-2)\,h)$$

$$= nh\,x^{(n-2)}\,(n-1)\,h$$

$$= n\,(n-1)h^2\,x^{(n-2)}$$

Proceeding in the same way, we get

$$\Delta^n\, x^{(n)} = n\,(n-1)\,(n-2) \,......\, 1\; h^n\, x^{(n-n)} = n\,!\; h^n$$

Again $\Delta^{n+1}\, x^{(n)} = \Delta(\Delta^n\, x^n) = \Delta(n\,!\; h^n) = 0$

1.10. TO SHOW THAT $f(a + nh) = f(a) + {}^nC_1\,\Delta f(a) + {}^nC_2\,\Delta^2 f(a) + \,.....\, + {}^nC_n\,\Delta^n f(a)$

We shall prove this by the method of mathematical induction

We have $\Delta f(a) = f(a+h) - f(a)$

$\therefore$ $f(a+h) = \Delta f(a) + f(a) = f(a) + \Delta f(a)$ it is true for $n = 1$

Again $\Delta f\,(a+h) = f(a+2h) - f(a+h)$

$\therefore$ $f(a+2h) = \Delta f(a+h) + f(a+h)$

$$= \Delta\,[\Delta\, f(a) + f(a)\,] + \Delta f(a) + f(a)$$

$$= f(a) + 2\Delta f(a) + \Delta^2 f(a)$$

$$f\,(a+2h) = f(a) + {}^2C_1\,\Delta\, f(a) + \Delta^2 f(a)$$

It is true for $n = 2$

Similarly $f\,(a+3h) = \Delta f\,(a+2h) + f\,(a+2h)$

$$= \Delta\,[f(a) + 2\Delta f(a) + \Delta^2 f(a)] + [f(a) + 2\Delta f(a) + \Delta^2 f\,(a)]$$

$$= f(a) + 3\Delta\, f(a) + 3\Delta^2 f(a) + \Delta^3 f(a)$$

$$f(a+3h) = f(a) + {}^3C_1\,\Delta\, f(a) + {}^3C_2\,\Delta^2 f(a) + \Delta^3 f(a)$$

It is true for $n = 3$

Now Assume that it is true for $n = k$ then

$$f(a+kh) = f(a) + {}^kC_1\,\Delta f(a) + {}^kC_2\,\Delta^2 f(a) + \,......\, + {}^kC_k\,\Delta^k f(a)$$

Now we shall show that this result is true for $n = k + 1$

Now $f(a + (k+1)\,h) = f(a+kh) + \Delta f\,(a+kh)$

$$= [f(a) + {}^kC_1\,\Delta f\,(a) + {}^kC_2\,\Delta^2 f(a) + \,......\, + {}^kC_k\,\Delta^k f(a)]$$

$$+ \Delta\,[\,f(a) + {}^kC_1\,\Delta f\,(a) + {}^kC_2\,\Delta^2 f\,(a) \,......\, {}^kC_k\,\Delta^k\; f(a)]$$

$$= f(a) + [{}^kC_1 + 1]\,\Delta f(a) + [{}^kC_2 + {}^kC_1]\,\Delta^2 f(a)$$
$$+ [{}^kC_3 + {}^kC_2]\,\Delta^3 f(a) + \ldots\ldots + \Delta^{k+1} f(a).$$
$$f(a + (k+1)h) = f(a) + {}^{k+1}C_1\,\Delta f(a) + {}^{k+1}C_2\Delta^2 f(a) + {}^{k+1}C_3\Delta^3 f(a) + \ldots\ldots\ldots + \Delta^{k+1} f(a)$$

Hence the result is true for $n = k + 1$ $[\because\ {}^kC_r + {}^kC_{r+1} = {}^{k+1}C_{r+1}]$

So by the principle of mathematical induction it is true for all n, we have

$$f(a + nh) = f(a) + {}^nC_1\,\Delta f(a) + {}^nC_2\Delta^2 f(a) + \ldots\ldots\ldots + {}^nC_n\Delta^n f(a)$$

SOLVED EXAMPLES

Example 1. *Prove that* $\Delta^3 \equiv E^3 - 3E^2 + 3E - 1$.

Solution. By the definition we have

$$\Delta f(x) = f(x+h) - f(x),\ E f(x) = f(x+h)$$

and $$E^n f(x) = f(x + nh) \qquad \ldots(i)$$

$\therefore$ $$\Delta^2 f(x) = \Delta[f(x+h) - f(x)] = f(x+2h) - 2f(x+h) + f(x)$$

and $$\Delta^3 f(x) = \Delta[f(x+2h) - 2f(x+h) + f(x)]$$
$$= f(x+3h) - 3f(x+2h) + 3f(x+h) - f(x)$$
$$= E^3 f(x) - 3E^2 f(x) + 3E f(x) - f(x) \qquad \text{[Using } (i)]$$
$$\Delta^3 f(x) = (E^3 - 3E^2 + 3E - 1) f(x)$$

or $$\Delta^3 \equiv E^3 - 3E^2 + 3E - 1.$$ **Hence proved.**

Example 2. *Evaluate*

(i) $\Delta \cosh(a + bx)$ *(ii)* $\Delta \tan^{-1} ax$

Solution. (*i*) By the definition of Δ, we have

$$\Delta \cosh(a + bx) = \cosh(a + b(x+h)) - \cosh(a + bx)$$
$$= 2\sinh \frac{a + b(x+h) + a + bx}{2} \sinh \frac{a + b(x+h) - a - bx}{2}$$
$$= 2\sinh\left(a + bx + \frac{bh}{2}\right)\sinh\left(\frac{bh}{2}\right) \quad \textbf{Ans.}$$

(*ii*) By the definition of Δ, we have

$$\Delta \tan^{-1} ax = \tan^{-1} a(x+h) - \tan^{-1} ax$$
$$= \tan^{-1}\frac{a(x+h) - ax}{1 + a(x+h)ax} = \tan^{-1}\left(\frac{ah}{1 + a^2x^2 + a^2xh}\right). \quad \textbf{Ans.}$$

Example 3. *Evaluate:*

(i) $\Delta^2(3e^x)$ *(ii)* $\Delta\,[\text{Sin}\,(ax + b)]$.

Solution. (*i*) $\Delta^2 [3e^x] = 3\Delta^2 [e^x] = 3\ \Delta\ .\ \Delta\ [e^x] = 3\Delta\ [e^{x+1} - e^x]$

$$= 3[e^{x+2} - e^{x+1} - e^{x+1} + e^x]$$

$$= 3[e^2 - 2e + 1]e^x = 3(e - 1)^2\ e^x$$

(*ii*) By the definition of Δ, we have

$$\Delta[\sin (ax + b)] = \sin (a\ (x + h) + b) - \sin (ax + b)$$

$$= 2 \cos \left(\frac{a(x+h)+b+ax+b}{2}\right) \times \sin \left(\frac{a(x+h)+b-ax+b}{2}\right)$$

$$= 2 \cos \left(ax + b + \frac{ah}{2}\right) \sin \frac{ah}{2}.$$

Example 4. *Evaluate* $\Delta\ (3x + e^{2x} + \sin x)$.

Solution. By the definition of Δ, we have

$$\Delta(3x + e^{2x} + \sin x) = [3(x + h) + e^{2(x+h)} + \sin (x + h)] - [3x + e^{2x} + \sin x]$$

$$= 3h + e^{2x}\ [(e^{2h} - 1) + 2 \cos \left(\frac{x+h+x}{2}\right) \sin \frac{x+h-x}{2}$$

$$= 3h + e^{2x}(e^{2h} - 1) + 2 \cos \left(x + \frac{h}{2}\right) \sin \left(\frac{h}{2}\right).$$

Example 5. *Evaluate* $\Delta\ [e^{2x} \log 3x]$.

Solution. We have $\Delta\ [f(x)\ g(x)] = f(x + h)\ \Delta g(x) + g(x)\ \Delta f(x)$

Take $f(x) = e^{2x},\ g(x) = \log 3x$

then $\Delta\ (e^{2x} \log 3x) = e^{2(x+h)}\ \Delta \log 3x + \log 3x\ \Delta\ e^{2x}$

$$= e^{2(x+h)}\ [\ \log 3(x + h) - \log 3x] + \log 3x\ .\ (e^{2(x+h)} - e^{2x})$$

$$= e^{2x} \left[e^{2h} \log \left(1 + \frac{h}{x}\right) + (e^{2h} - 1) \log 3x\right].$$

Example 6. *Evaluate* $\Delta^n\ (e^{ax+b}),\ h = 1$.

Solution. We have $\Delta f(x) = f(x + 1) - f(x)$

$\therefore$ $\Delta e^{ax+b} = e^{a(x+1)+b} - e^{ax+b} = e^{ax+b}\ (e^a - 1)$

Again $\Delta^2\ e^{ax+b} = \Delta(\Delta e^{ax+b}) = \Delta\ [e^{ax+b}\ (e^a - 1)]$

$$= (e^a - 1)\ \Delta e^{ax+b} = (e^a - 1)\ e^{ax+b}\quad (e^a - 1)$$

$$= (e^a - 1)^2\ e^{ax+b}$$

Proceeding in the same way, we get

$$\Delta^n(e^{ax+b}) = (e^a - 1)^n\ e^{ax+b}.$$

Example 7. *Show that* $\Delta^r y_x = \nabla^r y_{x+r}$.

Solution. $\nabla^r\ y_{x+r} = (1 - E^{-1})^r\ y_{x+r}$ $\qquad [\because\ \ \nabla \equiv 1 - E^{-1}]$

$$= \left(\frac{E-1}{E}\right)^r y_{x+r} = (E-1)^r E^{-r} y_{x+r}$$

$$= (E-1)^r y_x \qquad [\because \quad \Delta = E - 1]$$

$$= \Delta^r y_x. \quad \textbf{Hence proved.}$$

Example 8. *Prove that* $\Delta \log f(x) = \log\left[1 + \frac{\Delta f(x)}{f(x)}\right]$.

Solution. By the definition of Δ, we have

$$\Delta \log f(x) = \log f(x+h) - \log f(x)$$

$$= \log \frac{f(x+h)}{f(x)} = \log\left[\frac{E\, f(x)}{f(x)}\right]$$

$$= \log\left[\frac{(1+\Delta)\, f(x)}{f(x)}\right] \qquad [\because \quad E = 1 + \Delta]$$

$$= \log\left[\frac{f(x) + \Delta f(x)}{f(x)}\right] = \log\left[1 + \frac{\Delta f(x)}{f(x)}\right] \quad \textbf{Hence proved.}$$

Example 9. *Evaluate* $\left(\frac{\Delta^2}{E}\right)x^3$, $h = 1$.

Solution. We have $\left(\frac{\Delta^2}{E}\right)x^3 = \left[\frac{(E-I)^2}{E}\right]x^3 = \left[\frac{E^2 + I - 2E}{E}\right]x^3$

$$= [E + E^{-1} - 2I]\, x^3 = (x+1)^3 + (x-1)^3 - 2x^3$$

$$= x^3 + 3x^2 + 3x + 1 + x^3 - 3x^2 + 3x - 1 - 2x^3 = 6x.$$

Example 10. *Evaluate* $\Delta^3 (1-x)(1-2x)(1-3x)$, $h = 1$.

Solution. Here $\quad f(x) = (1-x)(1-2x)(1-3x)$

$$= 1 - 6x + 11x^2 - 6x^3$$

which is a polynomial of degree 3 in x

$$\therefore \qquad \Delta^3 f(x) = \Delta^3(1 - 6x + 11x^2 - 6x^3)$$

$$= 0 - 6.0 + 11.0 - 6\,.\,3! = -36.$$

Example 11. *If*

x	1	2	3	4	5
y	2	5	10	20	30

find by forward difference table $\Delta^4 y(1)$.

Solution. First, we form forward difference table

x	$y = f(x)$	Δy	$\Delta^2 y$	$\Delta^3 y$	$\Delta^4 y$
1	2				
		3			
2	5		2		
		5		3	
3	10		5		– 8
		10		– 5	
4	20		0		
		10			
5	30				

By above we observe that $\Delta^4 y(1) = -8$.

Example 12. *Represent the function $f(x) = x^4 - 12x^3 + 42x^2 - 30x + 9$ and its successive differences into factorial notation.*

Solution. Let $x^4 - 12x^3 + 42x^2 - 30x + 9 = Ax^{(4)} + Bx^{(3)} + Cx^{(2)} + Dx^{(1)} + E$

$$= Ax(x-1)(x-2)(x-3) + Bx(x-1)(x-2) + Cx(x-1) + Dx + E \qquad ...(i)$$

where A, B, C, D and E are constants . Now, we will find the value of these constants

Putting $x = 0$ in (*i*) we get, $E = 9$

Again putting $x = 1$ in (*i*), we get $1 - 12 + 42 - 30 + 9 = D + E$

$\Rightarrow \qquad D = 1$

Putting $x = 2$ in (*i*), we get $16 - 12 \times 8 + 42 \times 4 - 30 \times 2 + 9 = 2C + 2D + E$

$\Rightarrow \qquad C = 13$

Putting $x = 3$ in (*i*), we get $81 - 12 \times 27 + 42 \times 9 - 30 \times 3 + 9 = 6B + 6C + 3D + E$

$\Rightarrow \qquad B = -6$

Equating the coefficient of x^4 on both sides, we get $A = 1$. Putting the values of A, B, C, D, E in (*i*), we get

$$f(x) = x^4 - 12x^3 + 42x^2 - 30x + 9 = x^{(4)} - 6x^{(3)} + 13x^{(2)} + x^{(1)} + 9$$

Now
$$\Delta f(x) = 4x^{(3)} - 18x^{(2)} + 26x^{(1)} + 1$$
$$\Delta^2 f(x) = 12x^{(2)} - 36x^{(1)} + 26$$
$$\Delta^3 f(x) = 24x^{(1)} - 36$$
$$\Delta^4 f(x) = 24$$
$$\Delta^5 f(x) = 0$$

Aliter: Let
$$f(x) = x^4 - 12x^3 + 42x^2 - 30x + 9$$
$$= Ax^{(4)} + Bx^{(3)} + Cx^{(2)} + Dx^{(1)} + E$$

Now, we obtain the values of A, B, C, D, E by synthetic division. The procedure is as follows:

1	1	– 12	42	– 30	9 = E
	0	1	– 11	31	
2	1	– 11	31	1 = D	
	0	2	– 18		
3	1	– 9	13 = C		
	0	3			
4	1	– 6 = B			
	0				
	1 = A				

Hence $$f(x) = x^{(4)} - 6x^{(3)} + 13x^{(2)} + x^{(1)} + 9.$$

Example 13. *Find the function whose first difference is $9x^2 + 11x + 5$.*

Solution. Let $f(x)$ be the required function then $\Delta f(x) = 9x^2 + 11x + 5$

First, we change $\Delta f(x)$ in factorial notation

Let $$f(x) = 9x^2 + 11x + 5 = Ax^{(2)} + Bx^{(1)} + C$$
$$= Ax(x-1) + Bx + C \qquad ...(i)$$

Putting $x = 0$ we get $C = 5$

Putting $x = 1$ we get $9 + 11 + 5 = B + C \Rightarrow C = 20$

On comparing like term in (*i*) we get $A = 9$

On putting in (*i*), we get

$$\Delta f(x) = 9x^{(2)} + 20x^{(1)} + 5$$

Integrating, we get $f(x) = \dfrac{9x^{(3)}}{3} + \dfrac{20x^{(2)}}{2} + 5x + C_1$ where C_1 is constant of Integration

$$= 3x(x-1)(x-2) + 10x(x-1) + 5x + C_1$$
$$f(x) = 3x^3 + x^2 + x + C.$$

Example 14. *Find the function whose first difference is e^{ax+b}.*

Solution. Let $f(x)$ be the required function

Then $$\Delta f(x) = e^{ax+b} \qquad ...(i)$$

Let us consider $$f(x) = Ae^{ax+b}$$

so that $$\Delta f(x) = \Delta[Ae^{ax+b}] = A\Delta\, e^{ax+b}$$
$$= A\,[e^{a(x+1)+b} - e^{ax+b}]$$
$$= Ae^{ax+b}\,[e^a - 1] \qquad ...(ii)$$

On comparing (*i*) and (*ii*) we get

$$A = \frac{1}{e^a - 1}$$

$$\therefore \qquad f(x) = \frac{e^{ax+b}}{e^a - 1}.$$

Example 15. *What is the lowest degree polynomial which takes the following values*

x	*0*	*1*	*2*	*3*	*4*	*5*
f(x)	*0*	*3*	*8*	*15*	*24*	*35*

Solution. First we prepare the forward difference table

x	$f(x)$	$\Delta f(x)$	$\Delta^2 f(x)$	$\Delta^3 f(x)$
0	0			
		3		
1	3		2	
		5		0
2	8		2	
		7		0
3	15		2	
		9		0
4	24		2	
		11		
5	35			

We know that $f(a + nh) = f(a) + {}^nC_1\, \Delta f(a) + {}^nC_2\, \Delta^2 f(a) + {}^nC_3\, \Delta^3 f(a) + \ldots..$

Putting $a = 0$, $h = 1$, $n = x$, we get

$$f(x) = f(0) + {}^xC_1\, \Delta f(0) + {}^xC_2\, \Delta^2 f(0) + {}^xC_3\, \Delta^3 f(0) + \ldots.. \qquad \ldots(1)$$

Putting the value of $f(0)$, $\Delta f(0)$, $\Delta^2 f(0)$ and $\Delta^3 f(0)$ in (1) from Difference table, we get

$$f(x) = 0 + x\,.\,3 + \frac{x(x-1)}{2!}\,2 + \frac{x\,(x-1)\,(x-2)}{3!}\,.\,0 + 0$$

$$= 3x + x\,(x-1) = x^2 + 2x.$$

Example 16. *A second degree polynomial passes through the points (0, 1) (1, 3), (2, 7), and (3, 13). Find the polynomial.*

Solution. First we prepare the forward difference table

x	$f(x)$	$\Delta f(x)$	$\Delta^2 f(x)$	$\Delta^3 f(x)$
0	1			
		2		
1	3		2	
		4		0
2	7		2	
		6		
3	13			

We know that

$$f(a + nh) = f(a) + {}^nC_1 \, \Delta f(a) + {}^nC_2 \, \Delta^2 f(a) + {}^nC_3 \, \Delta^3 f(a) \ldots\ldots \, {}^nC_n \, \Delta^n f(a)$$

Putting $a = 0$, $h = 1$, $n = x$, we get

$$f(x) = f(0) + {}^xC_1 \, \Delta f(0) + {}^xC_2 \, \Delta^2 f(0) + {}^xC_3 \, \Delta^3 f(0) + \ldots \qquad \ldots(1)$$

Putting the values of $f(0)$, $\Delta f(0)$, $\Delta^2 f(0)$ and $\Delta^3 f(0)$ from difference table in (1) we get

$$f(x) = 1 + x \,.\, 2 + \frac{x(x-1)}{2!} 2 + \frac{x\,(x-1)\,(x-2)}{3!} \,.\, 0$$

$$= 1 + 2x + x\,(x - 1) = x^2 + x + 1.$$

Example 17. *Given $u_0 = 1$, $u_1 = 11$, $u_2 = 21$, $u_3 = 28$ and $u_4 = 29$ find $\Delta^4 u_0$ without forming difference table.*

Solution. We have $\Delta^4 u_0 = (E - I)^4 u_0 = (E^4 - {}^4C_1 E^3 + {}^4C_2 E^2 - {}^4C_3 E + I) \, u_0$

$$= E^4 u_0 - 4E^3 u_0 + 6E^2 u_0 - 4E \, u_0 + u_0$$

$$= u_4 - 4u_3 + 6u_2 - 4u_1 + u_0$$

$$= 29 - 4 \times 28 + 6 \times 21 - 4 \times 11 + 1$$

$$= 29 - 112 + 126 - 44 + 1 = 0.$$

Example 18. *Prove that*

(i) $f(4) = f(3) + \Delta f(2) + \Delta^2 f(1) + \Delta^3 f(1)$

(ii) $f(4) = f(0) + 4\Delta f(0) + 6\Delta^2 f(-1) + 10\Delta^3 f(-1)$

as for as third difference.

Solution. (*i*) We have

$$\Delta f\,(3) = f(4) - f(3)$$

or

$$f(4) = f(3) + \Delta f(3)$$

$$= f(3) + \Delta[f(2) + \Delta f(2)] \qquad [\because \ \Delta f(2) = f(3) - f(2)]$$

$$= f(3) + \Delta f(2) + \Delta^2 f(2)$$

$$= f(3) + \Delta f(2) + \Delta^2[f(1) + \Delta f(1)] \qquad [\because \ \Delta f(1) = f(2) - f(1)]$$

$$f(4) = f(3) + \Delta f(2) + \Delta^2 f(1) + \Delta^3 f(1)$$ **Hence proved.**

(*ii*) We have

$$f(4) = f(-1+5) = E^5 f(-1) = (1+\Delta)^5 f(-1)$$
$$= (1 + {}^5C_1\,\Delta + {}^5C_2\,\Delta^2 + {}^5C_3\,\Delta^3 + {}^5C_4\,\Delta^4 + {}^5C_5\,\Delta^5) f(-1)$$
$$= f(-1) + 5\Delta f(-1) + 10\,\Delta^2 f(-1) + 10\,\Delta^3 f(-1) \qquad \text{taking upto 3rd difference}$$
$$= [f(-1) + \Delta f(-1)] + 4\,[\Delta f(-1) + \Delta^2 f(-1)] + 6\,\Delta^2 f(-1) + 10\Delta^3 f(-1)$$
$$= [f(-1) + \Delta f(-1)] + 4\Delta[f(-1) + \Delta f(-1)] + 6\Delta^2 f(-1) + 10\Delta^3 f(-1)$$
$$= f(0) + 4\Delta f(0) + 6\Delta^2 f(-1) + 10\Delta^3 f(-1) \qquad [\because\ \Delta f(-1) = f(0) - f(-1)]$$

$\Rightarrow\quad f(4) = f(0) + 4\Delta f(0) + 6\Delta^2 f(-1) + 10\Delta^3 f(-1)$ **Hence proved.**

Example 19. *Prove that* $u_x = u_{x-1} + \Delta u_{x-2} + \Delta^2 u_{x-3} + \ldots\ldots + \Delta^{n-1} u_{x-n} + \Delta^n u_{x-n}$.

Solution. Consider $u_x - \Delta^n u_{x-n} = u_x - \Delta^n E^{-n} u_x = \left(1 - \dfrac{\Delta^n}{E^n}\right) u_x = \left(\dfrac{E^n - \Delta^n}{E^n}\right) u_x$

$$= \frac{1}{E^n} \frac{(E-\Delta)\,[E^{n-1} + E^{n-2}\,\Delta + E^{n-3}\,\Delta^2 + \ldots\ldots + \Delta^{n-1}]u_x}{(E-\Delta)} \qquad (\because\ E = 1+\Delta)$$
$$= (E^{-1} + \Delta E^{-2} + \Delta^2 E^{-3} + \ldots\ldots + \Delta^{n-1} E^{-n}) u_x$$
$$u_x - \Delta^n u_{x-n} = u_{x-1} + \Delta u_{x-2} + \Delta^2 u_{x-3} + \ldots\ldots + \Delta^{n-1} u_{x-n}$$

$\therefore\qquad u_x = u_{x-1} + \Delta u_{x-2} + \Delta^2 u_{x-3} + \ldots\ldots + \Delta^{n-1} u_{x-n} + \Delta^n u_{x-n}.$ **Hence proved.**

Example 20. *Prove that* $u_1 x + u_2 x^2 + u_3 x^3 + \ldots\ldots$

$$= \frac{x}{1-x} u_1 + \frac{x^2}{(1-x)^2} \Delta u_1 + \frac{x^3}{(1-x)^3} \Delta^2 u_1 + \ldots\ldots \text{ and } 0 < x < 1.$$

Solution. R.H.S. $= \dfrac{x}{1-x} u_1 + \dfrac{x^2}{(1-x)^2} \Delta u_1 + \dfrac{x^3}{(1-x)^3} \Delta^2 u_1 + \ldots\ldots$

$$= \frac{x}{1-x} u_1 + \frac{x^2}{(1-x)^2} (E-1) u_1 + \frac{x^3}{(1-x)^3} (E-1)^2 u_1 + \ldots\ldots$$
$$= \left(\frac{x}{1-x} - \frac{x^2}{(1-x)^2} + \frac{x^3}{(1-x)^3} \ldots\ldots\right) u_1 + \left(\frac{x^2}{(1-x)^2} - \frac{2x^3}{(1-x)^3} + \ldots\ldots\right) E u_1$$
$$+ \left(\frac{x^3}{(1-x)^3} + \ldots\ldots\right) E^2 u_1 + \ldots\ldots$$
$$= \frac{x}{1-x}\left(1 + \frac{x}{(1-x)}\right)^{-1} u_1 + \frac{x^2}{(1-x)^2}\left(1 + \frac{x}{1-x}\right)^{-2} u_2$$
$$+ \frac{x^3}{(1-x)^3}\left(1 + \frac{x}{1-x}\right)^{-3} u_3 + \ldots\ldots$$
$$= u_1 x + u_2 x^2 + u_3 x^3 + \ldots\ldots.$$

Example 21. *Prove that* $u_0 + u_1 + u_2 + \ldots\ldots + u_n =$

$$^{n+1}C_1\, u_0 + {}^{n+1}C_2\, \Delta\, u_0 + {}^{n+1}C_3 \Delta^2\, u_0 + \ldots\ldots + \Delta^n\, u_0.$$

Solution. We have $u_0 + u_1 + u_2 + \ldots\ldots + u_n = u_0 + Eu_0 + E^2\, u_0 + \ldots\ldots + E^n\, u_0$

$$= (1 + E + E^2 + \ldots\ldots + E^n)\, u_0 = \frac{E^{n+1} - 1}{E - 1}\, u_0 \qquad \text{[Sum of } n \text{ term in G.P.]}$$

$$= \frac{(1+\Delta)^{n+1} - 1}{\Delta} u_0$$

$$= \frac{1}{\Delta}[1 + {}^{n+1}C_1\, \Delta + {}^{n+1}C_2\, \Delta^2 + {}^{n+1}C_3\, \Delta^3 + \ldots\ldots + {}^{n+1}C_{n+1}\, \Delta^{n+1} - 1]\, u_0$$

$$= \frac{1}{\Delta}\, [^{n+1}\, C_1\, \Delta u_0 + {}^{n+1}C_2 \Delta^2\, u_0 + {}^{n+1}C_3\, \Delta^3\, u_0 + \ldots\ldots + {}^{n+1}C_{n+1}\;\; \Delta^{n+1}\, u_0]$$

$$= {}^{n+1}C_1 u_0 + {}^{n+1}C_2\, \Delta\, u_0 + {}^{n+1}C_3\, \Delta^2\, u_0 + \ldots\ldots + \Delta^n\, u_0 = \text{R.H.S.}$$

Example 22. *Prove that* $u_0 + {}^nC_1\, u_1 x + {}^nC_2\, u_2 x^2 + \ldots\ldots + u_n x^n$

$$= (1 + x)^n\, u_0 + {}^nC_1\, (1 + x)^{n-1}\, x\, \Delta u_0 + {}^nC_2\, (1 + x)^{n-2}\, x^2\, \Delta^2\, u_0 + \ldots\ldots + x^n\, \Delta^n\, u_0$$

Solution. R.H.S. $(1 + x)^n\, u_0 + {}^nC_1\, (1 + x)^{n-1}\, x\Delta u_0 + {}^nC_2\, (1 + x)^{n-2}\, x^2 \Delta^2\, u_0 + \ldots\ldots + x^n \Delta^n\, u_0$

$$= ((1 + x) + x\Delta)^n\, u_0 = (1 + x\, (\, 1 + \Delta))^n\, u_0 = (1 + xE)^n\, u_0$$

$$= (1 + {}^nC_1\, xE + {}^nC_2\, x^2\, E^2 + {}^nC_3\, x^3\, E^3 + \ldots\ldots + x^n\, E^n\,)\, u_0$$

$$= u_0 + {}^nC_1\, u_1\, x + {}^nC_2\, u_2\, x^2 + {}^nC_3\, u_3\, x^3 + \ldots\ldots + x^n\, u_n = \text{L.H.S.}$$

Example 23. $\Delta^n\, u_x = u_{x+n} - {}^nC_1 u_{x+n-1} + {}^nC_2\, u_{x+n-2} - \ldots\ldots + (-1)^n\, u_x.$

Solution. R.H.S. $u_{x+n} - {}^nC_1\, u_{x+n-1} + {}^nC_2\, u_{x+n-2} - \ldots\ldots + (-1)^n\, u_x$

$$= (E^n - {}^nC_1\, E^{n-1} + {}^nC_2\, E^{n-2} - \ldots\ldots + (-1)^n)\, u_x$$

$$= (E - 1)^n u_x = \Delta^n y_x = \text{L.H.S.}$$

EXERCISE 1.1

1. Prove that $\Delta^2 \equiv E^2 - 2E + 1$.
2. Prove that if $f(x)$ and $g(x)$ are the function of x then

   (*i*) $\Delta[f(x) + g(x)] = \Delta f(x) + \Delta g(x)$ (*ii*) $\Delta[af(x)] = a\, \Delta f(x)$

   (*iii*) $\Delta[f(x)\, g(x)] = f(x)\, \Delta\, g(x) + g(x + 1)\, f(x) = f(x + 1)\, \Delta\, g(x) + g(x)\, \Delta f(x)$

   (*iv*) $\Delta\left[\dfrac{f(x)}{g(x)}\right] = \dfrac{g(x)\, \Delta\, f(x) - f(x)\, \Delta\, g(x)}{g(x)\, g(x+h)}$.
3. Evaluate

   (*i*) $\Delta[\sinh\, (a + bx)]$ (*ii*) $\Delta[\tan\, ax]$

   (*iii*) $\Delta\, [\cot\, 2^x]$ (*iv*) $\Delta\, (x + \cos\, x)$

(v) $\Delta\,(x^2 + e^x + 2)$ (vi) $\Delta\,[\log x]$

(vii) $\Delta\,[e^{ax}\log bx]$ (viii) $\Delta\left[\dfrac{x^2}{\cos 2x}\right]$

4. Evaluate

(i) $\Delta^2 \cos 2x$ (ii) $\Delta^2\,(ab^{cx})$

(iii) $\left(\dfrac{\Delta^2}{E}\right)x^3$ (iv) $\Delta^2\left[\dfrac{5x+12}{x^2+5x+6}\right]$.

5. Evaluate

(i) $\Delta^n\left(\dfrac{1}{x}\right)$ (ii) $\Delta^n\,[\sin(ax+b)]$

(iii) $\Delta^6\,(ax-1)(bx^2-1)(cx^3-1)$ (iv) $\Delta^n\,[ax^n + bx^{n-1}]$

6. Prove that $e^x = \left(\dfrac{\Delta^2}{E}\right)e^x \cdot \dfrac{Ee^x}{\Delta^2 e^x}$; the interval of differencing being h.

7. Prove that $\nabla^n y_x = \Delta^n y_{x-n}$.

8. Evaluate

(i) $(2\Delta^2 + \Delta - 1)(x^2 + 2x + 1)$ (ii) $(\Delta + 1)(2\Delta - 1)(x^2 + 2x + 1)$

(iii) $(E+2)(E+1)(2^{x+h} + x)$ (iv) $(E^2 + 3E + 2)\,2^{x+h} + x$

9. Write down the polynomial of lowest degree which satisfies the following set of number 0, 7, 26, 63, 124, 215, 342, 511.

10. A third degree polynomial passes through the points (0, – 1) (1, 1) (2, 1) and (3, – 2). Find the polynomial.

11. Construct a forward difference table for

x	0	5	10	15	20	25
$f(x)$	7	11	14	18	24	32

12. If $f(0) = -3, f(1) = 6, f(2) = 8, f(3) = 12$ prepare forward difference table.

13. Given $f(0) = 3, f(1) = 12, f(2) = 81, f(3) = 200, f(4) = 100$ and $f(5) = 8$. Form a difference table and find $\Delta^5 f(0)$.

14. Given $u_0 = 3, u_1 = 12, u_2 = 81, u_3 = 200, u_4 = 100, u_5 = 8$ find $\Delta^5 u_0$ without forming difference table.

15. If $f(0) = -3, f(1) = 6, f(2) = 8, f(3) = 12$ and the third difference being constant, find $f(6)$.

16. Represent the function $f(x) = 2x^3 - 3x^2 + 3x - 10$ and its successive differences into factorial notation.

17. Find the function whose first difference is $x^3 + 3x^2 + 5x + 12$.

18. Obtain the function whose first difference is:

(i) e^x (ii) $x(x-1)$ (iii) a

(iv) $x^{(2)} + 5x$ (v) $\sin x$ (vi) 5^x.

19. Prove that $u_0 + {}^xC_1\,\Delta u_1 + {}^xC_2\,\Delta^2 u_2 + \ldots = u_x + {}^xC_1\,\Delta^2 u_{x-1} + {}^xC_2\,\Delta^4 u_{x-2} + \ldots$

20. Prove that $u_{x+n} = u_n + {}^xC_1\,\Delta u_{x-1} + {}^{x+1}C_2\,\Delta^2 u_{x-2} + {}^{x+2}C_3\Delta^3 u_{x-3} + \ldots$

21. Prove that $\Delta^n u_{x-n} = u_x - {}^nC_1\,u_{x-1} + {}^nC_2\,u_{x-2} - {}^nC_3\,u_{x-3} + \ldots$

22. Prove that $u_0 + \dfrac{u_1 x}{1!} + \dfrac{u_2 x^2}{2!} + \dfrac{u_3 x^3}{3!} + \ldots = e^x\left[u_0 + x\,\Delta u_0 + \dfrac{x^2}{2!}\Delta^2 u_0 + \dfrac{x^3}{3!}\Delta^3 u_0 + \ldots\right]$

ANSWERS

3. (*i*) $2\sinh\left(\dfrac{b}{2}\right)\cosh\left(a + \dfrac{b}{2} + bx\right)$ (*ii*) $\dfrac{\sin a}{\cos ax \cos a(x+1)}$

(*iii*) $-\operatorname{cosec} 2^{x+1}$ (*iv*) $h - 2\sin\left(x + \dfrac{h}{2}\right)\sin\left(\dfrac{h}{2}\right)$

(*v*) $2hx + h^2 + e^x(e^h - 1)$ (*vi*) $\log\left(1 + \dfrac{h}{x}\right)$

(*vii*) $e^{ax}\left[e^{ah}\log\left(1 + \dfrac{h}{x}\right) + (e^{ah} - 1)\log bx\right]$ (*viii*) $\dfrac{h(2x+h)\cos 2x + 2x^2 \sinh h \sin(2x+h)}{\cos(2x+2h)\cos 2x}$

4. (*i*) $-4\sin^2 h\cos(2x + 2h)$ (*ii*) $(b^c - 1)^2\, ab^{cx}$

(*iii*) $6x$ (*iv*) $\dfrac{4}{(x+2)(x+3)(x+4)} + \dfrac{6}{(x+3)(x+4)(x+5)}$

5. (*i*) $\dfrac{(-1)^n\, n!}{x(x+1)(x+2)\ldots\ldots(x+n)}$ (*ii*) $\left(2\sin\dfrac{a}{2}\right)^n \sin\left(ax + b + n\left(\dfrac{a+\pi}{2}\right)\right)$

(*iii*) $720\,abc$ (*iv*) $a\,(n\,!)$.

8. (*i*) $5h^2 + 2hx + 2h - x^2 - 2x - 1$ (*ii*) $5h^2 + 2hx + 2h - x^2 - 2x - 1$

(*iii*) h (*iv*) h

9. $x^3 + 3x^2 + 3x$ **10.** $-\dfrac{1}{6}(x^3 + 3x^2 - 16x + 6)$

13. 755 **14.** 755

15. 126 **16.** $2x^{(3)} + 3x^{(2)} + 2x^{(1)} - 10,\ 6x^{(2)} + 6x^{(1)} + 2,\ 12x + 6,\ 12$

17. $\dfrac{1}{4}x^{(4)} + 2x^{(3)} + \dfrac{9}{2}x^{(2)} + 12x^{(1)} + C$

18. (*i*) $\dfrac{e^x}{(e^h - 1)} + C$ (*ii*) $\dfrac{x^{(3)}}{3} + C$ (*iii*) $ax + C$

(*iv*) $\dfrac{x^{(3)}}{3} + \dfrac{5x^{(2)}}{2} + C$ (*v*) $-\dfrac{1}{2}\sin x$ (*vi*) $\dfrac{1}{4}5^x$.

CHAPTER 2

Interpolation

INTRODUCTION

Suppose $y = f(x)$ be a function of x and $y_0, y_1, y_2, \ldots\ldots, y_n$ are the values of the function $f(x)$ at $x_0, x_1, x_2, \ldots., x_n$ respectively, then the method to obtaining the value of $f(x)$ at point $x = x_i$ which lie between x_0 and x_n is called interpolation.

Thus, interpolation is the technique of computing the value of the function outside the given interval.

If $x = x_i$ does not lie between x_0 and x_n then computing the value of $f(x)$ at this point is called the extra polation.

The study of interpolation depends on the calculus of finite difference.

In this chapter, we shall discuss Newton-Gregory forward and backward interpolation, Lagrange's, Stirling's interpolation formula and method of finding the missing one and more term.

2.1. TO FIND ONE MISSING TERM

Method 1. Suppose one value of $f(x)$ be missing from the set of $(n + 1)$ values (*i.e.*, n values are given) of x, the values of x being equidistant. Let the unknown value be X. Now construct the difference table. We can assume $y = f(x)$ to be a polynomial of degree $(n - 1)$ in x, since n values of y are given. Now equating to zero the nth difference, we get the value of X.

Method 2. Suppose one value of $f(x)$ be missing from the set of $(n + 1)$ values (*i.e.*, n values are given) of x, the values of x being equidistant. Then we can assume $y = f(x)$ to be a polynomial of degree $(n - 1)$ in x

$$\therefore \quad \Delta^n f(x) = 0$$

or $$(E - I)^n f(x) = 0$$

or $$(E^n - {}^nC_1 E^{n-1} + {}^nC_2 E^{n-2} - \ldots\ldots + (-1)^n I) f(x) = 0$$

or $$E^n f(x) - {}^nC_1 E^{n-1} f(x) + {}^nC_2 E^{n-2} f(x) - \ldots\ldots + (-1)^n f(x) = 0$$

or $$f(x+nh) - {}^nC_1 f(x+(n-1)h) + {}^nC_2 f(x+(n-2)h) - \ldots\ldots + (-1)^n f(x) = 0 \qquad \ldots(2.1)$$

If $x = x_0$ is the first value of x then putting $x = x_0$ in (2.1) and solving we get the missing term.

To find two missing term. Suppose two value X_1 and X_2 of $f(x)$ be missing from the set of $(n+2)$ values (*i.e.*, n values are given) of x, the values of x being equidistant. Then, we can assume $y = f(x)$ be a polynomial of degree $(n-1)$ in x

$$\therefore \qquad \Delta^n f(x) = 0$$

or $$f(x+nh) - {}^nC_1 f(x+(n-1)h) + {}^nC_2 f(x+(n-2)h) - \ldots\ldots + (-1)^n f(x) = 0 \qquad \ldots(2.2)$$

If $x = x_0$ is the first value of x then putting $x = x_0$, and $x = x_1$ successively in (2.2), we get two equation in missing X_1 and X_2. On solving we get X_1 and X_2.

2.2. NEWTON-GREGORY'S FORMULA FOR FORWARD INTERPOLATION WITH EQUAL INTERVALS

Let $y = f(x)$ be a function which assumes the values $f(a)$, $f(a+h)$, $f(a+2h)$, ..., $f(a+nh)$ for $x = a$, $a+h$, $a+2h$, ..., $a+nh$ respectively where h is the difference of the arguments. Let $f(x)$ be a polynomial in x of degree n. So $f(x)$ can be written as

$$f(x) = a_0 + a_1(x-a) + a_2(x-a)(x-a-h) + a_3(x-a)(x-a-h)(x-a-2h)$$
$$+ \ldots\ldots + a_n(x-a)(x-a-h)\ldots\ldots(x-a-(n-1)h) \qquad \ldots(2.3)$$

where $a_0, a_1, a_2, \ldots\ldots, a_n$ are constants

Putting successively the values $x = a, a+h, a+2h, \ldots\ldots, a+nh$ in (2.3), we get

$$f(a) = a_0 \quad \text{or} \quad a_0 = f(a)$$

$$f(a+h) = a_0 + a_1(a+h-a)$$

or $$f(a+h) = a_0 + a_1 h \Rightarrow a_1 = \frac{f(a+h)-a_0}{h} = \frac{f(a+h)-f(a)}{h} = \frac{\Delta f(a)}{h}$$

i.e., $$a_1 = \frac{\Delta f(a)}{h}$$

$$f(a+2h) = a_0 + a_1(a+2h-a) + a_2(a+2h-a)(a+2h-a-h)$$
$$= a_0 + a_1 2h + a_2 2h.h$$
$$2h^2 a_2 = f(a+2h) - 2ha_1 - a_0$$
$$a_2 = \frac{f(a+2h) - 2(f(a+h)-f(a)) - f(a)}{2!h^2}$$
$$= \frac{f(a+2h) - 2f(a+h) + f(a)}{2!h^2} = \frac{\Delta^2 f(a)}{2!h^2}$$

Proceeding in the same way, we get

$$a_3 = \frac{\Delta^3 f(a)}{3!h^3} \;......$$

$$a_n = \frac{\Delta^n f(a)}{n!h^n}$$

Putting the values $a_0, a_1, a_2, ..., a_n$ into (1), we get

$$f(x) = f(a) + (x-a)\frac{\Delta f(a)}{h} + (x-a)(x-a-h)\frac{\Delta^2 f(a)}{2!h^2}$$

$$+ (x-a)(x-a-h)(x-a-2h)\frac{\Delta^3 f(a)}{3!h^n} + ...$$

$$+ (x-a)(x-a-h)(x-a-2h) + ... + (x-a-(n-1)h)\frac{\Delta^n f(a)}{n!h^n!} \qquad ...(2.4)$$

This is Newton-Gregory formula for forward interpolation putting $x = a + hu$ in (2.4), we get

$$f(a+hu) = f(a) + u\Delta f(a) + \frac{u(u-1)}{2!}\Delta^2 f(a) + ... + \frac{u(u-1)(u-2).....(u-(n-1))}{n!}\Delta^n f(a).$$

2.3. NEWTON-GREGORY'S FORMULA FOR BACKWARD INTERPOLATION WITH EQUAL INTERVALS

Let $y = f(x)$ be a function which assumes the values $f(a), f(a+h), f(a+2h), ..., f(a+nh)$ for $x = a, a+h, a+2h, ..., a+nh$, respectively where h is the difference of arguments. Let $f(x)$ be a polynomials in x of degree n. So $f(x)$ can be written as

$$f(x) = a_0 + a_1(x-a-nh) + a_2(x-a-nh)(x-a-(n-1)h)$$
$$+ a_3(x-a-nh)(x-a-(n-1)h)(x-a-(n-2)h) + ...$$
$$+ a_n(x-a-nh)(x-a-(n-1)h)\;.....\;(x-a-h) \qquad ...(2.5)$$

where $a_0, a_1, a_2, ..., a_n$ are constants

Putting successively the values $x = a + nh, a + (n-1)h, a + (n-2)h, ..., a + h$ in (2.5), we get

$$f(a+nh) = a_0 \quad \Rightarrow \quad a_0 = f(a+nh)$$

$$f(a+(n-1)h) = a_0 + a_1(a+(n-1)h - a - nh)$$

$$a_1 = \frac{f(a+nh) - f(a+(n-1)h)}{h} = \frac{\nabla f(a+nh)}{h}$$

$$f(a+(n-2)h) = a_0 + a_1(a+(n-2)h - a - nh)$$
$$+ a_2(a+(n-2)h - a - nh).(a+(n-2)h - a - (n-1)h)$$

$$a_2 = \frac{\nabla^2 f(a+nh)}{2!h^2}$$

Proceeding in the same way, we get

$$a_3 = \frac{\nabla^3 f(a+nh)}{3!\,h^3} \quad \ldots\ldots$$

$$a_n = \frac{\nabla^n f(a+nh)}{n!\,h^n}$$

Putting the values $a_0, a_1, a_2, \ldots\ldots, a_n$ into (2.5), we get

$$f(x) = f(a+nh) + (x-a-nh)\frac{\nabla f(a+nh)}{h} + (x-a-nh)(x-a-(n-1)h)\frac{\nabla^2 f(a+nh)}{2!\,h^2}$$

$$+ \ldots\ldots + (x-a-nh)(x-a-(n-1)h)\ldots\ldots(x-a-h)\frac{\nabla^n f(a+nh)}{n!\,h^n} \quad \ldots(2.6)$$

This is the Newton-Gregory's formula for backward interpolation putting $x = a + nh + hu$ in (2.6), we get

$$f(a+nh+hu) = f(a+nh) + u\nabla f(a+nh) + \frac{u(u+1)}{2!}\nabla^2 f(a+nh)$$

$$+ \ldots\ldots + \frac{u(u+1)(u+2)\ldots\ldots(u+n-1)}{n!}\nabla^n f(a+nh).$$

2.4. LAGRANGE'S INTERPOLATION FORMULA FOR UNEQUAL INTERVALS

Let $y_0, y_1, y_2, \ldots, y_n$ be the values of function $y = f(x)$ corresponding to the arguments $x_0, x_1, x_2, \ldots., x_n$ not necessarily equally spaced. Since there are $(n+1)$ values of $f(x)$ so $(n+1)$th difference is zero. Thus $f(x)$ is supposed to be polynomial in x of degree n.

Then $\quad y = f(x) = a_0(x-x_1)(x-x_2)\ldots..(x-x_n) + a_1(x-x_0)(x-x_2)\ldots.(x-x_n)$

$$+ a_2(x-x_0)(x-x_1)\ldots..(x-x_n)\ldots.. + a_n(x-x_0)(x-x_1)\ldots..(x-x_{n-1}) \quad \ldots(2.7)$$

where $a_0, a_1, a_2, \ldots\ldots, a_n$ are constants.

To determine a_0 put $x = x_0$ and $y = y_0$ in (2.7), we get

$$y_0 = a_0(x_0-x_1)(x_0-x_2)\ldots\ldots(x_0-x_n)$$

$$\Rightarrow \quad a_0 = \frac{y_0}{(x_0-x_1)(x_0-x_2)\ldots\ldots(x_0-x_n)}$$

Similarly to determine a_1 put $x = x_1$ and $y = y_1$ in (2.7), we get

$$y_1 = a_1(x_1-x_0)(x_1-x_2)\ldots\ldots(x_1-x_n)$$

$$\Rightarrow \quad a_1 = \frac{y_1}{(x_1-x_0)(x_1-x_2)\ldots\ldots(x_1-x_n)}$$

Proceeding in this way, we get

$$a_n = \frac{y_n}{(x_n-x_0)(x_n-x_1)\ldots\ldots(x_n-x_{n-1})}$$

Putting these values of $a_1, a_2, \ldots\ldots, a_n$ in (2.7), we get

$$y = f(x) = \frac{(x-x_1)(x-x_2)\ldots\ldots(x-x_n)}{(x_0-x_1)(x_0-x_2)\ldots\ldots(x_0-x_n)}\,y_0$$

$$+\frac{(x-x_0)(x-x_2)(x-x_3)\ldots\ldots(x-x_n)}{(x_1-x_0)(x_1-x_2)(x_1-x_3)\ldots\ldots(x_1-x_n)}\,y_1 + \ldots\ldots$$

$$+\frac{(x-x_0)(x-x_1)(x-x_2)\ldots\ldots(x-x_{n-1})}{(x_n-x_0)(x_n-x_1)(x_n-x_2)\ldots\ldots(x_n-x_{n-1})}\,y_n$$

which is the Lagrange's interpolation formula.

2.5. STIRLING'S DIFFERENCE FORMULA

The mean of Gauss's forward difference formula and Gauss's backward difference formula gives Stirling's difference formula

We have Gauss's forward difference formula is

$$y_u = y_0 + u\,\Delta y_0 + \frac{u(u-1)}{2!}\Delta^2 y_{-1} + \frac{(u+1)\,u\,(u-1)}{3!}\Delta^3 y_{-1}$$

$$+\frac{(u+1)\,u\,(u-1)(u-2)}{4!}\Delta^4 y_{-2} + \ldots\ldots \qquad \ldots(2.8)$$

and Gauss's backward difference formula is

$$y_u = y_0 + u\Delta y_{-1} + \frac{u\,(u+1)}{2!}\Delta^2 y_{-1} + \frac{(u+1)\,u\,(u-1)}{3!}\Delta^3 y_{-2} + \frac{(u+2)(u+1)\,u\,(u-1)}{4!}\Delta^4 y_{-2} + \ldots\ldots \qquad \ldots(2.9)$$

The mean of (2.8) and (2.9) is

$$y_u = y_0 + u\,\frac{(\Delta y_0 + \Delta y_{-1})}{2} + \frac{u^2}{2!}\Delta^2 y_{-1} + \frac{u(u^2-1)}{3!}\,\frac{(\Delta^3 y_{-1} + \Delta^3 y_{-2})}{2} + \frac{u^2(u^2-1)}{4!}\Delta^4 y_{-2} + \ldots\ldots$$

This formula is called the Stirling's difference formula.

SOLVED EXAMPLES

Example 1. *Given $u_0 = 580$, $u_1 = 556$, $u_2 = 520$, $u_3 =$ —, $u_4 = 384$, find u_3.*

Solution. Let the missing term $u_3 = X$

$\therefore$ The forward difference table is

x	u_x	Δu_x	$\Delta^2 u_x$	$\Delta^3 u_x$	$\Delta^4 u_x$
0	580				
		-24			
1	556		-12		
		-36		$X-472$	
2	520		$X-484$		$1860-4X$
		$X-520$		$1388-X$	
3	X		$904-2X$		
		$384-X$			
4	384				

Here four values of u_x are given. Therefore, we can assume u_x to be a polynomial of degree 3 in x

$$\therefore \quad \Delta^4 u_x = 0 \quad \text{or} \quad 1860 - 4X = 0$$

$$\text{or} \quad X = 465.$$

Aliter: Here four values of u_x are given. Therefore, we can assume u_x to be a polynomial of degree 3 in x

$$\therefore \quad \Delta^4 u_x = 0$$

$$\text{or} \quad (E - I)^4 u_x = 0$$

$$\text{or} \quad (E^4 - {}^4C_1 E^3 + {}^4C_2 E^2 - {}^4C_3 E + I)\, u_x = 0$$

$$\text{or} \quad E^4 u_x - 4E^3 u_x + 6E^2 u_x - 4Eu_x + u_x = 0$$

$$\text{or} \quad u_{x+4h} - 4u_{x+3h} + 6u_{x+2h} - 4u_{x+h} + u_x = 0$$

Putting $x = 0$ and $h = 1$, we get

$$u_4 - 4u_3 + 6u_2 - 6u_1 + u_0 = 0$$

$$\text{or} \quad 384 - 4X + 6 \times 520 - 6 \times 556 + 580 = 0$$

$$\text{or} \quad 1860 - 4X = 0 \quad \Rightarrow \quad X = 465.$$

Example 2. *Estimate the missing term in the following:*

x	1	2	3	4	5	6	7
y	2	4	8	—	32	64	128

Solution. Let X be the missing term. Since there are 6 values of y are given, then we have $\Delta^6 y = 0$

$$\text{or} \quad (E - I)^6 y_x = 0$$

$$\text{or} \quad (E^6 - {}^6C_1 E^5 + {}^6C_2 E^4 - {}^6C_3 E^3 + {}^6C_4 E^2 - {}^6C_5 E + I)\, y_x = 0$$

$$\text{or} \quad y_{x+6h} - 6y_{x+5h} + 15y_{x+4h} - 20y_{x+3h} + 15y_{x+2h} - 6y_{x+h} + y_x = 0$$

Putting $x = 1$ and $h = 1$ in above, we get

$$y_7 - 6y_6 + 15y_5 - 20y_4 + 15y_3 - 6y_2 + y_1 = 0$$

or $$128 - 6 \times 64 + 15 \times 32 - 20X + 15 \times 8 - 6 \times 4 + 2 = 0$$

or $$128 - 384 + 480 - 20X + 120 - 24 + 2 = 0$$

or $$322 - 20X = 0$$

$$X = \frac{322}{20} = 16.1.$$

Example 3. *Obtain the missing terms in the following table:*

x	1	2	3	4	5	6	7	8
$f(x)$	1	8	—	64	—	216	343	512

Solution. Let X_1 and X_2 are the missing term. Here six values of $f(x)$ are given. Therefore, we can assume $f(x)$ to be a polynomial of degree 5 in x

$\therefore$ $$\Delta^6 f(x) = 0$$

or $$(E - I)^6 f(x) = 0$$

or $$(E^6 - {}^6C_1 E^5 + {}^6C_2 E^4 - {}^6C_3 E^3 + {}^6C_4 E^2 - {}^6C_5 E + I) f(x) = 0$$

or $$f(x+6h) - 6f(x+5h) + 15f(x+4h) - 20f(x+3h) + 15f(x+2h) - 6f(x+h) + f(x) = 0 \quad ...(i)$$

Putting $h = 1$ and $x = 1$, and 2 successively in (i), we get

$$f(7) - 6f(6) + 15f(5) - 20f(4) + 15f(3) - 6f(2) + f(1) = 0$$

and $$f(8) - 6f(7) + 15f(6) - 20f(5) + 15f(4) - 6f(3) + f(2) = 0$$

or $$343 - 6 \times 216 + 15X_2 - 20 \times 64 + 15X_1 - 6 \times 8 + 1 = 0$$

and $$512 - 6 \times 343 + 15 \times 216 - 20X_2 + 15 \times 64 - 6X_1 + 8 = 0$$

or $$15X_2 + 15X_1 = 2280$$

and $$20X_2 + 6X_1 = 2662$$

On solving, we have $X_1 = 27$, $X_2 = 125$

i.e., $f(3) = 27$ and $f(5) = 125$.

Example 4. *Find the first term of the series whose second and subsequent terms are 8, 3, 0, –1, 0.*

Solution. Given

x	0	1	2	3	4	5
$f(x)$	—	8	3	0	– 1	0

There are 5 values of $f(x)$ are given

Then we have $$\Delta^5 f(x) = 0$$

or $$(E - I)^5 f(x) = 0$$

or $$(E^5 - {}^5C_1\, E^4 + {}^5C_2\, E^3 - {}^5C_3\, E^2 + {}^5C_4\, E - {}^5C_5\, I)\, f(x) = 0$$

or $$E^5 f(x) - 5E^4 f(x) + 10E^3 f(x) - 10E^2 f(x) + 5E f(x) - f(x) = 0$$

or $$f(x+5) - 5f(x+4) + 10f(x+3) - 10f(x+2) + 5f(x+1) - f(x) = 0$$

Putting $x = 0$, we get

$$f(5) - 5f(4) + 10f(3) - 10f(2) + 5f(1) - f(0) = 0$$

or $$0 - 5(-1) + 10 \times 0 - 10 \times 3 + 5 \times 8 - f(0) = 0 \quad \text{or} \quad f(0) = 15.$$

Example 5. *Given that $u_0 + u_8 = 1.9243$, $u_1 + u_7 = 1.9590$,*

$u_2 + u_6 = 1.9823$, and $u_3 + u_5 = 1.9956$. Find u_4.

Solution. Since there are 8 values of u_x are given.

Then, we have $\quad \Delta^8 u_0 = 0 \quad$ or $\quad (E - I)^8 u_0 = 0$

or $$(E^8 - {}^8C_1\, E^7 + {}^8C_2\, E^6 - {}^8C_3\, E^5 + {}^8C_4\, E^4 - {}^8C_5\, E^3 + {}^8C_6\, E^2 - {}^8C_7\, E + {}^8C_8\, I)u_0$$

or $$E^8u_0 - 8E^7u_0 + 28E^6u_0 - 56E^5u_0 + 70E^4u_0 - 56E^3u_0 + 28E^2u_0 - 8Eu_0 + u_0 = 0$$

or $$u_8 - 8u_7 + 28u_6 - 56u_5 + 70\, u_4 - 56u_3 + 28u_2 - 8u_1 + u_0 = 0$$

or $$(u_8 + u_0) - 8\,(u_7 + u_1) + 28\,(u_6 + u_2) - 56\,(u_5 + u_3) + 70u_4 = 0$$

or $$1.9243 - 8 \times 1.9590 + 28 \times 1.9823 - 56 \times 1.9956 + 70\, u_4 = 0$$

or $$70u_4 = 69.9969 \quad \text{or} \quad u_4 = 0.9999.$$

Example 6. *From the following table, find the number of students who obtain less than 45 marks.*

Marks	*30–40*	*40–50*	*50–60*	*60–70*	*70–80*
No. of students	*31*	*42*	*51*	*35*	*31*

Solution. The difference table of the given data is as under

Marks	*No. of students*	$\Delta f(x)$	$\Delta^2 f(x)$	$\Delta^3 f(x)$	$\Delta^4 f(x)$
Below 40	31				
		42			
Below 50	73		9		
		51		– 25	
Below 60	124		– 16		37
		35		12	
Below 70	159		– 4		
		31			
Below 80	190				

Here $h = 10$, $a = 40$, $x = 45$, $u = \dfrac{x - a}{h} = \dfrac{45 - 40}{10} = 0.5$

By Newton-Gregory forward interpolation formula

$$f(x) = f(a) + u\,\Delta f(a) + \frac{u(u-1)}{2!}\Delta^2 f(a) + \frac{u(u-1)(u-2)}{3!}\Delta^3 f(a) + \ldots\ldots$$

$$f(45) = f(40) + 0.5\Delta f(40) + \frac{(0.5)(0.5-1)}{2!}\Delta^2 f(40) + \frac{(0.5)(0.5-1)(0.5-2)}{3!}\Delta^3 f(40)$$

$$+ \frac{(0.5)(0.5-1)(0.5-2)(0.5-3)}{4!}\Delta^4 f(40)$$

$$= 31 + 0.5 \times 42 + \frac{(0.5)(-0.5)}{2}9 + \frac{(0.5)(-0.5)(-1.5)}{6}(-25) + \frac{(0.5)(-0.5)(1.5)(-2.5)}{24}37$$

$$= 31 + 21 - 1.125 - 1.563 - 1.445$$

$f(45) = 47.867.$

Example 7. *Using Newton's forward difference formula find the value of f(1.6) if*

x	*1*	*1.4*	*1.8*	*2.2*
f (x)	*3.49*	*4.82*	*5.96*	*6.5*

Solution. The difference table of the given data is as under:

x	$f(x)$	$\Delta f(x)$	$\Delta^2 f(x)$	$\Delta^3 f(x)$
1	3.49			
		1.33		
1.4	4.82		– 0.19	
		1.14		– 0.41
1.8	5.96		– 0.6	
		0.54		
2.2	6.5			

Here $a = 1, h = 0.4, \quad x = 1.6$ then $u = \frac{1.6-1}{0.4} = \frac{0.6}{0.4} = \frac{3}{2} = 1.5$

By Newton-Gregory forward formula

$$f(x) = f(a) + u\,\Delta f(a) + \frac{u(u-1)}{2!}\Delta^2 f(a) + \frac{u(u-1)(u-2)}{3!}\Delta^3 f(a) + \ldots.$$

$$f(1.6) = f(1) + 1.5\,\Delta f(1) + \frac{1.5(1.5-1)}{2!}\Delta^2 f(1) + \frac{1.5(1.5-1)(1.5-2)}{3!}\Delta^3 f(1)$$

$$= 3.49 + 1.5 \times 1.33 + \frac{1.5 \times 0.5}{2!}(-0.19) + \frac{1.5 \times 0.5 \times (-0.5)}{3!}(-0.41)$$

$$= 3.49 + 1.995 - 0.07125 + 0.025625$$

$$f(1.6) = 5.439375$$

Solution by using of Backward Interpolation Formula

The backward difference table of the given data is as under

x	$f(x)$	$\nabla f(x)$	$\nabla^2 f(x)$	$\nabla^3 f(x)$
1	3.49			
		1.33		
1.4	4.82		– 0.19	
		1.14		– 0.41
1.8	5.96		– 0.6	
		0.54		
2.2	6.5			

Here $x = 1.6$, $a + nh = 2.2$, $h = 0.4$ then $u = \dfrac{x-(a+nh)}{h} = \dfrac{1.6-2.2}{0.4} = \dfrac{-0.6}{0.4} = -1.5$

By Newton-Gregory backward formula

$$f(a+nh+hu) = f(a+nh) + u\nabla f(a+nh) + \frac{u(u+1)}{2!}\nabla^2 f(a+nh)$$

$$+ \frac{u(u+1)(u+2)}{3!}\nabla^3 f(a+nh) + \ldots\ldots$$

$$f(1.6) = f(2.2) + (-1.5)\times 0.54 + \frac{(-1.5)(-1.5+1)}{2!}(-0.6) + \frac{(-1.5)(-1.5+1)(-1.5+2)}{3!}(-0.41)$$

$$= 6.5 - 0.81 - 0.225 - 0.025625$$

$$f(1.6) = 5.439375.$$

Example 8. *The population of a town in the decennial census were as under estimate the population for the year 1895 and 1925.*

Year x	*1891*	*1901*	*1911*	*1921*	*1931*
Population f(x) (In thousands)	*46*	*66*	*81*	*93*	*101*

Solution. The difference table of the given data is as under:

x	$y = f(x)$	$\Delta f(x)$	$\Delta^2 f(x)$	$\Delta^3 f(x)$	$\Delta^4 f(x)$
1891	46				
		20			
1901	66		– 5		
		15		2	
1911	81		– 3		– 3
		12		– 1	
1921	93		– 4		
		8			
1931	101				

(*i*) Here $a = 1891, h = 10, x = 1895, u = \dfrac{1895 - 1891}{10} = \dfrac{4}{10} = 0.4$

By Newton-Gregory forward formula,

$$f(x) = f(a) + u\Delta f(a) + \frac{u(u-1)}{2!}\Delta^2 f(a) + \frac{u(u-1)(u-2)}{3!}\Delta^3 f(a) + \frac{u(u-1)(u-2)(u-3)}{4!}\Delta^4 f(a) + \ldots\ldots \quad \ldots(i)$$

$$f(1995) = f(1891) + 0.4\Delta f(1891) + \frac{(0.4)(0.4-1)}{2!}\Delta^2 f(1891) + \frac{(0.4)(0.4-1)(0.4-2)}{3!}\Delta^3 f(1891) + \frac{(0.4)(0.4-1)(0.4-2)(0.4-3)}{4!}\Delta^4 f(1891)$$

$$= 46 + (0.4) \times 20 + \frac{(0.4)(-0.6)}{2!}(-5) + \frac{(0.4)(-0.6)(-1.6)}{3!}(2) + \frac{(0.4)(-0.6)(-1.6)(-2.6)}{4!}(-3)$$

$$= 46 + 8 + 0.6 + 0.128 + 0.1248 = 54.8528$$

(*ii*) Again here $a = 1891, h = 10, x = 1925, \quad u = \dfrac{1925 - 1891}{10} = 3.4$

Put these values in (*i*), we get

$$f(1925) = f(1891) + 3.4\,\Delta f(1891) + \frac{(3.4)(3.4-1)}{2!}\Delta^2 f(1891)$$

$$+ \frac{(3.4)(3.4-1)(3.4-2)}{3!} \Delta^3 f(1891) + \frac{(3.4)(3.4-1)(3.4-2)(3.4-3)}{4!} \Delta^4 f(1891)$$

$$= 46 + 3.4 \times 20 + \frac{3.4 \times 2.4}{2!} \times (-5) + \frac{3.4 \times 2.4 \times 1.4}{3!}(2) + \frac{3.4 \times 2.4 \times 1.4 \times 0.4}{4!}(-3)$$

$$= 46 + 68 - 20.4 + 3.808 - 0.5712 = 96.8368.$$

Solution by using Backward Interpolation Formula

The backward difference table of the given data is as under

x	$y = f(x)$	$\nabla f(x)$	$\nabla^2 f(x)$	$\nabla^3 f(x)$	$\nabla^4 f(x)$
1891	46				
		20			
1901	66		– 5		
		15		2	
1911	81		– 3		– 3
		12		– 1	
1921	93		– 4		
		8			
1931	101				

(*i*) Here $(n + hu) = 1931$, $h = 10$, $x = 1895$ then $u = \dfrac{1895 - 1931}{10} = \dfrac{-36}{10} = -3.6$

By Newton-Gregory backward formula, we have

$$f(a + nh + hu) = f(a + nh) + u\, \nabla f(a + nh) + \frac{u(u+1)}{2!} \nabla^2 f(a + nh)$$

$$+ \frac{u(u+1)(u+2)}{3!} \nabla^3 f(a + nh) + \frac{u(u+1)(u+2)(u+3)}{4!} \nabla^4 f(a + nh) + \ldots..$$

$$f(1995) = f(1931) + (-3.6) \times \nabla f(1931) + \frac{(-3.6)(-3.6+1)}{2!} \nabla^2 f(1931)$$

$$+ \frac{(-3.6)(-3.6+1)(-3.6+2)}{3!} \nabla^3 f(1931)$$

$$+ \frac{(-3.6)(-3.6+1)(-3.6+2)(-3.6+3)}{4!} \nabla^4 f(1931)$$

$$= 101 + (-3.6) \times 8 + \frac{(-3.6)\times(-2.6)}{2!}(-4) + \frac{(-3.6)(-2.6)(-1.6)}{3!}(-1)$$

$$+ \frac{(-3.6)(-2.6)(-1.6)(-0.6)}{4!}(-3)$$

$$= 101 - 28.8 - 18.72 + 2.496 - 1.1232 = 54.8528$$

$f(1995) = 54.8528.$

(*ii*) Again here $\quad n + hu = 1931, h = 10, x = 1925$

Then $$u = \frac{x-(a+nh)}{h} = \frac{1925-1931}{10} = \frac{-6}{10} = -0.6$$

By Newton-Gregory formula, we have

$$f(1925) = f(1931) + (-0.6)\,\nabla f(1931) + \frac{(-0.6)(-0.6+1)}{2!}\nabla^2 f\,(1931)$$

$$+ \frac{(-0.6)(-0.6+1)(-0.6+2)}{3!}\nabla^3 f(1931)$$

$$+ \frac{(-0.6)(-0.6+1)(-0.6+2)(-0.6+3)}{4!}\nabla^4 f(1931)$$

$$= 101 + (-0.6)\times 8 + \frac{(-0.6)\times(0.4)}{2!}(-4) + \frac{(-0.6)(0.4)(1.4)}{3!}(-1)$$

$$+ \frac{(-0.6)(0.4)(1.4)(2.4)}{4!}\times(-3)$$

$$= 101 - 4.8 + 0.48 + 0.056 + 0.1008 = 96.8368.$$

Example 9. *From the following table, find the form of the function f (x).*

x	*0*	*1*	*2*	*3*	*4*
f(x)	*3*	*6*	*11*	*18*	*27*

Solution. The difference table of the given data is as under:

x	$f(x)$	$\Delta f(x)$	$\Delta^2 f(x)$	$\Delta^3 f(x)$	$\Delta^4 f(x)$
0	3				
		3			
1	6		2		
		5		0	
2	11		2		0
		7		0	
3	18		2		
		9			
4	27				

Here $a = 0, h = 1, u = \dfrac{x-0}{1} = x$

By Newton-Gregory formula

$$f(x) = f(a) + u\,\Delta f(a) + \frac{u(u-1)}{2!}\,\Delta^2 f(a) + \frac{u(u-1)(u-2)}{3!}\,\Delta^3 f(a) + \ldots..$$

$$= f(0) + x\,\Delta f(0) + \frac{x(x-1)}{2!}\,\Delta^2 f(0) + \frac{x(x-1)(x-2)}{3!}\,\Delta^3 f(0) + \ldots..$$

$$= 3 + x\,.\,3 + \frac{x(x-1)}{2!}\,2$$

$$f(x) = x^2 + 2x + 3.$$

Example 10. *Use Lagrange's interpolation formula to find y when x = 2 given.*

x	0	1	3	4
y	5	6	50	105

Solution. Here $x_0 = 0,\quad x_1 = 1,\quad x_2 = 3,\quad x_3 = 4$

and $y_0 = 5,\quad y_1 = 6,\quad y_2 = 50,\quad y_3 = 105$

Putting $x = 2$ and above values in the Lagrange's formula, we get

$$y(2) = \frac{(2-1)(2-3)(2-4)}{(0-1)(0-3)(0-4)} \times 5 + \frac{(2-0)(2-3)(2-4)}{(1-0)(1-3)(1-4)} \times 6 + \frac{(2-0)(2-1)(2-4)}{(3-0)(3-1)(3-4)} \times 50$$

$$+ \frac{(2-0)(2-1)(2-3)}{(4-0)(4-1)(4-3)} \times 105$$

$$= \frac{1 \times (-1) \times (-2)}{(-1) \times (-3) \times (-4)} \times 5 + \frac{2 \times (-1) \times (-2)}{1 \times (-2) \times (-3)} \times 6 + \frac{2 \times 1 \times (-2)}{3 \times 2 \times (-1)} \times 50 + \frac{2 \times 1 \times (-1)}{4 \times 3 \times 1} \times 105$$

$$= -\frac{10}{12} + 4 + \frac{100}{3} - \frac{105}{6} = \frac{-10 + 48 + 400 - 210}{12} = 19.$$

Example 11. *The value of x and y are given as below:*

x	0	1	2	5
y	2	5	7	8

Find the value of y when x = 4.

Solution. Here $x_0 = 0,\quad x_1 = 1,\quad x_2 = 2,\quad x_3 = 5$

and $y_0 = 2,\quad y_1 = 5,\quad y_2 = 7,\quad y_3 = 8$

Putting $x = 4$ and the above values in Lagrange's formula, we get

$$f(4) = \frac{(4-1)(4-2)(4-5)}{(0-1)(0-2)(0-5)} \times 2 + \frac{(4-0)(4-2)(4-5)}{(1-0)(1-2)(1-5)} \times 5 + \frac{(4-0)(4-1)(4-5)}{(2-0)(2-1)(2-5)} \times 7 + \frac{(4-0)(4-1)(4-2)}{(5-0)(5-1)(5-2)} \times 8$$

$$= \frac{3\times 2\times(-1)}{(-1)(-2)(-5)} \times 2 + \frac{4\times 2\times(-1)}{1\times(-1)\times(-4)} \times 5 + \frac{4\times 3\times(-1)}{2\times 1\times(-3)} \times 7 + \frac{4\times 3\times 2}{5\times 4\times 3} \times 8$$

$$= \frac{6}{5} + (-10) + 14 + \frac{16}{5} = \frac{6+20+16}{5} = \frac{42}{5} = 8.4$$

$f(4) = 8.4.$

Example 12. *Find the value of y at x = 5 given that:*

x	*1*	*3*	*4*	*8*	*10*
y	*8*	*15*	*19*	*32*	*40*

Solution. Here $x_0 = 1,\ x_1 = 3,\ x_2 = 4,\ x_3 = 8,\ x_4 = 10$

and $y_0 = 8,\ y_1 = 15,\ y_2 = 19,\ y_3 = 32,\ y_4 = 40$

Putting $x = 5$ and the above values in Lagrange's formula, we get

$$f(5) = \frac{(5-3)(5-4)(5-8)(5-10)}{(1-3)(1-4)(1-8)(1-10)} \times 8 + \frac{(5-1)(5-4)(5-8)(5-10)}{(3-1)(3-4)(3-8)(3-10)} \times 15 + \frac{(5-1)(5-3)(5-8)(5-10)}{(4-1)(4-3)(4-8)(4-10)} \times 19 + \frac{(5-1)(5-3)(5-4)(5-10)}{(8-1)(8-3)(8-4)(8-10)} \times 32 + \frac{(5-1)(5-3)(5-4)(5-8)}{(10-1)(10-3)(10-4)(10-8)} \times 40$$

$$= \frac{2\times 1\times(-3)\times(-5)}{(-2)\times(-3)\times(-7)\times(-9)} \times 8 + \frac{4\times 1\times(-3)\times(-5)}{2\times(-1)\times(-5)\times(-7)} \times 15 + \frac{4\times 2\times(-3)\times(-5)}{3\times 1\times(-4)\times(-6)} \times 19 + \frac{4\times 2\times 1\times(-5)}{7\times 5\times 4\times(-2)} \times 32 + \frac{4\times 2\times 1\times(-3)}{9\times 7\times 6\times 2} \times 40$$

$$= \frac{40}{63} - \frac{90}{7} + \frac{95}{3} + \frac{32}{7} - \frac{80}{63} = -\frac{40}{63} - \frac{58}{7} + \frac{95}{5}$$

$f(5) = 22.746.$

Example 13. *Apply Lagrange's formula to find the cubic polynomial which includes the following values of x and y.*

x	*0*	*1*	*4*	*6*
y_x	*1*	*− 1*	*1*	*− 1*

Solution. Here $x_0 = 0, \quad x_1 = 1, \quad x_2 = 4, \quad x_3 = 6$

and $y_0 = 1, \quad y_1 = -1, \quad y_2 = 1, \quad y_3 = -1$

Putting the above values in Lagrange's formula, we get

$$f(x) = \frac{(x-1)(x-4)(x-6)}{(0-1)(0-4)(0-6)} 1 + \frac{(x-0)(x-4)(x-6)}{(1-0)(1-4)(1-6)} (-1) + \frac{(x-0)(x-1)(x-6)}{(4-0)(4-1)(4-6)} 1 + \frac{(x-0)(x-1)(x-4)}{(6-0)(6-1)(6-4)} (-1)$$

$$= -\frac{1}{24}[x^3 - 11x^2 + 34x - 24] - \frac{1}{15}[x^3 - 10x^2 + 24x] - \frac{1}{24}[x^3 - 7x^2 + 6x] - \frac{1}{60}[x^3 - 5x^2 + 4x]$$

$$= -\frac{1}{6}x^3 + \frac{3}{2}x^2 - \frac{10}{3}x + 1.$$

Example 14. *Using Lagrange's method, prove that*

$$y_3 = 0.05\,(y_0 + y_6) - 0.3(y_1 + y_5) + 0.75(y_2 + y_4).$$

Solution. Here $x_0 = 0, x_1 = 1, x_2 = 2, x_3 = 4, x_4 = 5, x_5 = 6$

and their corresponding values of function are given by y_0, y_1, y_2, y_4, y_5 and y_6

Now using Lagrange's formula, we have

$$y_x = \frac{(x-1)(x-2)(x-4)(x-5)(x-6)}{(0-1)(0-2)(0-4)(0-5)(0-6)} y_0 + \frac{(x-0)(x-2)(x-4)(x-5)(x-6)}{(1-0)(1-2)(1-4)(1-5)(1-6)} y_1$$

$$+ \frac{(x-0)(x-1)(x-4)(x-5)(x-6)}{(2-0)(2-1)(2-4)(2-5)(2-6)} y_2 + \frac{(x-0)(x-1)(x-2)(x-5)(x-6)}{(4-0)(4-1)(4-2)(4-5)(4-6)} y_4$$

$$+ \frac{(x-0)(x-1)(x-2)(x-4)(x-6)}{(5-0)(5-1)(5-2)(5-4)(5-6)} y_5 + \frac{(x-0)(x-1)(x-2)(x-4)(x-5)}{(6-0)(6-1)(6-2)(6-4)(6-5)} y_6$$

To find y_3, so putting $x = 3$ in above, we get

$$y_3 = \frac{12}{240} y_0 - \frac{18}{60} y_1 + \frac{36}{48} y_2 + \frac{36}{48} y_4 - \frac{18}{60} y_5 + \frac{12}{240} y_6$$

$$= \frac{1}{20}(y_0 + y_6) - \frac{3}{10}(y_1 + y_5) + \frac{3}{4}(y_2 + y_4)$$

$= 0.05\,(y_0 + y_6) - 0.3\,(y_1 + y_5) + 0.75(y_2 + y_4)$. **Hence proved.**

Example 15. *Use Stirling formula to compute $u_{12.2}$ from the following data.*

x	10	11	12	13	14
$10^5 u_x$	23967	28060	31788	35209	38368

Solution. Here $h = 1$. Now taking 12 as origin, the required value of u is $u = \dfrac{12.2 - 12}{1} = 0.2$

The difference table is as under:

x	u	y_u	Δy_u	$\Delta^2 y_u$	$\Delta^3 y_u$	$\Delta^4 y_u$
10	– 2	23967				
			4093			
11	– 1	28060		– 365		
			3728		58	
12	0	31788		– 307		– 13
			3421		45	
13	1	35209		– 262		
			3159			
14	2	38368				

Stirling formula is

$$y_u = y_0 + u\,\frac{(\Delta y_0 + \Delta y_{-1})}{2} + \frac{u^2}{2!}\,\Delta^2 y_{-1} + \frac{u(u^2-1)}{3!}(\Delta^3 y_{-1} + \Delta^3 y_{-2}) + \frac{u^2(u^2-1)}{4!}\,\Delta^4 y_{-2} + \ldots..$$

$$= 31788 + \frac{0.2(3421 + 3728)}{2} + \frac{(0.2)^2}{2!}(-307) + \frac{(0.2)((0.2)^2 - 1)}{3!}(45 + 58)$$

$$+ \frac{(0.2)^2((0.2)^2 - 1)}{4!}(-13)$$

$$= 31788 + 714.9 - 6.14 - 3.296 + 0.0208$$

$$= 32493.4848.$$

Example 16. *Use Stirling formula, to compute log 337.5 from the following data.*

x	310	320	330	340	350	360
$\log_{10} x$	2.4913617	2.5051500	2.5185139	2.5314789	2.5440680	2.5563025

Solution. Here $h = 10$. Now taking 330 as origin, the required value of u is

$$u = \frac{337.5 - 330}{10} = 0.75$$

The difference table is as under:

x	u	y_u	Δy_u	$\Delta^2 y_u$	$\Delta^3 y_u$	$\Delta^4 y_u$	$\Delta^5 y_u$
310	– 2	2.4913617					
			0.0137883				
320	– 1	2.5051500		– 0.0004244			
			0.0133639		0.0000255		
330	0	2.5185139		– 0.0003989		– 0.0000025	
			0.0129650		0.0000230		0.0000008
340	1	2.5314789		– 0.0003759		– 0.0000017	
			0.0125891		0.0000213		
350	2	2.5440680		– 0.0003546			
			0.0122345				
360	3	2.5563025					

Stirling formula is

$$y_u = y_0 + u\,\frac{(\Delta y_0 + \Delta y_{-1})}{2} + \frac{u^2}{2!}\,\Delta^2 y_{-1} + \frac{u(u^2-1)}{3!}\,\frac{(\Delta^3 y_{-1} + \Delta^3 y_{-2})}{2} + \frac{u^2(u^2-1)}{4!}\,\Delta^4 y_{-2} + \ldots\ldots$$

$$= 2.5185139 + (0.75)\,\frac{(0.0133639 + 0.0129650)}{2} + \frac{(0.75)^2}{2!}(-\,0.0003989)$$

$$+ \frac{(0.75)((0.75)^2-1)}{3!}\,\frac{(0.0000255 + 0.0000230)}{2} + \frac{(0.75)^2\,((0.75)^2-1)}{4!}\,(0.0000025)$$

$$= 2.52827374.$$

Example 17. *Use Stirling formula to find y_{32} from the following data*

$y_{20} = 14.035,\ y_{25} = 13.674,\ y_{30} = 13.275,\ y_{35} = 12.734,\ y_{40} = 12.089,\ y_{45} = 11.309.$

Solution. Here $h = 5$. Now taking 30 as origin, the required value of $u = \dfrac{32-30}{5} = 0.4$

The difference table is as under:

x	u	y_u	Δy_u	$\Delta^2 y_u$	$\Delta^3 y_u$	$\Delta^4 y_u$	$\Delta^5 y_u$
20	– 2	14.035					
			– 0.361				
25	– 1	13.674		– 0.038			
			– 0.399		– 0.104		
30	0	13.275		– 0.142		0.142	
			– 0.541		– 0.038		– 0.211
35	1	12.734		– 0.104		– 0.069	
			– 0.645		– 0.031		
40	2	12.089		– 0.135			
			– 0.780				
45	3	11.309					

Stirling formula is

$$y_u = y_0 + \frac{u(\Delta y_0 + \Delta y_{-1})}{2} + \frac{u^2}{2!}\Delta^2 y_{-1} + \frac{u(u^2-1)}{3!}\frac{(\Delta^3 y_{-1} + \Delta^3 y_{-2})}{2} + \frac{u^2(u^2-1)}{4!}\Delta^4 y_{-2} + ...$$

$$= 13.275 + (0.4)\frac{(-0.399 - 0.541)}{2} + \frac{(0.4)^2}{2!}(-0.142) + \frac{(0.4)((0.4)^2-1)}{3!}\left(\frac{0.038 - 0.104}{2}\right)$$

$$+ \frac{(0.4)^2((0.4)^2-1)}{4!}(0.142)$$

$$= 13.275 - 0.188 - 0.01136 + 0.001848 - 0.0007952$$

$$= 13.07669.$$

EXERCISE 2.1

1. Find the missing term in the following table:

x	1	2	3	4	5
y	2	5	7	—	32

2. Estimate the missing term in the following table.

x	0	1	2	3	4
$f(x)$	1	3	9	—	81

3. Given log 100 = 2, log 101 = 2.0043, log 103 = 2.0128, log 104 = 2.0170, find log 102.
4. Estimate the production of cotton in the year 1935 from the data given below (in millions of pales).

Year	1931	1932	1933	1934	1935	1936	1937
Production	17.1	13.0	14.0	9.6	—	12.4	18.2

5. Obtain the missing terms in the following table.

x	2	2.1	2.2	2.3	2.4	2.5	2.6
$f(x)$	0.135	—	0.111	0.100	—	0.082	0.024

6. The values of x and y are given as below:

x	5	6	9	11
y	12	13	14	16

Find the value of y when $x = 10$.

7. Find u_3 given $u_0 = 580$, $u_1 = 556$, $u_2 = 520$ and $u_4 = 385$.

8. The following table gives the normal weights of babies during the first 12 months life:

Age in months	0	2	5	8	10	12
Weight in lbs	7.5	10.25	15	16	18	21

Estimate the weight of the body at the age of 7 months.

9. Apply Lagrange's formula to find $f(x)$ at $x = 1.50$, using the following values of the function.

x	1.00	1.20	1.40	1.60	1.80	2.00
$f(x)$	0.2420	0.1942	0.1497	0.1109	0.0790	0.0540

10. Find the form of the function given by:

x	0	1	2	3	4
$f(x)$	3	6	11	18	27

11. Find the form of the function given by:

x	3	2	1	– 1
$f(x)$	3	12	15	– 21

12. Using Lagrange's formula, prove that $y_1 = y_3 - 0.3(y_5 - y_{-3}) + 0.2\,(y_{-3} - y_{-5})$.

13. Using Lagrange's formula, prove that $y_0 = \frac{1}{2}(y_1 + y_{-1}) - \frac{1}{8}\left[\frac{1}{2}(y_3 - y_1) - \frac{1}{2}(y_{-1} - y_{-3})\right]$.

14. Use Stirling's formula to find $f(28)$ given $f(20) = 49225$, $f(25) = 48316$, $f(30) = 47236$, $f(35) = 45926$, $f(40) = 44306$.

15. Use Stirling's formula to find $f(35)$ given $f(20) = 512$, $f(30) = 439$, $f(40) = 346$, $f(53) = 243$.

16. Find the value of $y(5)$ from the following table using Lagrange's interpolation formula.

x	1	2	3	4	7
y	2	4	8	16	128

17. Find $f(3)$ by Lagrange's interpolation formula from the following table.

x	0	1	2	4	5
$f(x)$	0	16	48	88	0

18. Find the value of sin 52° from the given table.

θ°	45°	50°	55°	60°
sin θ	0.7071	0.7660	0.8192	0.8660

19. Find the number of men getting wages between Rs 10 and Rs 15 from the following table.

Wages in Rs	0–10	10–20	20–30	30–40
Frequency	9	30	35	42

20. From the following table, find y when $x = 1.25$.

x	1	1.5	2.0	2.5
y	4.00	18.25	44.00	84.25

21. Find the value of the area of the circle of diameter 82 from the following data.

D(Diameter)	80	85	90	95	100
A(Area)	5026	5674	6362	7088	7854

22. Find the value of $f(1.5)$ and $f(7.5)$ from the given table.

x	1	2	3	4	5	6	7	8
$f(x)$	1	8	27	64	125	216	343	512

23. From the following table, find the form of the function $f(x)$.

x	3	5	7	9	11
$f(x)$	6	24	58	108	174

24. If l_x represents the number of persons living at age x in a life table, find as accurately as the data will permit the value of l_{47}. Given that $l_{20} = 512$, $l_{30} = 439$, $l_{40} = 346$, $l_{50} = 243$.

ANSWERS

1. 14 **2.** 31 **3.** 2.0086 **4.** 6.609 **5.** 0.123, 0.090

6. 14.67 **7.** 465.25 **8.** 15.7 lbs **9.** 0.1295 **10.** $x^2 + 2x + 3$

11. $x^3 - 9x^2 + 17x + 6$ **14.** 47692 **15.** 395 **16.** 32 **17.** 84

18. 0.788003 **19.** 15 **20.** 9.875 **21.** 5280

22. 3.375, 421.87 **23.** $2x^2 - 7x + 9$ **24.** 274.

CHAPTER 3

Solution of Linear Simultaneous Equations

INTRODUCTION

Simultaneous linear equations occur in the field of science and engineering like as analysis of a network under sinusoidal steady-state condition, determination of the output of a chemical plant and finding the cost of reaction, the analysis of electronic circuits having a number of invariant element etc.

We solved the system of simultaneous linear equation by matrix method or by Cramer's rule. But these methods are fail for large system. In this chapter we discuss some direct and iterative method of solutions.

3.1. LINEAR EQUATIONS

Let us consider m first degree equations in n variables

$$\left.\begin{array}{l} a_{11}x_1 + a_{12}x_2 + \ldots\ldots + a_{1n}x_n = b_1 \\ a_{21}x_1 + a_{22}x_2 + \ldots\ldots + a_{2n}x_n = b_2 \\ \vdots \qquad \vdots \qquad\qquad \vdots \qquad \vdots \\ a_{m1}x_1 + a_{m2}x_2 + \ldots\ldots + a_{mn}x_n = b_m \end{array}\right\}$$

The system of above equations can be written in the matrix form as follows

$$\begin{bmatrix} a_{11} & a_{12} & \ldots\ldots & a_{1n} \\ a_{21} & a_{22} & \ldots\ldots & a_{2n} \\ \vdots & & & \\ \vdots & & & \\ a_{m1} & a_{m2} & \ldots\ldots & a_{mn} \end{bmatrix} \begin{bmatrix} x_1 \\ x_2 \\ \vdots \\ \vdots \\ x_n \end{bmatrix} = \begin{bmatrix} b_1 \\ b_2 \\ \vdots \\ \vdots \\ b_m \end{bmatrix}$$

i.e., $$AX = B$$

where $$A = [a_{ij}]_{m \times n},\ X = \begin{bmatrix} x_1 \\ x_2 \\ \vdots \\ \vdots \\ x_n \end{bmatrix},\ B = \begin{bmatrix} b_1 \\ b_2 \\ \vdots \\ \vdots \\ b_m \end{bmatrix}$$

The system of equations given above is said to be homogeneous if all the b_i ($i = 1, 2 \ldots m$) are zero. Otherwise, it is called as non-homogeneous system. The solution of such equations can be obtained by

1. Determinant method
2. Matrix inversion method
3. Direct methods
   (*i*) Gauss elimination method (*ii*) Gauss-Jordan method (*iii*) Triangularization method
4. Indirect methods
   (*i*) Tacobi iterative method (*ii*) Gauss-Seidel iterative method (*iii*) Relaxation method

But in the present chapter we shall discuss only Gauss elimination and Gauss-Seidel method.

3.2. GAUSS ELIMINATION METHOD

In this method, the unknowns from the system of equations are eliminated successively such that system of equations is reduced to an upper triangular system from which the unknowns are determined by back substitution. We proceed stepwise as follows.

Consider the system of equations

$$\left.\begin{aligned} a_{11}x_1 + a_{12}x_2 + \ldots\ldots + a_{1n}x_n &= b_1 \\ a_{21}x_1 + a_{22}x_2 + \ldots\ldots + a_{2n}x_n &= b_2 \\ \vdots \quad\quad \vdots \quad\quad\quad\quad \vdots \quad &\quad \vdots \\ a_{m1}x_1 + a_{m2}x_2 + \ldots\ldots + a_{mn}x_n &= b_m \end{aligned}\right\} \quad \ldots(3.1)$$

Step 1. To eliminate x_1 from the second, third .... nth equation

Assuming $a_{11} \neq 0$. The variable x_1 eliminated from second equation by subtracting $\frac{a_{21}}{a_{11}}$ times the first equation from the second equation, similarly we eliminate x_1 from third equation by subtracting $\frac{a_{31}}{a_{11}}$ times the first equation from the third equation, etc. We get new system of equation as

$$\left.\begin{aligned} a_{11}x_1 + a_{12}x_2 + \ldots\ldots + a_{1n}x_n &= b_1 \\ b_{22}x_2 + \ldots\ldots + b_{2n}x_n &= b_2' \\ b_{32}x_2 + \ldots\ldots + b_{3n}x_n &= b_3' \\ \vdots \quad\quad\quad\quad \vdots \quad &\quad \vdots \\ b_{m2}x_2 + \ldots\ldots + b_{mn}x_n &= b_m' \end{aligned}\right\} \quad \ldots(3.2)$$

Step 2. To eliminate x_2 from the third, fourth ...... nth equation

Assuming $b_{22} \neq 0$. The variable x_2 eliminated from third equation by subtracting $\frac{b_{32}}{b_{22}}$ times the second equation from the third equation, similarly we eliminate x_2 from fourth equation by subtracting $\frac{b_{42}}{b_{22}}$ times the second equation from the fourth equation etc. We get new system of equation as

$$\left.\begin{aligned} a_{11}x_1 + a_{12}x_2 + a_{13}x_3 + \ldots\ldots + a_{1n}x_n &= b_1 \\ b_{22}x_2 + b_{23}x_3 + \ldots\ldots + b_{2n}x_n &= b_2' \\ c_{33}x_3 + \ldots\ldots + c_{3n}x_n &= b_3'' \\ \vdots \qquad \vdots \quad &\vdots \\ c_{m3}x_3 + \ldots\ldots + c_{mn}x_n &= b_m'' \end{aligned}\right\} \quad \ldots(3.3)$$

Proceeding in the same way we eliminate x_3 in third step, we eliminate x_4 in fourth step and so on. We get new system of equation as

$$\left.\begin{aligned} a_{11}x_1 + a_{12}x_2 + a_{13}x_3 + \ldots\ldots + a_{1n}x_n &= b_1 \\ b_{22}x_2 + b_{23}x_3 + \ldots\ldots + b_{2n}x_n &= b_2' \\ c_{33}x_3 + \ldots\ldots + c_{3n}x_n &= b_3'' \\ \vdots \quad &\vdots \\ d_{nn}x_n &= b_m^{(m-1)'} \end{aligned}\right\} \quad \ldots(3.4)$$

To evaluate the unknown

The value of $x_1\ x_2 \ldots\ldots x_n$ are given by (3.4) by back substitution.

3.3. GAUSS-SEIDEL METHOD

Let us consider a system of n equation in n variables in which $a_{ii} \neq 0$

$$\left.\begin{aligned} a_{11}x_1 + a_{12}x_2 + \ldots\ldots + a_{1n}x_n &= b_1 \\ a_{21}x_1 + a_{22}x_2 + \ldots\ldots + a_{2n}x_n &= b_2 \\ a_{31}x_1 + a_{32}x_2 + \ldots\ldots + a_{3n}x_n &= b_3 \\ \vdots \qquad \vdots \qquad\quad \vdots \quad &\vdots \\ a_{n1}x_1 + a_{n2}x_2 + \ldots\ldots + a_{nn}x_n &= b_n \end{aligned}\right\} \quad \ldots(3.5)$$

Above equation can be written as

$$\left.\begin{aligned} x_1 &= \frac{1}{a_{11}}[b_1 - a_{12}x_2 - a_{13}x_3 \ldots\ldots a_{1n}x_n] \\ x_2 &= \frac{1}{a_{22}}[b_2 - a_{21}x_1 - a_{23}x_3 \ldots\ldots a_{2n}x_n] \\ x_3 &= \frac{1}{a_{33}}[b_3 - a_{31}x_1 - a_{32}x_2 \ldots\ldots a_{3n}x_n] \\ &\vdots \\ x_n &= \frac{1}{a_{nn}}[b_n - a_{n1}x_1 - a_{n2}x_2 \ldots\ldots a_{n(n-1)}x_{n-1} \end{aligned}\right\} \quad \ldots(3.6)$$

Put the first approximations $x_1^{(1)}, x_2^{(1)}, \ldots\ldots, x_n^{(1)}$ in R.H.S. of first equation of (3.6), we get

$$x_1^{(2)} = \frac{1}{a_{11}}[b_1 - a_{12}x_2^{(1)} - a_{13}x_3^{(1)} \ldots\ldots a_{1n}x_n^{(1)}]$$

Now put $x_1^{(2)}\ x_2^{(1)}\ x_3^{(1)} \ldots\ldots x_n^{(1)}$ in R.H.S. of second equation of (3.6), we get

$$x_2^{(2)} = \frac{1}{a_{22}}[b_2 - a_{21}x_1^{(2)} - a_{23}x_3^{(1)} \ldots\ldots a_{2n}x_n^{(1)}]$$

Now put $x_1^{(2)}, x_2^{(2)}, x_3^{(1)}, \ldots\ldots, x_n^{(1)}$ in R.H.S of third equation of (3.6), we get

$$x_3^{(2)} = \frac{1}{a_{33}}[b_3 - a_{31}x_1^{(2)} - a_{32}x_2^{(2)} \ldots\ldots a_{3n}x_n^{(1)}]$$

Proceeding in the same way put $x_1^{(2)}, x_2^{(2)}, x_3^{(2)}, \ldots\ldots, x_n^{(1)}$ in the last equation of (3.6), we get

$$x_n^{(2)} = \frac{1}{a_{nn}}[b_n - a_{n1}x_1^{(2)} - a_{n2}x_2^{(2)} \ldots\ldots a_{n(n-1)}x_{n-1}^{(2)}]$$

This is the first stage of iteration

The whole process is repeated till the values of $x_1, x_2, \ldots\ldots, x_n$ are obtained to desired accuracy

Gauss-Seidel method is also known as a method of successive displacement.

SOLVED EXAMPLES

Example 1. *Solve by Gauss's elimination method the following:*

$$x_1 + x_2 + 2x_3 = 4 \quad \ldots(i)$$

$$3x_1 + x_2 - 3x_3 = -4 \quad \ldots(ii)$$

$$2x_1 - 3x_2 - 5x_3 = -5 \quad \ldots(iii)$$

Solution. Eliminating x_1 from second and third equation by subtracting 3 and 2 times of first equation respectively, we get

$$x_1 + x_2 + 2x_3 = 4 \quad \ldots(iv)$$

$$2x_2 + 9x_3 = 16 \quad \ldots(v)$$

$$5x_2 + 9x_3 = 13 \quad \ldots(vi)$$

Again eliminating x_2 from (*vi*) with the help of (*v*). Divide (*v*) by 2 and then this equation is subtracted after multiplies by 5 from (*vi*), we get

$$x_1 + x_2 + 2x_3 = 4 \quad \ldots(vii)$$

$$x_2 + \frac{9}{2}x_3 = 8 \quad \ldots(viii)$$

$$-\frac{27}{2}x_3 = -27$$

or $$x_3 = 2$$

Substitute the value of x_3 into (*viii*), we get

$$x_2 = 8 - \frac{9}{2} \times 2 = -1$$

and again substitute the values of x_2 and x_3 into (vii), we get

$$x_1 = 4 + 1 - 4 = 1$$

Hence, the solutions of the equation are

$$x_1 = 1, x_2 = -1, x_3 = 2.$$

Example 2. *Solve* $2x_1 + 4x_2 + x_3 = 3$

$$3x_1 + 2x_2 - 2x_3 = -2$$

$$x_1 - x_2 + x_3 = 6$$

by Gauss's elimination method.

Solution. We can write the given equations in the following order

$$x_1 - x_2 + x_3 = 6 \quad ...(i)$$

$$2x_1 + 4x_2 + x_3 = 3 \quad ...(ii)$$

$$3x_1 + 2x_2 - 2x_3 = -2 \quad ...(iii)$$

Eliminating x_1 from (ii) and (iii) equation by subtracting 2 and 3 times of first equation respectively, we get

$$x_1 - x_2 + x_3 = 6 \quad ...(iv)$$

$$6x_2 - x_3 = -9 \quad ...(v)$$

$$5x_2 - 5x_3 = -20 \quad ...(vi)$$

Again eliminating x_2 from (vi) with the help of (v). Divide (v) by 6 and then this equation is subtracted after multiplies by 5 from (vi), we get

$$x_1 - x_2 + x_3 = 6 \quad ...(vii)$$

$$6x_2 - x_3 = -9 \quad ...(viii)$$

$$x_3 = 3$$

Substitute the value of x_3 into $(viii)$, we get

$$x_2 = \frac{-9+3}{6} = -1$$

and again substitute the values of x_2 and x_3 into (vii), we get

$$x_1 = 6 - 1 + 3 = 2$$

Hence, the solutions of the equation are

$$x_1 = 2, x_2 = -1, x_3 = 3.$$

Example 3. *Solve* $6x + 3y + 2z = 6$

$$6x + 4y + 3z = 0$$

$$20x + 15y + 12z = 0$$

by Gauss's elimination method.

Solution. First, divide first equation by 6, we get

$$x + \frac{1}{2} y + \frac{1}{3} z = 1 \quad ...(i)$$

$$6x + 4y + 3z = 0 \quad ...(ii)$$

$$20x + 15y + 12z = 0 \quad ...(iii)$$

Eliminating x from (*ii*) and (*iii*) equation by subtracting 6 and 20 times of first equation respectively, we get

$$x + \frac{1}{2} y + \frac{1}{3} z = 1 \quad ...(iv)$$

$$y + z = -6 \quad ...(v)$$

$$5y + \frac{16}{3} z = -20 \quad ...(vi)$$

Now eliminating y from (*vi*) by subtracting 5 times of (*v*), we get

$$x + \frac{1}{2} y + \frac{1}{3} z = 1 \quad ...(vii)$$

$$y + z = -6 \quad ...(viii)$$

$$\frac{1}{3} z = 10$$

or

$$z = 30$$

Substitute the value of z into (*viii*), we get

$$y = -6 - 30 = -36$$

and again substitute the values of y and z into (*vii*), we get

$$x = 1 - \frac{1}{2} (-36) - \frac{1}{3} (80)$$

or

$$x = 9$$

Hence the solution of the equations are

$$x = 9, y = -36, z = 30$$

Example 4. *Solve the system of equation by Gauss-Seidel iteration method*

$$83x + 11y - 4z = 95$$

$$7x + 52y + 13z = 104$$

$$3x + 8y + 29z = 71$$

Solution. The given equation can be written in the iteration form as

$$x = \frac{1}{83} (95 - 11y + 4z) \quad ...(i)$$

$$y = \frac{1}{52} (104 - 7x - 13z) \quad (ii)$$

$$z = \frac{1}{29} (71 - 3x - 8y) \quad ...(iii)$$

Taking first $x^{(1)} = 0$, $y^{(1)} = 0$, $z^{(1)} = 0$ and put these values in (*i*), we get

$$x^{(2)} = \frac{1}{83}(95 - 11y^{(1)} - 4z^{(1)})$$

$$= \frac{1}{83}(95 - 11 \times 0 + 4 \times 0) = \frac{95}{83} = 1.14$$

Put $x^{(2)} = 1.14$, $y^{(1)} = 0$, $z^{(1)} = 0$, in (*ii*), we get

$$y^{(2)} = \frac{1}{52}(104 - 7x^{(2)} - 13z^{(1)}) = \frac{1}{52}(104 - 7 \times 1.14 - 13 \times 0)$$

$$= \frac{96.02}{52} = 1.85$$

Put $x^{(2)} = 1.14$, $y^{(2)} = 1.85$, $z^{(1)} = 0$ in (*iii*), we get

$$z^{(2)} = \frac{1}{29}(71 - 3x^{(2)} - 8y^{(2)})$$

$$= \frac{1}{29}(71 - 3 \times 1.14 - 8 \times 1.85) = \frac{52.78}{29} = 1.82$$

Now put $x^{(2)} = 1.14$, $y^{(2)} = 1.85$, $z^{(2)} = 1.82$ in (*i*), we get

$$x^{(3)} = \frac{1}{83}(95 - 11\,y^{(2)} + 4z^{(2)}]$$

$$x^{(3)} = \frac{1}{83}[95 - 11 \times 1.85 + 4 \times 1.82]$$

$$= \frac{81.93}{83} = 0.99$$

Put $x^{(3)} = 0.99$, $y^{(2)} = 1.85$, $z^{(2)} = 1.82$ in (*ii*), we get

$$y^{(3)} = \frac{1}{52}[104 - 7x^{(3)} - 13\,z^{(3)}] = \frac{1}{52}[104 - 7 \times 0.99 - 13 \times 1.82]$$

$$= \frac{73.41}{52} = 1.41.$$

Put $x^{(3)} = 0.99$, $y^{(3)} = 1.41$, $z^{(2)} = 1.82$ in (*iii*), we get

$$z^{(3)} = \frac{1}{29}(71 - 3x^{(3)} - 8y^{(3)})$$

$$= \frac{1}{29}(71 - 3 \times 0.99 - 8 \times 1.41)$$

$$= \frac{56.75}{29} = 1.95$$

Now put $x^{(3)} = 0.99$, $y^{(3)} = 1.41$, $z^{(3)} = 1.95$ in (*i*), we get

$$x^{(4)} = \frac{1}{83}(95 - 11y^{(3)} + 4z^{(3)})$$

$$= \frac{1}{83}(95 - 11 \times 1.41 + 4 \times 1.95)$$

$$= \frac{1}{83}(87.29) = 1.05$$

Put $x^{(4)} = 1.05,\ y^{(3)} = 1.41,\ z^{(3)} = 1.95$ in (*ii*), we get

$$y^{(4)} = \frac{1}{52}(104 - 7x^{(3)} - 13z^{(3)})$$

$$= \frac{1}{52}(104 - 7 \times 1.05 - 13 \times 1.95)$$

$$= \frac{71.3}{52} = 1.37$$

Put $x^{(4)} = 1.05,\ y^{(4)} = 1.37,\ z^{(3)} = 1.95$ in (*iii*), we get

$$z^{(4)} = \frac{1}{29}(71 - 3x^{(4)} - 8y^{(4)})$$

$$= \frac{1}{29}(71 - 3 \times 1.05 - 8 \times 1.37)$$

$$= \frac{56.89}{29} = 1.96$$

Here $\quad x^{(4)} = 1.05,\ y^{(4)} = 1.37,\ z^{(4)} = 1.96$

The values are sufficiently close to $x^{(3)},\ y^{(3)},\ z^{(3)}$ respectively. Hence the solution is

$$x = 1.05,\ y = 1.37,\ z = 1.96.$$

Example 5. *Solve the following system of equation by Gauss-Seidel iteration method*

$$10x + 2y + z = 9$$

$$2x + 20y - 2z = -44$$

$$-2x + 3y + 10z = 22$$

Solution. The given equation can be written in the iteration form as

$$x = \frac{1}{10}(9 - 2y - z) \qquad ...(i)$$

$$y = \frac{1}{20}(-44 - 2x + 2z) \qquad ...(ii)$$

$$z = \frac{1}{10}(22 + 2x - 3y) \qquad ...(iii)$$

Taking first $x^{(1)} = 0,\ y^{(1)} = 0,\ z^{(1)} = 0$ and put these values in (*i*), we get

$$x^{(2)} = \frac{1}{10}(9 - 2y^{(1)} - z^{(1)}) = \frac{1}{10}(9 - 2 \times 0 - 0) = \frac{9}{10} = 0.9$$

Put $x^{(2)} = 0.9,\ y^{(1)} = 0,\ z^{(1)} = 0$ in (*ii*), we get

$$y^{(2)} = \frac{1}{20}(-44 - 2x^{(2)} + 2z^{(1)}) = \frac{1}{20}(-44 - 2 \times 0.9 + 2 \times 0)$$

$$= -\frac{45.8}{20} = -2.29$$

Put $x^{(2)} = 0.9,\ y^{(2)} = -2.29,\ z^{(1)} = 0$ in (*iii*), we get

$$z^{(2)} = \frac{1}{10}(22 + 2 \times 0.9 - 3 \times (-2.29)) = \frac{1}{10}(30.67) = 3.067$$

Now put $x^{(2)} = 0.9,\ y^{(2)} = -2.29,\ z^{(2)} = 3.067$ in (*i*), we get

$$x^{(3)} = \frac{1}{10}(9 - 2y^{(2)} - z^{(2)}) = \frac{1}{10}(9 - 2 \times (-2.29) - 3.067)$$

$$= \frac{10.513}{10} = 1.051$$

Put $x^{(3)} = 1.051,\ y^{(2)} = -2.29,\ z^{(2)} = 3.067$ in (*ii*), we get

$$y^{(3)} = \frac{1}{20}(-44 - 2 \times x^{(3)} + 2z^{(2)}) = \frac{1}{20}(-44 - 2 \times 1.051 + 2 \times 3.067)$$

$$= \frac{1}{20}(-39.968) = -1.99$$

Put $x^{(3)} = 1.051,\ y^{(3)} = -1.99,\ z^{(2)} = 3.067$ in (*iii*), we get

$$z^{(3)} = \frac{1}{10}(22 + 2x^{(3)} - 3y^{(3)})$$

$$= \frac{1}{10}(22 + 2 \times 1.051 - 3 \times (-1.99))$$

$$= \frac{1}{10}(30.072) = 3.007$$

Now put $x^{(3)} = 1.051,\ y^{(3)} = -1.99,\ z^{(3)} = 3.007$ in (*i*), we get

$$x^{(4)} = \frac{1}{10}(9 - 2y^{(3)} - z^{(3)})$$

$$= \frac{1}{10}(9 - 2 \times (-1.99) - 3.007) = \frac{9.973}{10} = 0.997$$

Put $x^{(4)} = 0.997,\ y^{(3)} = -1.99,\ z^{(3)} = 3.007$ in (*ii*), we get

$$y^{(4)} = \frac{1}{20}(-44 - 2x^{(4)} + 2z^{(3)})$$

$$= \frac{1}{20}(-44 - 2 \times 0.997 + 2 \times 3.007) = \frac{1}{20}(39.98) = -1.99$$

Put $x^{(4)} = 0.997$, $y^{(4)} = -1.99$, $z^{(3)} = 3.007$ in (*iii*), we get

$$z^{(4)} = \frac{1}{10}(22 + 2x - 3y) = \frac{1}{10}(22 + 2 \times 0.997 - 3x(-1.99))$$

$$= \frac{1}{10}(29.964) = 2.99$$

Here $\quad x^{(4)} = .997,\ y^{(4)} = -1.99,\ z^{(4)} = 2.99$

The values are sufficiently close to $x^{(3)}$, $y^{(3)}$, $z^{(3)}$ respectively. Hence solution is

$$x = 0.997 \approx 1,\ 2y = -1.99 \approx 2,\ z = 2.99 \approx 3.$$

EXERCISE 3.1

Apply Gauss's elimination method to solve the following system of equations:

1. $4x_1 + x_2 + 3x_3 = 11$
$3x_1 + 4x_2 + 2x_3 = 11$
$2x_1 + 3x_2 + x_3 = 7$

2. $5x - y - 2z = 142$
$x - 3y - z = -30$
$2x - y - 3z = -50$

3. $2x + y + z = 10$
$3x + 2y + 3z = 18$
$x + 4y + 9z = 16$

4. $x + 4y - z = -5$
$x + y - 6z = -12$
$3x - y - z = 4$

5. $2x + 3y - z = 5$
$4x + 4y - 3z = 3$
$2x - 3y + 2z = 2$

6. $x_1 + x_2 + x_3 = 10$
$2x_1 + x_2 + 2x_3 = 17$
$3x_1 + 2x_2 + x_3 = 17$

Apply Gauss-seidel iteration method to solve the following system of equation:

7. $27x + 6y - z = 85$
$6x + 15y + 2z = 72$
$x + y + 54z = 110$

8. $2x + 4y + z = 3$
$3x + 2y - 2z = -2$
$x - y + z = 6$

9. $20x + y - 2z = 17$
$3x + 20y - z = -18$
$2x - 3y + 20z = 25$

10. $17x_1 + 65x_2 - 13x_3 + 50x_4 = 84$
$12x_1 + 16x_2 + 37x_3 + 18x_4 = 25$
$56x_1 + 23x_2 + 11x_3 - 19x_4 = 36$
$3x_1 - 5x_2 + 47x_3 + 10x_4 = 18$

ANSWERS

1. $x_1 = 1,\ x_2 = 1,\ x_3 = 2$

2. $x = 41.07,\ y = 15.77,\ z = 23.79$

3. $x = 7,\ y = -9,\ z = 5$

4. $x = 1.647,\ y = -1.140,\ z = 2.084$

5. $x = 1,\ y = 2,\ z = 3$

6. $x_1 = 2,\ x_2 = 3,\ x_3 = 5$

7. $x = 2.43,\ y = 3.57,\ z = 1.92$

8. $x = 2,\ y = -1,\ z = 3$

9. $x = 1,\ y = -1,\ z = 1$

10. $x_1 = 5.34,\ x_2 = -5.24,\ x_3 = -1.83,\ x_4 = 6.12$

CHAPTER 4

Solution of Algebraic and Transcendental Equations

4.1. ALGEBRAIC EQUATION

An expression of the form $f(x) \equiv a_0x^n + a_1x^{n-1} + a_2x^{n-2} + \ldots\ldots + a_{n-1}x + a_n = 0, a_0 \neq 0$ where all $a_0, a_1, \ldots\ldots, a_n$ are constants and n is a positive integer, called an algebraic equation of degree n, in x.

$4x^6 + 3x^5 + 9x^4 + x^3 + 3x - 6 = 0$, $x^3 - x - 1 = 0$ are the example of algebraic equation.

4.2. TRANSCENDENTAL EQUATION

An expression which contain some other functions such as exponential, trigonometric, logarithmic etc. are called transcendental equation $3x - \sqrt{1 + \sin x} = 0$, $x \log_{10} x = 72$, $\cos x = x e^x$ are the example of transcendental equation.

4.3. ROOT OF THE EQUATION

The value of x which satisfying the equation $f(x) = 0$ is called the root of the equation.

The roots of the linear, quadratic, cubic, or biquadratic equations are obtained by available methods, but for transcendental equation or higher degree equation can not solved by these methods easily. So those types of equation can be solved by numerical methods such as bisection, secant, Newton-Raphson, Regula-Falsi method etc.

In this chapter, we shall discuss only Regula-Falsi and Newton-Raphson method.

4.4. NEWTON-RAPHSON METHOD

Let $x = x_0$ be an approximate value of the roots of the equation $f(x) = 0$ which is algebraic or transcendental and let $x_0 + h$ be the correct value of the corresponding root where h be a real number sufficiently small. Then

$$f(x_0 + h) = 0 \qquad ...(4.1)$$

Expanding by Taylor's theorem

$$f(x_0 + h) = f(x_0) + \frac{h}{1!} f'(x_0) + \frac{h^2}{2!} f''(x_0) + \frac{h^3}{3!} f'''(x_0) + ... = 0$$

Since h is very small, so neglecting second and higher order terms, we get

$$f(x_0) + h f'(x_0) = 0$$

$$\Rightarrow \qquad h = -\frac{f(x_0)}{f'(x_0)} \text{ also } f'(x_0) \neq 0$$

Thus, the first approximation of the root is given by

$$x_1 = x_0 + h = x_0 - \frac{f(x_0)}{f'(x_0)}$$

Similarly taking x_1 as initial approximation, to be the better approximation of the root x_2 is obtained as

$$x_2 = x_1 - \frac{f(x_1)}{f'(x_1)} \qquad f'(x_0) \neq 0$$

Proceeding in the same way we get better approximation of the root is given by

$$\boxed{x_{n+1} = x_n - \frac{f(x_n)}{f'(x_n)}, \quad n = 0, 1, 2, 3 .....}$$

This is known as Newton-Raphson formula.

4.5. REGULA-FALSI METHOD

This is the oldest method for finding the real root of the equation $f(x) = 0$. In this method we take two points x_0 and x_1 such that $f(x_0)$ and $f(x_1)$ are of opposite signs *i.e.*, $f(x_0) f(x_1) < 0$. The root must lie in between x_0 and x_1 since the graph $y = f(x)$ crosses the x-axis between these two points.

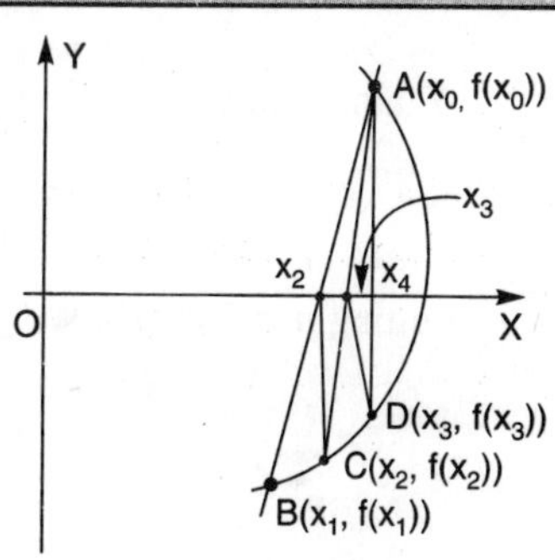

Now equation of the chord joining the two points $A[x_0, f(x_0)]$ and $B[x_1, f(x_1)]$ is

$$\frac{y - f(x_0)}{x - x_0} = \frac{f(x_1) - f(x_0)}{x_1 - x_0} \quad ...(4.2)$$

In this method the curve between the points $A\,[x_0, f(x_0)]$ and $B[x_1, f(x_1)]$ is replaced by the chord AB by joining the points A and B and taking the point of intersection of the chord with the x-axis as an approximation to the root which is given by putting $y = 0$ in (4.2). Thus, we have

$$x_2 = x_0 - \frac{(x_1 - x_0)}{f(x_1) - f(x_0)} f(x_0)$$

If now $f(x_0)$ and $f(x_2)$ are of opposite signs, then the root lies between x_0 and x_2. Then replace the part of curve between the points $A(x_0, f(x_0))$ and $C(x_2, f(x_2))$ by the chord joining these points and this chord intersect the x-axis then we get second approximation to the root which is given by

$$x_3 = x_0 - \frac{(x_2 - x_0)}{f(x_2) - f(x_0)} f(x_0)$$

The procedure is repeated till the root is found to desired accuracy.

SOLVED EXAMPLES

Example 1. *Find the real root of the equation $x^2 - 5x + 2 = 0$ by Newton-Raphson's method.*

Solution. Let $\quad f(x) = x^2 - 5x + 2 = 0$ and $f'(x) = 2x - 5 \quad ...(i)$

Now $\quad f(4) = 4^2 - 5 \times 4 + 2 = -2$

$\quad f(5) = 5^2 - 5 \times 5 + 2 = 2$

Therefore, one real root lies between 4 and 5

By Newton-Raphson's formula

$$x_{n+1} = x_n - \frac{f(x_n)}{f'(x_n)} = x_n - \frac{x_n^2 - 5x_n + 2}{2x_n - 5} = \frac{2x_n^2 - 5x_n - x_n^2 + 5x_n - 2}{2x_n - 5}$$

$$x_{n+1} = \frac{x_n^2 - 2}{2x_n - 5} \quad \text{where } n = 0, 1, 2, 3, ... \quad ...(ii)$$

Take $\quad x_0 = 4$

Putting $n = 0$ in (ii), we get first approximation

$$x_1 = \frac{x_0^2 - 2}{2x_0 - 5} = \frac{4^2 - 2}{2 \times 4 - 5} = \frac{16 - 2}{8 - 5} = \frac{14}{3} = 4.6667$$

Again, putting $n = 1$ in (ii), we get second approximation

$$x_2 = \frac{x_1^2 - 2}{2x_1 - 5} = \frac{(4.6667)^2 - 2}{2 \times 4.6667 - 5} = \frac{21.7780 - 2}{9.3334 - 5} = \frac{19.7780}{4.3334} = 4.5640$$

Putting $n = 2$ in (ii), we get third approximation

$$x_3 = \frac{x_2^2 - 2}{2x_2 - 5} = \frac{(4.5640)^2 - 2}{2 \times (4.5640) - 5} = \frac{20.8308 - 2}{9.1281 - 5} = \frac{18.8308}{4.1281} = 4.5616$$

Putting $n = 3$ in (*ii*), we get fourth approximation

$$x_4 = \frac{x_3^2 - 2}{2x_3 - 5} = \frac{(4.5616)^2 - 2}{2 \times 4.5616 - 5} = \frac{20.8084 - 2}{9.1232 - 5} = \frac{18.8084}{4.1232} = 4.5616$$

Here $x_3 = x_4$. Thus, the root of equation $x^2 - 5x + 2 = 0$ is 4.5616.

Example 2. *Find the real root of the equation $x^3 - 3x - 5 = 0$ by Newton-Raphson's method.*

Solution. Let $f(x) = x^3 - 3x - 5 = 0$ ...(*i*)

then $f'(x) = 3x^2 - 3$

Now $f(2) = 2^3 - 3 \times 2 - 5 = -3$

and $f(3) = 3^3 - 3 \times 3 - 5 = 13$

Therefore, one real root lies between 2 and 3

By Newton-Raphson's formula, we have

$$x_{n+1} = x_n - \frac{f(x_n)}{f'(x_n)} = x_n - \frac{x_n^3 - 3x_n - 5}{3x_n^2 - 3} = \frac{3x_n^3 - 3x_n - x_n^3 + 3x_n + 5}{3x_n^2 - 3}$$

$$x_{n+1} = \frac{2x_n^3 + 5}{3x_n^2 - 3} \quad \text{where } n = 0, 1, 2, 3, \ldots\ldots \qquad \text{...(ii)}$$

Take $x_0 = 2$

Putting $n = 0$ in (*ii*), we get first approximation

$$x_1 = \frac{2x_0^3 + 5}{3x_0^2 - 3} = \frac{2.2^3 + 5}{3.2^2 - 3} = \frac{21}{9} = 2.3333.$$

Again putting $n = 1$ in (*ii*), we get second approximation

$$x_2 = \frac{2 \,.\, x_1^3 + 5}{3x_1^2 - 3} = \frac{2 \,.\, (2.3333)^3 + 5}{3 \,.\, (2.3333)^2 - 3} = \frac{30.4063}{13.3333} = 2.2804$$

Putting $n = 2$ in (*ii*), we get third approximation

$$x_3 = \frac{2 \,.\, x_2^3 + 5}{3x_2^2 - 3} = \frac{2 \,.\, (2.2804)^3 + 5}{3 \,.\, (2.2804)^2 - 3} = \frac{28.7171}{12.6006} = 2.2790$$

Putting $n = 3$ in (*ii*), we get fourth approximation

$$x_4 = \frac{2x_3^3 + 5}{3x_3^2 - 3} = \frac{2.(2.2790)^3 + 5}{3.(2.2790)^2 - 3} = \frac{28.6735}{12.5815} = 2.2790$$

Here $x_3 = x_4$ therefore the root of the equation

$x^3 - 3x - 5 = 0$ is 2.2790.

Example 3. *Using Newton-Raphson's method find the cube root of 10 correct to four places of decimal.*

Solution. Let $f(x) = x^3 - 10 = 0$ ...(*i*)

then $f'(x) = 3x^2$

Now $f(2) = 2^3 - 10 = -2$ and $f(3) = 3^3 - 10 = 17$

Therefore, one real root lies between 2 and 3.

By Newton-Raphson's formula, we have

$$x_{n+1} = x_n - \frac{f(x_n)}{f'(x_n)} = x_n - \frac{x_n^3 - 10}{3x_n^2} = \frac{3x_n^3 - x_n^3 + 10}{3x_n^2}$$

$$x_{n+1} = \frac{2x_n^3 + 10}{3x_n^2} \qquad \text{where } n = 0, 1, 2, 3, \ldots\ldots \qquad \ldots(ii)$$

Take $x_0 = 2$

Putting $n = 0$ in (*ii*), we get first approximation

$$x_1 = \frac{2x_0^3 + 10}{3x_0^2} = \frac{2.2^3 + 10}{3.2^2} = \frac{26}{12} = 2.1666$$

Putting $n = 1$ in (*ii*), we get second approximation

$$x_2 = \frac{2x_1^3 + 10}{3x_1^2} = \frac{2(2.1666)^3 + 10}{3 \times (2.1666)^2} = \frac{30.3407}{14.0824} = 2.1545$$

Putting $n = 2$ in (*ii*), we get third approximation

$$x_3 = \frac{2x_2^3 + 10}{3x_2^2} = \frac{2.(2.1545)^3 + 10}{3.(2.1545)^2} = \frac{30.0018}{13.9256} = 2.1544$$

Putting $n = 3$ in (*ii*), we get fourth approximation

$$x_4 = \frac{2x_3^3 + 10}{3x_3^2} = \frac{2(2.1544)^3 + 10}{3.(2.1544)^2} = \frac{29.9990}{13.9247} = 2.1544$$

Here $x_3 = x_4$ therefore the cube root of 10 is 2.1544.

Example 4. *Find the real root of* $\tan x = 4x$ *by using Newton-Raphson's method.*

Solution. Let $f(x) = \tan x - 4x = 0$ ...(*i*)

then $f'(x) = \sec^2 x - 4 = 0$

Now $f(0) = 0$ and $f(1) = -3.982$

$\because$ $f(0) = 0$ so exact root of $f(x) = 0$ is 0

By Newton-Raphson's method, we have

$$x_{n+1} = x_n - \frac{f(x_n)}{f'(x_n)} = x_n - \frac{\tan x_n - 4x_n}{\sec^2 x_n - 4} = \frac{x_n \sec^2 x_n - \tan x_n}{\sec^2 x_n - 4} \quad ...(ii)$$

where $n = 0, 1, 2, 3, ...$

Take $x_0 = 0$.

Putting $n = 0$, we get first approximation

$$x_1 = \frac{x_0 \sec^2 x_0 - \tan x_0}{\sec^2 x_0 - 4} = \frac{0.1 - 0}{1 - 4} = 0$$

Hence the root of $\tan x - 4x = 0$ is 0.

Example 5. *Using Newton-Raphson's method find the solution of $e^x = 3x$.*

Solution. Let $f(x) = e^x - 3x = 0$ ...(i)

then $f'(x) = e^x - 3$

Now $f(0) = 1$ and $f(1) = -0.28$

Therefore root lies between 0 and 1

By Newton's formula, we have

$$x_{n+1} = x_n - \frac{f(x_n)}{f'(x_n)} = x_n - \frac{e^{x_n} - 3x_n}{e^{x_n} - 3} = \frac{x_n e^{x_n} - 3x_n - e^{x_n} + 3x_n}{e^{x_n} - 3}$$

$$= \frac{e^{x_n}(x_n - 1)}{e^{x_n} - 3}, \text{ where } n = 0, 1, 2, 3, ... \quad ...(ii)$$

Take $x_0 = 0$.

Putting $n = 0$ in (*ii*), we get first approximation

$$x_1 = \frac{e^{x_0}(x_0 - 1)}{e^{x_0} - 3} = \frac{e^0(0 - 1)}{e^0 - 3} = \frac{-1}{-2} = 0.5$$

Again putting $n = 1$ in (*ii*), we get second approximation

$$x_2 = \frac{e^{x_1}(x_1 - 1)}{e^{x_1} - 3} = \frac{e^{0.5}(0.5 - 1)}{e^{0.5} - 3} = \frac{-0.82436}{-1.35127} = 0.61006$$

Putting $n = 2$ in (*ii*), we get third approximation

$$x_3 = \frac{e^{x_2}(x_2 - 1)}{e^{x_2} - 3} = \frac{e^{0.61006}(0.61006 - 1)}{e^{0.61006} - 3} = \frac{-0.71770}{-0.15945} = 0.61900$$

Putting $n = 3$ in (*ii*), we get fourth approximation

$$x_4 = \frac{e^{x_3}(x_3 - 1)}{e^{x_3} - 3} = \frac{e^{0.61900}(0.61900 - 1)}{e^{0.61900} - 3} = \frac{-0.70754}{-1.14293} = 0.61905$$

Here $x_3 = x_4$. Therefore, the root of equation $e^x = 3x$ is 0.6190 **Ans.**

Example 6. *Find the real root of the equation 3x = cos x + 1 by Newton-Raphson's method*

Solution. Let $f(x) = 3x - \cos x - 1 = 0$ ...(*i*)

then $f'(x) = 3 + \sin x = 0$

Now $f(0) = -2$ and $f(1) = 3 - \cos 1 - 1 = 2 - 0.5403 = 1.4597$

Therefore, root lies between 0 and 1

By Newton-Raphson's formula, we have

$$x_{n+1} = x_n - \frac{f(x_n)}{f'(x_n)} = x_n - \frac{3x_n - \cos x_n - 1}{3 + \sin x_n} = \frac{x_n \sin x_n + \cos x_n + 1}{3 + \sin x_n} \quad ...(ii)$$

Take $x_0 = 0.$

Putting $n = 0$ in (*ii*), we get first approximation

$$x_1 = \frac{x_0 \sin x_0 + \cos x_0 + 1}{3 + \sin x_0} = \frac{0 \sin 0 + \cos 0 + 1}{3 + \sin 0} = \frac{2}{3} = 0.6666$$

Putting $n = 1$ in (*ii*), we get second approximation

$$x_2 = \frac{x_1 \sin x_1 + \cos x_1 + 1}{3 + \sin x_1} = \frac{(0.6666) \sin (0.6666) + \cos (0.6666) + 1}{3 + \sin (0.6666)}$$

$$= \frac{0.6666 \times 0.6183 + 0.7859 + 1}{3 + 0.6183} = \frac{0.4122 + 0.7859 + 1}{3.6183}$$

$$= \frac{2.1481}{3.6183} = 0.6074$$

Putting $n = 2$ in (*ii*), we get third approximation

$$x_3 = \frac{x_2 \sin x_2 + \cos x_2 + 1}{3 + \sin x_2} = \frac{(0.6074) \sin (0.6074) + \cos (0.6074) + 1}{3 + \sin (0.6074)}$$

$$= \frac{0.6074 \times 0.5707 + 0.8211 + 1}{3 + 0.5707} = \frac{0.3466 + 0.8211 + 1}{3.5707} = \frac{2.1677}{3.5707} = 0.6071$$

Putting $n = 3$ in (*ii*), we get fourth approximation

$$x_4 = \frac{x_3 \sin x_3 + \cos x_3 + 1}{3 + \sin x_3} = \frac{(0.6071) \sin (0.6071) + \cos (0.6071) + 1}{3 + \sin (0.6071)}$$

$$= \frac{0.6071 \times 0.5704 + 0.8213 + 1}{3 + 0.5704} = \frac{0.3463 + 0.8213 + 1}{3.5704} = \frac{2.1676}{3.5704} = 0.6071.$$

Here $x_3 = x_4$ therefore, the root of $3x - \cos x - 1 = 0$ is 0.6071.

Example 7. *Use the method of false position, find the real root of the equation* $x^3 - 2x - 5 = 0$.

Solution. We have $f(x) = x^3 - 2x - 5 = 0$ ...(*i*)

Now $f(2) = 2^3 - 2 \times 2 - 5 = -1$

and $f(3) = 3^3 - 2 \times 3 - 5 = 16$

Therefore, one real root lies between 2 and 3

$\therefore$ Taking $x_0 = 2$ and $x_1 = 3$ $\Rightarrow$ $f(x_0) = f(2) = -1,\ f(x_1) = f(3) = 16$

Then by Regula-Falsi method, we get

$$x_2 = x_0 - \frac{x_1 - x_0}{f(x_1) - f(x_0)} f(x_0) = 2 - \frac{3-2}{16-(-1)}(-1)$$

$$= 2 + \frac{1}{17} = \frac{35}{17} = 2.0588$$

Now $f(x_2) = f(2.0588) = (2.0588)^3 - 2 \times (2.0588) - 5$ by (*i*)

$$= -0.3910$$

Therefore, one real root lies between 2.0588 and 3

$\therefore$ Taking $x_0 = 2.0588,\ x_1 = 3$ $\Rightarrow$ $f(x_0) = f(2.0588) = -0.3910$

$f(x_1) = f(3) = 16$

Then by Regula-Falsi method, we get

$$x_3 = x_0 - \frac{x_1 - x_0}{f(x_1) - f(x_0)} f(x_0) = 2.0588 - \frac{3-2.0588}{16-(-0.3910)} \times (-0.3910)$$

$$= 2.0588 + \frac{0.9412}{16.391} \times (0.3910) = 2.0588 + 0.02245 = 2.0812$$

Now $f(x_3) = f(2.0812) = (2.0812)^3 - 2 \times (2.0812) - 5 = -0.1479$

Therefore, one root lies between 2.0812 and 3

$\therefore$ Taking $x_0 = 2.0812,\ x_1 = 3$

$\Rightarrow$ $f(x_0) = f(2.0812) = -0.1479, f(x_1) = f(3) = 16$

Then by Regula-Falsi method, we get

$$x_4 = x_0 - \frac{x_1 - x_0}{f(x_1) - f(x_0)} f(x_0)$$

$$= 2.0812 - \frac{3 - 2.0812 \times (-0.1479)}{16-(-0.1479)} = 2.0812 + \frac{0.9188 \times (0.1479)}{16.1479}$$

$$= 2.0812 + 0.0084 = 2.0896$$

Now $f(x_4) = f(2.0896) = (2.0896)^3 - 2 \times (2.0896) - 5 = -0.0551$

Therefore, one root lies between 2.0896 and 3

$\therefore$ Taking $x_0 = 2.0896,\ x_1 = 3$

$f(x_0) = f(2.0896) = -0.0551,\ f(x_1) = f(3) = 16$

Then by Regula-Falsi method, we get

$$x_5 = x_0 - \frac{x_1 - x_0}{f(x_1) - f(x_0)} f(x_0) = 2.0896 - \frac{3 - 2.0896}{16 - (-0.0551)} (-0.0551)$$

$$= 2.0896 + \frac{0.9104}{16.0551} \times 0.0551 = 2.0896 + 0.0031 = 2.0927$$

Now $\qquad f(x_5) = f(2.0927) = (2.0927)^3 - 2 \times (2.0927) - 5 = -0.0206$

Therefore, one real root lies between 2.0927 and 3

$\therefore$ Taking $x_0 = 2.0927$, $x_1 = 3$

$$f(x_0) = f(2.0927) = -0.0206, f(x_1) = f(3) = 16$$

Then by Regula-Falsi method, we get

$$x_6 = x_0 - \frac{x_1 - x_0}{f(x_1) - f(x_0)} f(x_0) = 2.0927 - \frac{3 - 2.0927}{16 - (-0.0206)} \times (-0.0206)$$

$$= 2.0927 + \frac{0.9073}{16.0206} \times 0.0206 = 2.0927 + 0.0011 = 2.0938$$

Now $\qquad f(x_6) = f(2.0938) = (2.0938)^3 - 2 \times (2.0938) - 5 = -0.0083$

Therefore, one root lies between 2.0938 and 3

$\therefore$ Taking $x_0 = 2.0938$, $x_1 = 3$

$$f(x_0) = f(2.0938) = -0.0083$$

$$f(x_1) = f(3) = 16$$

Then by Regula-Falsi method, we get

$$x_7 = x_0 - \frac{x_1 - x_0}{f(x_1) - f(x_0)} \times f(x_0) = 2.0938 - \frac{3 - 2.0938}{16 - (-0.0083)} \times (-0.0083)$$

$$= 2.0938 + \frac{0.9062}{16.0083} \times 0.0083$$

$$= 2.0938 + 0.00046 = 2.0942$$

Now $\qquad f(2.0942) = (2.0942)^3 - 2 \times (2.0942) - 5 = -0.0030$

Therefore, root lies between 2.0942 and 3

$\therefore$ Taking $x_0 = 2.0942$, $x_1 = 3$

$\Rightarrow \qquad f(x_0) = f(2.0942) = -0.0030,\ f(x_1) = f(3) = 16$

Then by Regula-Falsi method, we get

$$x_8 = x_0 - \frac{x_1 - x_0}{f(x_1) - f(x_0)} f(x_0) = 2.0942 - \frac{3 - 2.0942}{16 - (-0.0030)} \times (-0.0030)$$

$$= 2.0942 + \frac{0.9052}{16.0030} \times 0.0030 = 2.0942 + 0.00016 = 2.0943$$

Here $x_7 = x_8$

Hence, the root is 2.094 correct to three decimal places.

Example 8. *Determine the real root of the equation $xe^x = 2$ using Regula-Falsi method correct to three decimal places.*

Solution. Let $f(x) = xe^x - 2 = 0$

Now $f(0) = 0 - 2 = -2$

and $f(0.5) = 0.5e^{0.5} - 2 = -1.1756$

$f(0.7) = -0.5903$

$f(0.9) = 0.2136$

Therefore, one real root lies between 0.7 and 0.9

$\therefore$ Taking $x_0 = 0.7, x_1 = 0.9$

$\Rightarrow$ $f(x_0) = f(0.7) = -0.5903, f(x_1) = f(0.9) = 0.2136$

Then by Regula-Falsi method, we get

$$x_2 = x_0 - \frac{x_1 - x_0}{f(x_1) - f(x_0)} f(x_0)$$

$$= 0.7 - \frac{0.9 - 0.7}{0.2136 - (-0.5903)} \times (-0.5903) = 0.7 + \frac{0.2}{0.8039} \times (0.5903)$$

$$= 0.8469$$

Now $f(x_2) = f(0.8469) = (0.8469)\, e^{0.8469} - 2 = -0.0247$

Therefore, root lies between 0.8469 and 0.9

$\therefore$ Taking $x_0 = 0.8469, x_1 = 0.9$

$\Rightarrow$ $f(x_0) = f(0.8469) = -0.0247, f(x_1) = f(0.9) = 0.2136$

Then by Regula-Falsi method, we get

$$x_3 = x_0 - \frac{x_1 - x_0}{f(x_1) - f(x_0)} f(x_0) = 0.7 - \frac{0.9 - 0.8469}{0.2136 - (-0.0247)} \times (-0.0247)$$

$$= 0.7 + \frac{0.0531}{0.2383} \times 0.0247 = 0.8524$$

Now $f(x_3) = f(0.8524) = (0.8524)\, e^{0.8524} - 2 = -0.00089$

Therefore, root lies between 0.8524 and 0.9

$\therefore$ Taking $x_0 = 0.8524, x_1 = 0.9$

$\Rightarrow$ $f(x_0) = f(0.8524) = -0.00089, f(x_1) = f(0.9) = 0.2136$

$$x_4 = x_0 - \frac{x_1 - x_0}{f(x_1) - f(x_0)} = 0.8524 - \frac{0.9 - 0.8524}{0.2136 - (-0.00089)} \times (-0.00089)$$

$$= 0.8524 + \frac{0.0476}{0.21449} \times (0.00089) = 0.8526$$

Here $x_3 = x_4$

Hence, the root is 0.852 correct to three decimal places.

EXERCISE 4.1

1. Find the real root of the equation $x^4 - x - 10 = 0$, correct to three decimal places by Newton-Raphson's method.
2. By using Newton-Raphson's method find the real root of the following equation:
   (*i*) $x^4 - x - 13 = 0$ (*ii*) $x^3 - 5x + 3 = 0$
   (*iii*) $x^3 - 4x + 1 = 0$
3. Using Newton-Raphson's method. Find the square root of 12 correct to three places of decimal.
4. Using Newton-Raphson's method obtain the real root of the following equation:
   (*i*) $x = \sqrt{29}$ (*ii*) $x^2 + 4 \sin x = 0$
   (*iii*) $x \sin x + \cos x = 0$ (*iv*) $\log x = \cos x$
5. Using Regula-Falsi method, obtain the real root of the following equation:
   (*i*) $x^3 - 9x + 1 = 0$ (*ii*) $x^2 - 2x - 1 = 0$
   (*iii*) $x^3 - x^2 - 2 = 0$
6. Find a real root of the equation $x^2 - \log_e x - 12 = 0$, using Regula-Falsi method correct to three decimal place.
7. Find real root of the following equation by using Regula-Falsi method:
   $xe^x - 3 = 0$.

ANSWERS

1. 1.85558 **2.** (*i*) 1.967 (*ii*) 0.6566 (*iii*) 0.254
3. 3.4641 **4.** (*i*) 5.3852 (*ii*) – 1.9338 (*iii*) 2.798 (*iv*) 1.3030
5. (*i*) 2.9428 (*ii*) 2.4141 (*iii*) 1.6955 **6.** 3.6461 **7.** 1.046

CHAPTER 5

Curve Fitting

INTRODUCTION

Curve fitting have a great importance in the field of statistics (not only theoretical but also practical) Engineering and Science.

5.1. SCATTER DIAGRAM

Let (x_i, y_i): $i = 1, 2 \ldots n$ be n sets of numerical values of two variables x and y. If we plot these n sets on the graph then we get a diagram. This resulting diagram is called the scatter diagram.

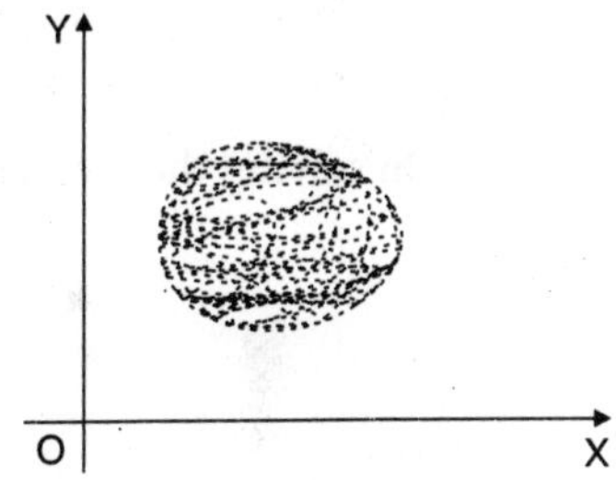

5.2. CURVE FITTING

Let (x_1, y_1) (x_2, y_2) ..... (x_n, y_n) are the n numerical values of two variables x (independent) and y (dependent). By the scatter diagram, we get an approximate relation between these two variables called empirical law. Curve fitting means that relationship between two variable in the form of equation of the curve from the given data. In other words, the method of obtaining an equation of best fit is called curve fitting.

Methods of Curve Fitting

Following are some methods for fitting a curve:

1. Graphical method
2. Method of least squares
3. Method of group averages
4. Method of moments.

But in this chapter, we shall discuss only method of least squares for straight line and parabola.

5.3. METHOD OF LEAST SQUARES

Let us consider a set of m observations $(x_i, y_i) : i = 1, 2, 3..... m$ of two variables x and y. We wish to fit a polynomial of degree n. Assume that such curve is

$$y = a + bx + cx^2 + ... qx^n \text{ of these values.} \quad ...(5.1)$$

In order to determine $a, b, c ..... q$ such that it represents the curve of best fit. Now we apply the principle of least squares so put $x = x_1, x_2 ..... x_m$ in (5.1), we get

$$\left.\begin{aligned} y'_1 &= a + bx_1 + cx_1^2 .... qx_1^n \\ y'_2 &= a + bx_2 + cx_2^2 + ... + qx_2^n \\ &\vdots \\ y'_m &= a + bx_m + cx_m^2 + ... + qx_m^n \end{aligned}\right\} \quad ...(5.2)$$

where $y'_1, y'_2 ..... y'_m$ are the expected value of y for $x = x_1, x_2 ..... x_m$ respectively, which are other than observed values $y_1, y_2 ..... y_m$ of y for $x = x_1, x_2 ..... x_m$ respectively. In general expected values and observed values are different, the difference $y_i - y'_i$; $i = 1, 2 ..... m$ are called residuals.

Let $A = \Sigma (y_i - y'_i)^2 = \Sigma [y_i - (a + bx_i + cx_i^2 ..... qx_i^n)^2]; i = 1, 2 ..... m.$

The constant $a, b, c ..... q$ are choosen as the sum of squares of $y_i - y'_i ; i = 1, 2 ..... m$ is minimum.

Now for maximum and minimum

$$\frac{\partial A}{\partial a} = 0, \frac{\partial A}{\partial b} = 0 ..... \frac{\partial A}{\partial q} = 0$$

On solving these equations, we get

$$\left.\begin{aligned} ma + b\Sigma x + c\Sigma x^2 + ...+ q\Sigma x^n &= \Sigma y \\ a\Sigma x + b\Sigma x^2 + c\Sigma x^3 + ...+ q\Sigma x^{n+1} &= \Sigma xy \\ a\Sigma x^2 + b\Sigma x^3 + c\Sigma x^4 + ... + q\Sigma x^{n+2} &= \Sigma x^2 y \\ &\vdots \\ a\Sigma x^n + b\Sigma x^{n+1} + c\Sigma x^{n+2} + ...+ q\Sigma x^{2n} &= \Sigma x^n y \end{aligned}\right\} \quad ...(5.3)$$

These equations are known as normal equations. On solving these normal equations, we get the values of constants $a, b, c ..... q$.

Note 1: If $n = 1$, then the curve to be fitted is a straight line $y = a + bx$ whose normal equations are

$$\Sigma y = ma + b\Sigma x$$

$$\Sigma xy = a\Sigma x + b\Sigma x^2$$

which can be solved for a and b.

Note 2: If $n = 2$, then the curve to be fitted is a parabola of second degree $y = a + bx + cx^2$ whose normal equations are

$$\Sigma y = ma + b\Sigma x + c\Sigma x^2$$

$$\Sigma xy = a\Sigma x + b\Sigma x^2 + c\Sigma x^3$$

$$\Sigma x^2 y = a\Sigma x^2 + b\Sigma x^3 + c\Sigma x^4$$

which can be solved for a, b and c.

5.4. WORKING RULE TO FIT A STRAIGHT LINE TO GIVEN DATA BY METHOD OF LEAST SQUARES

(1) Let the equation of straight line be $y = a + bx$.

(2) From the given data we calculate Σx, Σy, Σx^2, and Σxy.

(3) Putting above values in normal equations

$$ma + b\Sigma x = \Sigma y \qquad ...(i)$$

$$a\Sigma x + b\Sigma x^2 = \Sigma xy. \qquad ...(ii)$$

(4) Solving equations (*i*) and (*ii*) for a and b.

(5) Putting the values of a and b in $y = a + bx$ we get equation of straight line of best fit.

5.5. WORKING RULE TO FIT A PARABOLA TO THE GIVEN DATA BY METHOD OF LEAST SQUARES

(1) Let the equation of the parabola be $y = a + bx + cx^2$.

(2) From the given data we calculate Σx, Σy, Σx^2, Σx^3, Σx^4, Σxy and Σx^2y.

(3) Putting above values in normal equations

$$ma + b\Sigma x + c\Sigma x^2 = \Sigma y \qquad ...(i)$$

$$a\Sigma x + b\Sigma x^2 + c\Sigma x^3 = \Sigma xy \qquad ...(ii)$$

$$a\Sigma x^2 + b\Sigma x^3 + c\Sigma x^4 = \Sigma x^2y. \qquad ...(iii)$$

(4) Solving equations (*i*), (*ii*) and (*iii*) for a,b and c.

(5) Putting the values of a and b in $y = a + bx + cx^2$ we get equation of parabola of best fit.

SOLVED EXAMPLES

Example 1. *Fit a straight line $y = a + bx$, to the following data regarding x as the independent variable by the method of least squares.*

x	0	1	3	6	8
y	1	3	2	4	5

Solution. Let the equation of straight line be $y = a + bx$ and the normal equations are:

$$ma + b\Sigma x = \Sigma y$$

$$a\Sigma x + b\Sigma x^2 = \Sigma xy$$

Now

x	y	x^2	xy
0	1	0	0
1	3	1	3
3	2	9	6
6	4	36	24
8	5	64	40
$\Sigma x = 18$	$\Sigma y = 15$	$\Sigma x^2 = 110$	$\Sigma xy = 73$

Here $m = 5$, $\Sigma x = 18$, $\Sigma y = 15$, $\Sigma x^2 = 110$, $\Sigma xy = 73$.

Put these values in normal equations

$$5a + 18b = 15 \qquad ...(i)$$

$$18a + 110b = 73 \qquad ...(ii)$$

On solving (i) and (ii), we get

$$a = 1.488,\ b = 0.420$$

Thus, the equation of straight line $y = 1.488 + 0.42x$.

Example 2. *Fit a straight line to the following data regarding x as a independent variable.*

x	*1*	*2*	*3*	*4*	*5*
y	*5*	*7*	*9*	*10*	*11*

Solution. Let equation of straight line be $y = a + bx$ and the normal equations are:

$$ma + b\Sigma x = \Sigma y$$

$$a\Sigma x + b\Sigma x^2 = \Sigma xy$$

Now

x	y	x^2	xy
1	5	1	5
2	7	4	14
3	9	9	27
4	10	16	40
5	11	25	55
$\Sigma x = 15$	$\Sigma y = 42$	$\Sigma x^2 = 55$	$\Sigma xy = 141$

Here $m = 5$, $\Sigma x = 15$, $\Sigma y = 42$, $\Sigma x^2 = 55$, $\Sigma xy = 141$.

Put these values in normal equations, we have

$$5a + 15b = 42 \qquad ...(i)$$

$$15a + 55b = 141 \qquad ...(ii)$$

On solving (*i*) and (*ii*), we get

$$a = 3.9 \quad \text{and} \quad b = 1.5$$

Thus, the equation of straight line is $y = 3.9 + 1.5x$.

Example 3. *Find a least squares straight line for the following data.*

x	*1*	*2*	*3*	*4*
y	*3*	*7*	*13*	*21*

Solution. Let equation of straight line be $y = a + bx$ and the normal equations are:

$$ma + b\Sigma x = \Sigma y$$

$$a\Sigma x + b\Sigma x^2 = \Sigma xy$$

Now

x	y	x^2	xy
1	3	1	3
2	7	4	14
3	13	9	39
4	21	16	84
$\Sigma x = 10$	$\Sigma y = 44$	$\Sigma x^2 = 30$	$\Sigma xy = 140$

Here $m = 4$, $\Sigma x = 10$, $\Sigma y = 44$, $\Sigma x^2 = 30$, $\Sigma xy = 140$.

Put these values in normal equations

$$4a + 10b = 44 \quad \text{...(i)}$$

$$10a + 30b = 140 \quad \text{...(ii)}$$

On solving (*i*) and (*ii*), we get

$$a = -4 \quad \text{and} \quad b = 6$$

Thus, the equation of straight line is $y = -4 + 6x$.

Example 4. *Fit a straight line to the following data regarding x as the independent variable.*

x	*1*	*2*	*3*	*4*	*5*	*6*
y	*1200*	*900*	*600*	*200*	*110*	*50*

Solution. Let the equation of straight line be $y = a + bx$ and the normal equations are:

$$ma + b\Sigma x = \Sigma y$$

$$a\Sigma x + b\Sigma x^2 = \Sigma xy$$

Now

x	y	x^2	xy
1	1200	1	1200
2	900	4	1800
3	600	9	1800
4	200	16	800
5	110	25	550
6	50	36	300
$\Sigma x = 21$	$\Sigma y = 3060$	$\Sigma x^2 = 91$	$\Sigma xy = 6450$

Here $m = 6$, $\Sigma x = 21$, $\Sigma y = 3060$, $\Sigma x^2 = 91$, $\Sigma xy = 6450$.

Put these values in normal equations

$$6a + 21b = 3060 \qquad \text{...}(i)$$

$$21a + 91b = 6450 \qquad \text{...}(ii)$$

On solving (*i*) and (*ii*), we get

$$a = 1361.97 \quad \text{and} \quad b = -243.42$$

Thus, the equation of straight line is $y = 1361.97 - 243.42\,x$.

Example 5. *Fit a polynomial of the second degree to the data points given in the following table.*

x	0	1	2
y	1	6	17

Solution. Let $y = a + bx + cx^2$ be a parabola to be fitted for the given data and normal equations are:

$$ma + b\Sigma x + c\Sigma x^2 = \Sigma y$$

$$a\Sigma x + b\Sigma x^2 + c\Sigma x^3 = \Sigma xy$$

$$a\Sigma x^2 + b\Sigma x^3 + c\Sigma x^4 = \Sigma x^2 y$$

Now

x	y	x^2	x^3	x^4	xy	x^2y
0	1	0	0	0	0	0
1	6	1	1	1	6	6
2	17	4	8	16	34	68
$\Sigma x = 3$	$\Sigma y = 24$	$\Sigma x^2 = 5$	$\Sigma x^3 = 9$	$\Sigma x^4 = 17$	$\Sigma xy = 40$	$\Sigma x^2y = 74$

Here $m = 3$, $\Sigma x = 3$, $\Sigma y = 24$, $\Sigma x^2 = 5$, $\Sigma x^3 = 9$, $\Sigma x^4 = 17$, $\Sigma xy = 40$, $\Sigma x^2y = 74$.

Put these values in normal equations, we have

$$3a + 3b + 5c = 24 \quad ...(i)$$

$$3a + 5b + 9c = 40 \quad ...(ii)$$

$$5a + 9b + 17c = 74 \quad ...(iii)$$

On solving (*i*), (*ii*) and (*iii*), we get $a = 1, b = 2, c = 3$

Thus, the equation of required parabola is $y = 1 + 2x + 3x^2$.

Example 6. *Find the values of a, b and c so that* $y = a + bx + cx^2$ *is the best fit to the data.*

x	0	1	2	3	4
y	1	0	3	10	21

Solution. Let $y = a + bx + cx^2$ be a parabola to be fitted for the given data and normal equations are

$$ma + b\Sigma x + c\Sigma x^2 = \Sigma y$$

$$a\Sigma x + b\Sigma x^2 + c\Sigma x^3 = \Sigma xy$$

$$a\Sigma x^2 + b\Sigma x^3 + c\Sigma x^4 = \Sigma x^2 y$$

Now

x	y	x^2	x^3	x^4	xy	x^2y
0	1	0	0	0	0	0
1	0	1	1	1	0	0
2	3	4	8	16	6	12
3	10	9	27	81	30	90
4	21	16	64	256	84	336
$\Sigma x = 10$	$\Sigma y = 35$	$\Sigma x^2 = 30$	$\Sigma x^3 = 100$	$\Sigma x^4 = 354$	$\Sigma xy = 120$	$\Sigma x^2y = 438$

Here $m = 5, \Sigma x = 10, \Sigma y = 35, \Sigma x^2 = 30, \Sigma x^3 = 100, \Sigma x^4 = 354, \Sigma xy = 120$ and $\Sigma x^2y = 438$

Put these values in normal equations, we have

$$5a + 10b + 30c = 35 \quad ...(i)$$

$$10a + 30b + 100c = 120 \quad ...(ii)$$

$$30a + 100b + 354c = 438 \quad ...(iii)$$

On solving (*i*), (*ii*) and (*iii*), we get $a = 1, b = -3, c = 2$ and equation of parabola

$$y = 1 - 3x + 2x^2.$$

Example 7. *Fit a second degree parabola to the following data x as the independent variable.*

x	1	2	3	4	5	6	7	8	9
y	2	6	7	8	10	11	11	10	9

Solution. Let equation of second degree parabola be $y = a + bx + cx^2$ and the normal equations are

$$ma + b\Sigma x + c\Sigma x^2 = \Sigma y$$

$$a\Sigma x + b\Sigma x^2 + c\Sigma x^3 = \Sigma xy$$

$$a\Sigma x^2 + b\Sigma x^3 + c\Sigma x^4 = \Sigma x^2 y$$

Now

x	y	x^2	x^3	x^4	xy	x^2y
1	2	1	1	1	2	2
2	6	4	8	16	12	24
3	7	9	27	81	21	63
4	8	16	64	256	32	128
5	10	25	125	625	50	250
6	11	36	216	1296	66	396
7	11	49	343	2401	77	539
8	10	64	512	4096	80	640
9	9	81	729	6561	81	729
$\Sigma x = 45$	$\Sigma y = 74$	$\Sigma x^2 = 285$	$\Sigma x^3 = 2025$	$\Sigma x^4 = 15333$	$\Sigma xy = 421$	$\Sigma x^2y = 2771$

Here $m = 9$, $\Sigma x = 45$, $\Sigma y = 79$, $\Sigma x^2 = 285$, $\Sigma x^3 = 2025$, $\Sigma x^4 = 15333$, $\Sigma xy = 421$ and $\Sigma x^2y = 2771$

Put these values in normal equations, we have

$$9a + 45b + 285c = 74 \quad ...(i)$$

$$45a + 285b + 2025c = 421 \quad ...(ii)$$

$$285a + 2025b + 15333c = 2771 \quad ...(iii)$$

On solving (*i*), (*ii*) and (*iii*), we get $a = -0.923$, $b = 3.52$, $c = -0.267$

Thus, the equation of parabola is $y = -0.923 + 3.52x - 0.267x^2$.

EXERCISE 5.1

1. Find a least squares straight line for the following data:

x	1	2	3	4	5	6
y	6	4	3	5	4	2

2. Find a least squares straight line for the following data:

x	0	1	2	3	4
y	1.0	2.9	4.8	6.7	8.6

3. Show that the line of best fit to the following data is given by

x	6	7	7	8	8	8	9	9	10
y	5	5	4	5	4	3	4	3	3

4. Find the best values of a and b so that $y = a + bx$ fits the data given in the table.

x	0	1	2	3	4
y	1	1.8	3.3	4.5	6.3

5 Obtain the least squares straight line fit to the following data:

x	0.2	0.4	0.6	0.8	1
$f(x)$	0.447	0.632	0.775	0.894	1

6. Fit a linear curve to the data $\{(x, y): (1, 1)\ (2, 5)\ (3, 11)\ (4, 8)\ (5, 14)\}$.

7. Fit a second degree parabola to the following data taking x as independent and y as dependent variable.

x	1	2	3	4	5	6	7	8	9
y	2	6	7	8	10	11	11	10	9

8. Fit a parabolic curve to the following data x as the independent variable.

x	1	2	3	4	5
y	6	17	34	57	86

9. Find the least squares approximation of second degree for the discrete data.

x	−2	−1	0	1	2
$f(x)$	15	1	1	3	19

10. Fit a parabola $y = a + bx + cx^2$ in least square sense to the data.

x	10	12	15	23	20
y	14	17	23	25	21

ANSWERS

1. $y = 5.799 - 0.514x$ **2.** $y = 1 + 1.9x$ **3.** $y = 8 - 0.5x$ **4.** $y = 0.72 + 1.33x$

5. $f(x) = 0.3392 + 0.684x$ **6.** $y = -\dfrac{9}{10} + \dfrac{29}{10}x$ **7.** $y = -1 + 3.55x - 0.27x^2$

8. $y = 1 + 2x + 3x^2$ **9.** $f(x) = \dfrac{0.37 + 35x + 155x^2}{35}$

10. $y = -8.89 + 3.03x - 0.07x^2$.

UNIT II

NUMERICAL ANALYSIS-II

In this unit, we shall discuss numerical differentiation, numerical integration and numerical solution of ordinary differential equation of first order.

The chapter first deals with differentiation of function is solved by first approximation with the help of interpolation formula and differentiating this formula as many times as required.

The chapter second deals with integration of function by Trapezoidal rule, Simpson's "1/3" rule and Simpson's "3/8" rule.

Chapter third deals with solution of ordinary differential equation of first order by Euler's method, Euler's modified method, Picard's method, Milne's method, and Runge-Kutta method.

CHAPTER 1

Numerical Differentiation

INTRODUCTION

Numerical differentiation is the process of obtaining the value of the derivative of a function from a set of numerical values of that function

1. If the argument are equally spaced.
   (*a*) We will use Newton-Gregory forward formula. If we desire to find the derivative of the function at a point near to beginning.
   (*b*) If we desire to find the derivative of the function at a point near to end then we will use Newton-Gregory backard formula.
   (*c*) If the derivative at a point is near the middle of the table we apply stirling difference formula.
2. In case the argument are unequally spaced then we should use Newton's divided differenc formula.

1.1. DERIVATIVES USING FORWARD DIFFERENCE FORMULA

We have Newton's Gregory forward difference formula

$$y = y_0 + u\,\Delta y_0 + \frac{u(u-1)}{2!}\Delta^2 y_0 + \frac{u(u-1)(u-2)}{3!}\Delta^3 y_0 + \ldots\ldots \qquad \ldots(1.1)$$

where $u = \dfrac{x-a}{h}$. $\qquad \ldots(1.2)$

Differentiating both sides of (1.1) w.r.t. x, we get

$$\frac{dy}{dx} = \frac{d}{dx}\left[y_0 + u\,\Delta y_0 + \frac{u(u-1)}{2!}\Delta^2 y_0 + \frac{u(u-1)(u-2)}{3!}\Delta^3 y_0 + \ldots..\right]$$

$$= \left[\Delta y_0 + \frac{(2u-1)}{2!}\Delta^2 y_0 + \frac{3u^2-6u+2}{3!}\Delta^3 y_0 + \ldots..\right]\frac{du}{dx} \qquad \ldots(1.3)$$

But by (1.2) $$\frac{du}{dx} = \frac{1}{h}$$

Putting in (1.3), we get

$$\frac{dy}{dx} = \frac{1}{h}\left[\Delta y_0 + \frac{2u-1}{2!}\Delta^2 y_0 + \frac{3u^2-6u+2}{3!}\Delta^3 y_0 + ...\right] \quad ...(1.4)$$

At $x = a \Rightarrow u = 0$ then from (1.4), we get

$$\left(\frac{dy}{dx}\right)_{x=a} = \frac{1}{h}\left[\Delta y_0 - \frac{1}{2}\Delta^2 y_0 + \frac{1}{3}\Delta^3 y_0 - \frac{1}{4}\Delta^4 y_0 + ...\right] \quad ...(1.5)$$

Now differentiating both sides (1.4) w.r.t. x, we get

$$\frac{d^2y}{dx^2} = \frac{1}{h}\left[\Delta^2 y_0 + \frac{6u-6}{3!}\Delta^3 y_0 + \frac{12u^2-36u+22}{4!}\Delta^4 y_0 + ...\right]\frac{du}{dx}$$

$$= \frac{1}{h^2}\left[\Delta^2 y_0 + \frac{6u-6}{3!}\Delta^3 y_0 + \frac{12u^2-36u+22}{4!}\Delta^4 y_0 .....\right]$$

$$\left[\because \frac{du}{dx} = \frac{1}{h}\right] \quad ...(1.6)$$

At $x = a$, $u = 0$ then from (1.6), we get

$$\left(\frac{d^2y}{dx^2}\right)_{x=a} = \frac{1}{h^2}\left[\Delta^2 y_0 - \Delta^3 y_0 + \frac{11}{12}\Delta^4 y_0 .....\right]$$

Proceeding in the same way, we get successive differentiations at the required points as

$$\left(\frac{d^3y}{dx^3}\right)_{x=a} = \frac{1}{h^3}\left[\Delta^3 y_0 - \frac{3}{2}\Delta^4 y_0 + ...\right]$$

$$\vdots \qquad\qquad \vdots$$

1.2. DERIVATIVES USING BACKWARD DIFFERENCE FORMULA

We have Newton's Gregory backward difference formula

$$y = y_n + u\,\nabla y_n + \frac{u(u+1)}{2!}\nabla^2 y_n + \frac{u(u+1)(u+2)}{3!}\nabla^3 y_n + ... \quad ...(1.7)$$

where $$u = \frac{x - x_n}{h}. \quad ...(1.8)$$

Differentiating both sides of (1.7) w.r.t. x, we get

$$\frac{dy}{dx} = \frac{d}{dx}\left[y_n + u\nabla y_n + \frac{u(u+1)}{2!}\nabla^2 y_n + \frac{u(u+1)(u+2)}{3!}\nabla^3 y_n + ...\right]$$

$$= \left[\nabla y_n + \frac{2u+1}{2!}\nabla^2 y_n + \frac{3u^2+6u+2}{3!}\nabla^3 y_n + ...\right]\frac{du}{dx} \qquad ...(1.9)$$

But by (1.8) $\frac{du}{dx} = \frac{1}{h}$ putting in (1.9), we get

$$\frac{dy}{dx} = \frac{1}{h}\left[\nabla y_n + \frac{2u+1}{2!}\nabla^2 y_n + \frac{3u^2+6u+2}{3!}\nabla^3 y_n + ...\right] \qquad ...(1.10)$$

At $x = x_n \Rightarrow u = 0$ then from (1.10), we get

$$\left(\frac{dy}{dx}\right)_{x=x_n} = \frac{1}{h}\left[\nabla y_n + \frac{1}{2!}\nabla^2 y_n + \frac{1}{3!}\nabla^3 y_n + ...\right] \qquad ...(1.11)$$

Now differentiating both sides of (1.10) w.r.t. x, we get

$$\frac{d^2y}{dx^2} = \frac{1}{h}\left[\nabla^2 y_n + \frac{6u+6}{3!}\nabla^3 y_n + \frac{12u^2+36u+22}{4!}\nabla^4 y_n + ...\right]\frac{du}{dx}$$

$$= \frac{1}{h^2}\left[\nabla^2 y_n + \frac{6u+6}{3!}\nabla^3 y_n + \frac{12u^2+36u+22}{4!}\nabla^4 y_n + ...\right] \qquad ...(1.12)$$

At $x = x_n$, $u = 0$ then from (1.12), we get

$$\left(\frac{d^2y}{dx^2}\right)_{x=x_n} = \frac{1}{h^2}\left[\nabla^2 y_n + \nabla^3 y_n + \frac{11}{12}\nabla^4 y_n + ...\right] \qquad ...(1.13)$$

Proceeding in the same way, we get successive differentiation at the required point

$$\left(\frac{d^3y}{dx^3}\right)_{x=x_n} = \frac{1}{h^3}\left[\nabla^3 y_n + \frac{3}{2}\nabla^4 y_n + ...\right].$$

1.3. DERIVATIVES USING STIRLING DIFFERENCE FORMULA

We have Stirling difference formula

$$y = y_0 + u\left(\frac{\Delta y_0 + \Delta y_1}{2}\right) + \frac{u^2}{2!}\Delta^2 y_{-1} + \frac{u(u^2-1)}{3!}\left(\frac{\Delta^3 y_{-1} + \Delta^3 y_{-2}}{2}\right)$$

$$+ \frac{u^2(u^2-1)}{4!}\Delta^4 y_{-2} + ... \qquad ...(1.14)$$

where $$u = \frac{x - x_0}{h} \qquad ...(1.15)$$

Now differentiating both sides of (1.14) w.r.t. x, we get

$$\frac{dy}{dx} = \frac{d}{dx}\left[y_0 + u\left(\frac{\Delta y_0 + \Delta y_{-1}}{2}\right) + \frac{u^2}{2!}\Delta^2 y_{-1} + \frac{u(u^2-1)}{3!}\left(\frac{\Delta^3 y_{-1} + \Delta^3 y_{-2}}{2}\right) + \frac{u^2(u^2-1)}{4!}\Delta^4 y_{-2} + \ldots\ldots\right]$$

$$= \left[\left(\frac{\Delta y_0 + \Delta y_{-1}}{2}\right) + u\Delta^2 y_{-1} + \frac{3u^2-1}{3!}\left(\frac{\Delta^3 y_{-1} + \Delta^3 y_{-2}}{2}\right) + \left(\frac{4u^3-2u}{4!}\right)\Delta^4 y_{-2} \ldots\ldots\right]\frac{du}{dx}$$

But $\frac{du}{dx} = \frac{1}{h}$ by (1.15) put in above, we get

$$\frac{dy}{dx} = \frac{1}{h}\left[\frac{\Delta y_0 + \Delta y_{-1}}{2} + u\Delta^2 y_{-1} + \frac{3u^2-1}{3!}\left(\frac{\Delta^3 y_{-1} + \Delta^3 y_{-2}}{2}\right) + \frac{4u^3-2u}{4!}\Delta^4 y_{-2} \ldots\ldots\right] \quad \ldots(1.16)$$

At $x = x_0$, $u = 0$ then from (1.16), we get

$$\left(\frac{dy}{dx}\right)_{x=x_0} = \frac{1}{h}\left[\frac{\Delta y_0 + \Delta y_{-1}}{2} - \frac{1}{6}\left(\frac{\Delta^2 y_{-1} + \Delta^3 y_{-2}}{2}\right) + \ldots\ldots\right]$$

Again differentiating (1.16) w.r.t. x on both sides, we get

$$\frac{d^2y}{dx^2} = \frac{1}{h}\left[\Delta^2 y_{-1} + \frac{6u}{3!}\left(\frac{\Delta^3 y_{-1} + \Delta^3 y_{-2}}{2}\right) + \left(\frac{12u^2-2}{4!}\right)\Delta^4 y_{-2} + \ldots\ldots\right]\frac{du}{dx}$$

$$= \frac{1}{h^2}\left[\Delta^2 y_{-1} + \frac{6u}{3!}\left(\frac{\Delta^3 y_{-1} + \Delta^3 y_{-2}}{2}\right) + \left(\frac{12u^2-2}{4!}\right)\Delta^4 y_{-2} + \ldots\ldots\right] \quad \left[\because \frac{du}{dx} = \frac{1}{h}\right] \quad \ldots(1.17)$$

At $x = x_0$, $u = 0$ then from (1.17), we get

$$\left(\frac{d^2u}{dx^2}\right)_{x=x_0} = \frac{1}{h^2}\left[\Delta^2 y_{-1} - \frac{1}{12}\Delta^4 y_{-2} + \ldots\ldots\right]$$

Proceeding in the same way, we get successive differentiation at the required point.

1.4. DERIVATIVES USING NEWTON'S DIVIDED DIFFERENCE FORMULA

We have Newton's divided difference formula

$$y = f(x_0) + (x - x_0)\,\Delta f(x_0) + (x - x_0)(x - x_1)\,\Delta^2 f(x_0) + (x - x_0)(x - x_1)(x - x_2)\,\Delta^3 f(x_0) + \ldots \quad \ldots(1.18)$$

Now, differentiating (1.18) w.r.t. x, we get

$$\frac{dy}{dx} = \Delta f(x_0) + [(x - x_1) + (x - x_0)]\ \Delta^2 f(x_0) + [(x - x_1)(x - x_2) + (x - x_0)(x - x_2) + (x - x_0)(x - x_1)]\ \Delta^3 f(x_0) \ldots\ldots \quad \ldots(1.19)$$

Putting $x = a$ in (1.18) we get value of first derivative at $x = a$

Again differentiating (1.19) w.r.t. x, we get

$$\frac{d^2y}{dx^2} = 2\Delta^2 f(x_0) + [2(x - x_0) + 2(x - x_1) + 2(x - x_2)]\ \Delta^2 f(x_0) + \ldots \quad \ldots(1.20)$$

Putting $x = a$ in (1.20) we get value of second derivative at $x = a$

Note 1. If we want to determine the value of the derivatives of the function near the beginning of arguments, we use Newton's forward formula.

2. If derivative required near the end of arguments, we use Newton's backward formula.

3. If derivative required at the middle of the given arguments, we apply and central difference formula.

4. And we use Newton's divided difference formula when argument are not equally spaced.

SOLVED EXAMPLES

Example 1. *Find* $\frac{dy}{dx}$ *and* $\frac{d^2y}{dx^2}$ *of* $y = x^{1/3}$ *at* $x = 50$ *from the following table:*

x	50	51	52	53	54	55	56
$y = x^{1/3}$	3.6840	3.7084	3.7325	3.7563	3.7798	3.8030	3.8259

Solution. Since $x = 50$ lies near the beginning of the table therefore in this case we shall use Newton's forward formula. The difference table is as below:

x	$y = x^{1/3}$	Δy	$\Delta^2 y$	$\Delta^3 y$
50	3.6840			
		0.0244		
51	3.7084		– 0.0003	
		0.0241		0
52	3.7325		– 0.0003	
		0.0238		0
53	3.7563		– 0.0003	
		0.0235		0
54	3.7798		– 0.0003	
		0.0232		0
55	3.8030		– 0.0003	
		0.0229		
56	3.8259			

Here $a = 50$, $h = 1$ then

$$\left(\frac{dy}{dx}\right)_{x=a} = \frac{1}{h}\left[\Delta y_0 - \frac{1}{2}\Delta^2 y_0 + \frac{1}{3}\Delta^3 y_0 \ldots\ldots\right]$$

$$\left(\frac{dy}{dx}\right)_{x=50} = \frac{1}{1}\left[0.0244 - \frac{1}{2}(-0.0003) + \frac{1}{3}(0)\right]$$

$$= 0.0244 + 0.00015 = 0.02455. \quad \textbf{Ans.}$$

and
$$\left(\frac{d^2y}{dx^2}\right)_{x=a} = \frac{1}{h^2}\left[\Delta^2 y_0 - \Delta^3 y_0 + \ldots\ldots\right]$$

$$\left(\frac{d^2y}{dx^2}\right)_{x=50} = \frac{1}{1^2}\left[(-0.0003)\right]$$

$$\left(\frac{d^2y}{dx^2}\right)_{x=50} = -0.0003. \quad \textbf{Ans.}$$

Example 2. *Find f ′(1.5) and f ″(1.5) from the following table:*

x	*1.5*	*2.0*	*2.5*	*3.0*	*3.5*	*4.0*
f(x)	*3.375*	*7.000*	*13.625*	*24.000*	*38.875*	*59.000*

Solution. Since $x = 1.5$ lies near the beginning of the table therefore in this case we shall use Newton's forward formula. The difference table is

x	$f(x)$	$\Delta f(x)$	$\Delta^2 f(x)$	$\Delta^3 f(x)$	$\Delta^4 f(x)$
1.5	3.375				
		3.625			
2.0	7.000		3.000		
		6.625		.75	
2.5	13.625		3.75		0
		10.375		.75	
3.0	24.000		4.5		0
		14.875		.75	
3.5	38.875		5.25		
		20.125			
4.0	59.000				

Here $a = 1.5$ and $h = 0.5$ then

$$\left(\frac{dy}{dx}\right)_{x=a} = \frac{1}{h}\left[\Delta y_0 - \frac{1}{2}\Delta^2 y_0 + \frac{1}{3}\Delta^3 y_0 \ldots\ldots\right]$$

or
$$\left(\frac{dy}{dx}\right)_{x=1.5} = \frac{1}{0.5}\left[3.625 - \frac{1}{2}\times 3.000 + \frac{1}{3}\times(0.75) - 0 \ldots\ldots\right]$$

or $$f'(1.5) = \frac{1}{0.5}[3.625 - 1.5 + 0.25] = \frac{1}{0.5} \times 2.375 = 4.75$$

and $$\left(\frac{d^2y}{dx^2}\right)_{x=a} = \frac{1}{h^2}\left[\Delta^2 y_0 - \Delta^3 y_0 + \frac{11}{12}\Delta^4 y_0 \ldots\ldots\right]$$

$$\left(\frac{d^2y}{dx^2}\right)_{x=1.5} = \frac{1}{(0.5)^2}\left[3.000 - 0.75 + \frac{11}{12}.0\ldots\ldots\right]$$

$$f''(1.5) = \frac{1}{0.25}[2.25] = 9$$

Hence $f'(1.5) = 4.75$ and $f''(1.5) = 9$ **Ans.**

Example 3. *Find the first and second derivatives of the functions tabulated below at the point 1.1*

x	1.0	1.2	1.4	1.6	1.8	2.0
$f(x)$	0	0.1280	0.5440	1.2960	2.4320	4.0000

Solution. Since $x = 1.1$ lies near the beginning of the table therefore in this case we shall use Newton's Gregory forward formula. The difference table is as below:

x	$f(x)$	$\Delta f(x)$	$\Delta^2 f(x)$	$\Delta^3 f(x)$	$\Delta^4 f(x)$
1.0	0				
		0.1280			
1.2	0.1280		0.2880		
		0.4160		0.0480	
1.4	0.5440		0.3360		0
		0.7520		0.0480	
1.6	1.2960		0.3840		0
		1.1360		0.0480	
1.8	2.4320		0.4320		
		1.5680			
2.0	4.0000				

Since 1.1 lies between given argument so we have

$$f(x) = f(x_0) + u\Delta f(x_0) + \frac{u(u-1)}{2!}\Delta^2 f(x_0) + \frac{u(u-1)(u-2)}{3!}\Delta^3 f(x_0) + \ldots\ldots \quad \ldots(i)$$

where $$u = \frac{x - x_0}{h} = \frac{x-1}{0.2} = 5(x-1) \quad \ldots(ii)$$

Differentiating (i) w.r.t. x on both sides, we get

$$f'(x) = \left[\Delta f(x_0) + \frac{2u-1}{2!}\Delta^2 f(x_0) + \frac{3u^2 - 6u + 2}{3!}\Delta^3 f(x_0)\ldots\ldots\right]\frac{du}{dx}$$

But by (*ii*) $\frac{du}{dx} = 5$

$$\therefore \quad f'(x) = 5\left[\Delta f(x_0) + \frac{2u-1}{2!}\Delta^2 f(x_0) + \frac{3u^2-6u+2}{3!}\Delta^3 f(x_0)\ldots\ldots\right] \quad \ldots(iii)$$

At $x = 1.1$, $u = 5(1.1 - 1) = 0.5$

$$\therefore \quad f'(1.1) = 5\left[0.128 + \frac{2(0.5)-1}{2}(0.2880) + \frac{3(0.5)^2-6(0.5)+2}{6}(0.048)\right]$$

$$= 5[0.128 + 0 - 0.002] = 0.63$$

$$f'(1.1) = 0.63$$

Differentiating (*iii*) again w.r.t. x, we get

$$f''(x) = 5\left[\Delta^2 f(x_0) + \frac{6u-6}{3!}\Delta^3 f(x_0)\ldots\ldots\right]\frac{du}{dx}$$

$$= 25\,[\Delta^2 f(x_0) + (u-1)\,\Delta^3 f(x_0) + \ldots\ldots] \qquad \left[\because \frac{du}{dx} = 5\right]$$

At $x = 1.1 \Rightarrow u = 5(1.1 - 1) = 0.5$

$$\therefore \quad f''(1.1) = 25\,[0.2880 + (0.5-1)(0.0480)] = 25[0.2880 - 0.024]$$

$$f''(1.1) = 6.6$$

Hence $f'(1.1) = 0.63$ and $f''(1.1) = 6.6$ **Ans.**

Example 4. *Given that*

$\theta°$	*0*	*10*	*20*	*30*	*40*
sin $\theta°$	*0.000*	*0.1736*	*0.3420*	*0.5000*	*0.6428*

Find cos θ *when* $\theta = 10°$.

Solution. Since $\theta = 10°$ lies near the beginning of the table therefore in this case we shall use Newton's forward formula. The difference table is as below:

$\theta°$	sin $\theta°$	$\Delta f(x)$	$\Delta^2 f(x)$	$\Delta^3 f(x)$	
0	0.000				
		0.1736			
10	0.1736		– 0.0052		
		0.1684		– 0.0052	
20	0.3420		– 0.0104		0.0004
		0.1580		– 0.0048	
30	0.5000		– 0.0152		
		0.1428			
40	0.6428				

Here $f(x) = \sin\theta$, $a = 10$, $h = 10° = 0.1745$ radian

$$\left(\frac{dy}{d\theta}\right)_{\theta=10°} = \frac{1}{h}\left[\Delta y_0 - \frac{1}{2}\Delta^2 y_0 + \frac{1}{3}\Delta^3 y_0 - \frac{1}{4}\Delta^4 y_0 \ldots\ldots\right]$$

$$(\cos\theta)_{\theta=10°} = \frac{1}{10}\left[0.1684 + \frac{1}{2}(0.0104) + \frac{1}{3}(-0.0048)\right]$$

$$= \frac{1}{0.1745} \times 0.1720$$

$$= 0.9856 \quad \textbf{Ans.}$$

Example 5. *From the following table find the values of* $\frac{dy}{dx}$ *and* $\frac{d^2y}{dx^2}$ *at* $x = 2.03$

x	1.96	1.98	2.00	2.02	2.04
y	0.7825	0.7739	0.7651	0.7563	0.7473

Solution.

x	y	∇y	$\nabla^2 y$	$\nabla^3 y$	$\nabla^4 y$
1.96	0.7825				
		– 0.0086			
1.98	0.7739		– 0.0002		
		– 0.0088		0.0002	
2.00	0.7651		0		– 0.0004
		– 0.0088		– 0.0002	
2.02	0.7563		– 0.0002		
		– 0.0090			
2.04	0.7473				

Here $x_n = 2.04$, $h = 0.02$, $x = 2.03$

$$\Rightarrow \qquad u = \frac{x - x_n}{h} = \frac{2.03 - 2.04}{0.02} = -\frac{0.01}{0.02} = -\frac{1}{2}$$

Now by derivative of Newton's backward formula, we have

$$\left(\frac{dy}{dx}\right) = \frac{1}{h}\left[\nabla y_n + \frac{2u+1}{2!}\nabla^2 y_n + \frac{3u^2+6u+2}{3!}\nabla^2 y_n + \frac{4u^3+18u^2+22u+6}{4!}\nabla^4 y_n + \ldots\ldots\right] \qquad \ldots(i)$$

$$\left(\frac{dy}{dx}\right)_{x=2.03} = \frac{1}{0.02}\left[-0.0090 + \frac{2\left(-\frac{1}{2}\right)+1}{2!}(-0.0002) + \frac{3\left(-\frac{1}{2}\right)^2+6\left(-\frac{1}{2}\right)+2}{3!}(-0.0002)\right.$$

$$\left. + \frac{4\left(-\frac{1}{2}\right)^3+18\left(-\frac{1}{2}\right)^2+22\left(-\frac{1}{2}\right)+6}{4!}(-0.0004)\right]$$

$$= \frac{1}{0.02}[-0.0090 + 0 + 0.000008 + 0.000017)$$

$$= -0.44875$$

Again differentiating (i), w.r.t. x, we get

$$\left(\frac{d^2y}{dx^2}\right) = \frac{1}{h^2}\left[\nabla^2 y_n + \frac{6u+6}{3!}\nabla^3 y_n + \frac{12u^2+36u+22}{4!}\nabla^4 y_n + \ldots\ldots\right]$$

$$\left(\frac{d^2y}{dx^2}\right)_{x=2.03} = \frac{1}{(0.02)^2}\left[-0.0002 + \frac{6\left(-\frac{1}{2}\right)+6}{6}(-0.0002)\right.$$

$$\left. + \frac{12\left(-\frac{1}{2}\right)^2+36\left(-\frac{1}{2}\right)+22}{24}\times(-0.0004)\right]$$

$$= \frac{1}{0.0004}[-0.0002 - 0.0001 - 0.00012]$$

$$\left(\frac{d^2y}{dx^2}\right)_{x=2.03} = -1.05$$

Hence $\left(\frac{dy}{dx}\right)_{x=2.03} = -0.44875$ and $\left(\frac{d^2y}{dx^2}\right)_{x=2.03} = -1.05$. **Ans.**

Example 6. *Find $f'(93)$ from the following table:*

x	60	75	90	105	120
$f(x)$	28.2	38.2	43.2	40.9	37.2

Solution. Since 93 lies near the central point of the table therefore in this case we shall use Stirling formula. The difference table is given by

u	x	$f(x)$	$\Delta f(x)$	$\Delta^2 f(x)$	$\Delta^3 f(x)$	$\Delta^4 f(x)$
– 2	60	28.2				
			10.0			
– 1	75	38.2		– 5		
			5		– 2.3	
0	90	43.2		– 7.3		8.2
			– 2.3		5.9	
+ 1	105	40.9		– 1.4		
			– 3.7			
2	120	37.2				

Here $x_0 = 90, x = 93, h = 15$

$$\therefore \qquad u = \frac{93-90}{15} = \frac{3}{15} = \frac{1}{5} = 0.2$$

Putting these values in Stirling formula for first derivative, we get

$$\frac{dy}{dx} = \frac{1}{h}\left[\frac{\Delta y_0 + \Delta y_{-1}}{2} + u\Delta^2 y_{-1} + \frac{3u^2-1}{3!}\left(\frac{\Delta^3 y_{-1} + \Delta^3 y_{-2}}{2}\right) + \frac{4u^3 - 2u}{4!}\Delta^4 y_{-2} \ldots\ldots\right]$$

$$\left(\frac{dy}{dx}\right)_{x=93} = \frac{1}{15}\left[\frac{5+(-2.3)}{2} + (0.2)\times(-7.3) + \frac{3(0.2)^2-1}{3!}\left(\frac{(-2.3)+(5.9)}{2} + \frac{4(0.2)^3 - 2(0.2)}{4!}(8.2)\right)\right]$$

$$f'(93) = \frac{1}{15}\left[\frac{2.7}{2} + (-1.46) - \frac{3.168}{3!\times 2} - \frac{3.0716}{4!}\right]$$

$$= \frac{1}{15}(1.35 - 1.46 - 0.26400 - 0.1257)$$

$f'(93) = -0.3331.$ **Ans.**

Example 7. *Find $f'(6)$ from the following table:*

x	0	1	3	4	5	7	9
$f(x)$	150	108	0	– 54	– 100	– 144	– 84

Solution. In this case the values of the arguments are not equally spaced. So we shall use Newton's divided difference formula. The divided difference table is given below.

x	$f(x)$	$\Delta f(x)$	$\Delta^2 f(x)$	$\Delta^3 f(x)$	$\Delta^4 f(x)$
0	150				
		$\frac{108-150}{1-0} = -42$			
1	108		$\frac{-54+42}{3-0} = -4$		
				$\frac{0+4}{4-0} = 1$	
		$\frac{0-108}{3-1} = -54$	$\frac{-54+54}{4-1} = 0$		$\frac{1-1}{5-0} = 0$
3	0			$\frac{4-0}{5-1} = 1$	
		$\frac{-54-0}{4-3} = -54$			
4	– 54		$\frac{-46+54}{5-3} = +4$		$\frac{1-1}{7-1} = 0$
		$\frac{-100+54}{5-4} = -46$		$\frac{8-4}{7-3} = 1$	
5	– 100		$\frac{-22+46}{7-4} = 8$		$\frac{1-1}{9-3} = 0$
		$\frac{-144+100}{7-5} = -22$		$\frac{13-8}{9-4} = 1$	
7	– 144		$\frac{30+22}{9-5} = 13$		
		$\frac{-84+144}{9-7} = 30$			
9	– 84				

By Newton's divided difference formula, we have

$$f(x) = f(x_0) + (x - x_0)\Delta f(x_0) + (x - x_0)(x - x_1)\ \Delta^2 f(x) + (x - x_0)(x - x_1)(x - x_2)\Delta^3 f(x) + \ldots\ldots \quad \ldots(i)$$

Differentiating (i) w.r.t. x, we get

$$f'(x) = \Delta f(x_0) + [(x - x_1) + (x - x_0)]\ \Delta^2 f(x) + [(x - x_1)(x - x_2) + (x - x_0)(x - x_2) + (x - x_0)(x - x_1)] \times \Delta^3 f(x) + \ldots\ldots$$

Here $x = 6, x_0 = 0, x_1 = 1, x_2 = 3, x_3 = 4$ ...... putting in above, we get

$$f'(6) = -42 + [(6 - 1) + (6 - 0)](-4) + [(6 - 1)(6 - 3) + (6 - 0)(6 - 3) + (6 - 0)(6 - 1)]1$$

$$= -42 + 11 \times (-4) + (15 + 18 + 30) \times 1$$

$$= -42 - 44 + 63 = -23$$

$f'(6) = -23$. **Ans.**

Example 8. *From the following table, find $f'(10)$.*

x	3	5	11	27	34
$f(x)$	− 13	23	899	17315	35606

Solution. In this case the values of the argument are not equally spaced. So we shall use Newton's divided difference formula. The divided difference table is given below.

x	$f(x)$	$\Delta f(x)$	$\Delta^2 f(x)$	$\Delta^3 f(x)$	$\Delta^4 f(x)$
3	− 13	$\frac{23-(-13)}{5-3}=18$			
			$\frac{146-18}{11-3}=16$		
5	23	$\frac{899-23}{11-5}=146$		$\frac{40-16}{27-3}=1$	
			$\frac{1026-146}{27-5}=40$		$\frac{1-1}{34-3}=0$
11	899	$\frac{17315-899}{27-11}=1026$		$\frac{69-40}{34-5}=1$	
			$\frac{2613-1026}{34-11}=69$		
27	17315	$\frac{35606-17315}{34-27}=2613$			
34	35606				

By Newton's divided difference formula, we have

$$f(x) = f(x_0) + (x - x_0)\, \Delta f(x_0) + (x - x_0)(x - x_1)\, \Delta^2 f(x_0) + (x - x_0)(x - x_1)(x - x_2)\, \Delta^2 f(x_0) + \ldots\ldots \quad \ldots(i)$$

Differentiating (i) w.r.t. x, we get

$$f'(x) = \Delta f(x_0) + ((x - x_1) + (x - x_0)\, \Delta^2 f(x_0) + [(x - x_1)(x - x_2) + (x - x_0)(x - x_2) + (x - x_0)(x - x_1)] \times \Delta^2 f(x_0) + \ldots\ldots$$

Here $x = 10, x_0 = 3, x_1 = 5, x_2 = 11, x_3 = 27, x_4 = 34.$

$$f'(10) = 18 + ((10 - 5) + (10 - 3))\, 16 + [(10 - 5)(10 - 11) + [(10 - 3)(10 - 11) + (10 - 3)(10 - 5)] \times 1$$

$$= 18 + 12 \times 16 + [(-5) + (-7) + 35)] \times 1$$

$$= 18 + 192 + 23$$

$f'(10) = 233.$ **Ans.**

EXERCISE 1.1

1. Find $f'(1)$ for $f(x) = \frac{1}{1+x^2}$ using the following data:

x	1.0	1.1	1.2	1.3	1.4
$f(x)$	0.2500	0.2268	0.2066	0.1890	0.1736

2. Find the first, second and third derivatives of the function tabulated below, at the point $x = 1.5$

x	1.5	2.0	2.5	3.0	3.5	4.0
$f(x)$	3.375	7.000	13.625	24.000	38.875	59.000

3. Find $y'(0)$ and $y''(0)$ from the following table:

x	0	1	2	3	4	5
y	4	8	15	7	6	2

4. Find the derivative of $f(x)$ at $x = 0.4$ from the following table:

x	0.1	0.2	0.3	0.4
$y = f(x)$	1.10517	1.22140	1.34986	1.49182

5. Find $f'(1.1)$ and $f''(1.1)$ from the following table:

x	1	1.1	1.2	1.3	1.4	1.5	1.6
y	7.989	8.403	8.781	9.129	9.451	9.750	10.031

6. Find $f'(1.5)$ from the following table:

x	0.0	0.5	1.0	1.5	2.0
$f(x)$	0.3989	0.3521	0.2420	0.1295	0.0540

7. Find $f'(0.6)$ and $f''(0.6)$ from the following table:

x	0.4	0.5	0.6	0.7	0.8
$f(x)$	1.5836	1.7974	2.0442	2.3275	2.6510

8. Find $f'(0.04)$ from the following table:

x	0.01	0.02	0.03	0.04	0.05	0.06
$f(x)$	0.1023	0.1047	0.1071	0.1096	0.1122	0.1148

9. Find $f'(7.50)$ from the following table:

x	7.47	7.48	7.49	7.50	7.51	7.52	7.53
$y = f(x)$	0.193	0.195	0.198	0.201	0.203	0.206	0.208

10. From the following table, find the first derivative at $x = 4$:

x	1	2	4	8	10
y	0	1	5	21	27

11. Find $f'(5)$ from the following table:

x	0	2	3	4	7	9
$f(x)$	4	26	58	112	466	922

ANSWERS

1. – 0.24833 **2.** 4.75, 9, 6 **3.** – 27.9 and 117.67 **4.** 1.49133

5. 3.9435 and – 3.545 **6.** – 0.1868 **7.** 2.6445 and 3.64833

8. 0.2561 **9.** 0.223 **10.** 2.8826 **11.** 84856

CHAPTER 2

Numerical Integration

INTRODUCTION

Numerical integration is the process of obtaining the value of a definite integral from a set of numerical values of the integrand. The process to finding the value of the definite integral $I = \int_a^b f(x)\, dx$ of a function of a single variable, is called as numerical quadrature. If we apply this for function of two variables it is called mechanical cubature.

The problem of numerical integration is solved by first approximating the function $f(x)$ by a interpolating polynomial and then integrating it between the desired limit.

Thus $$f(x) \approx P_n(x)$$

$$\int_a^b f(x)\, dx = \int_a^b P_n(x)\, dx.$$

2.1. A GENERAL QUADRATURE FORMULA FOR EQUALLY SPACED ARGUMENTS

Let $$I = \int_a^b f(x)\, dx$$

Further let $y = f(x)$, consider the values $y_0, y_1, y_2, \ldots\ldots, y_n$ for $x = x_0, x_0 + h, x_0 + 2h, \ldots\ldots, x_0 + nh$.

Let us divide the interval (a, b) into n sub-intervals of equal width h *i.e.*, $\frac{b-a}{n} = h$

Let $x_0 = a, x_1 = x_0 + h = a + h, x_2 = x_0 + 2h = a + 2h, \ldots\ldots, x_n = x_0 + nh = b$

Then $$I = \int_a^b f(x)\, dx = \int_{x_0}^{x_0+nh} f(x)\, dx \qquad \ldots(2.1)$$

Now by Newton's forward interpolation formula

$$y = f(x) = y_0 + u\,\Delta y_0 + \frac{u(u-1)}{2!}\Delta^2 y_0 + \frac{u(u-1)(u-2)}{3!}\Delta^3 y_0 + \ldots\ldots.$$

where $$u = \frac{x - x_0}{h}$$

$\therefore$ $$du = \frac{dx}{h} \;\Rightarrow\; dx = h\,du$$

Putting above relation in (2.1), we get

$$I = h\int_0^n \left[y_0 + u\Delta y_0 + \frac{u(u-1)}{2!}\Delta^2 y_0 + \frac{u(u-1)(u-2)}{3!}\Delta^3 y_0 \ldots\ldots. \right] du$$

$$= h\left[n y_0 + \frac{n^2}{2}\Delta y_0 + \left(\frac{n^3}{3} - \frac{n^2}{2}\right)\frac{\Delta^2 y_0}{2!} + \left(\frac{n^4}{4} - n^3 + n^2\right)\frac{\Delta^3 y_0}{3!} \ldots\ldots + \text{up to } (n+1) \text{ terms} \right]$$

$$\therefore \int_{x_0}^{x_0+nh} f(x)\,dx = h\left[n y_0 + \frac{n^2}{2}\Delta y_0 + \left(\frac{n^3}{3} - \frac{n^2}{2}\right)\frac{\Delta^2 y_0}{2!} + \left(\frac{n^4}{4} - n^3 + n^2\right)\frac{\Delta^3 y_0}{3!} \ldots\ldots + \text{up to } (n+1) \text{ terms} \right] \quad \ldots(2.2)$$

This is the general quadrature formula.

2.2. THE TRAPEZOIDAL RULE

Putting $n = 1$ in general quadrature formula (2.2) and neglecting all differences higher then first, we get

$$\int_{x_0}^{x_0+h} f(x)\,dx = h\left[y_0 + \frac{1}{2}\Delta y_0 \right] = h\left[y_0 + \frac{1}{2}(y_1 - y_0) \right] = h\frac{(y_0 + y_1)}{2}$$

Similarly $$\int_{x_0+h}^{x_0+2h} f(x)\,dx = h\frac{(y_1 + y_2)}{2}$$

$$\vdots \qquad\qquad \vdots$$

$$\int_{x_0+(n-1)h}^{x_0+nh} f(x)\,dx = h\frac{(y_{n-1} + y_n)}{2}$$

Adding these n integrals, we get

$$\int_{x_0}^{x_0+nh} f(x)\,dx = h\left[\frac{1}{2}(y_0 + y_n) + (y_1 + y_2 \ldots\ldots y_{n-1}) \right]$$

This rule is called the Trapezoidal rule.

2.3. SIMPSON'S ONE-THIRD RULE

Putting $n = 2$ in general quadrature formula (2.2) and neglecting all differences higher then second, we get

$$\int_{x_0}^{x_0+2h} f(x)\,dx = h\left[2y_0 + \frac{2^2}{2}\Delta y_0 + \left(\frac{2^3}{3} - \frac{2^2}{2}\right)\frac{\Delta^2 y_0}{2!}\right]$$

$$= h\left[2y_0 + 2(y_1 - y_0) + \frac{1}{3}(y_2 - 2y_1 + y_0)\right]$$

$$= \frac{h}{3}\,[y_0 + 4y_1 + y_2]$$

Similarly $\displaystyle\int_{x_0+2h}^{x_0+4h} f(x)\,dx = \frac{h}{3}\,[\,y_2 + 4y_3 + y_4\,]$

$\vdots \qquad\qquad \vdots$

$$\int_{x_0+(n-1)h}^{x_0+nh} f(x)\,dx = \frac{h}{3}\,(\,y_{n-2} + 4y_{n-1} + y_n)$$

Adding all these integrals, we get

$$\int_{x_0}^{x_0+nh} f(x)\,dx = \frac{h}{3}\,[y_0 + 2(y_2 + y_4 + y_6 \;......\; 4_{n-2}) + 4\,(y_1 + y_3 + y_5 \;......\; y_{n-1}) + y_n]$$

This is known as Simpson's $\frac{1}{3}$ rule.

2.4. SIMPSON'S THREE-EIGHTH RULE

Putting $n = 3$ in general quadrature formula (2.2) and neglecting all differences higher then third, we get

$$\int_{x_0}^{x_0+3h} f(x)\,dx = h\left[3y_0 + \frac{9}{2}\Delta y_0 + \left(\frac{3^3}{3} - \frac{3^2}{2}\right)\frac{\Delta^2 y_0}{2!} + \left(\frac{3^4}{4} - 3^3 + 3^2\right)\frac{\Delta^3 y_0}{3!}\right]$$

$$= h\left[3y_0 + \frac{9}{2}(y_1 - y_0) + \frac{9}{4}(y_2 - 2y_1 + y_0) + \frac{3}{8}(y_3 - 3y_2 + 3y_1 - y_0)\right]$$

$$= \frac{3h}{8}\,[y_0 + 3y_1 + 3y_2 + y_3]$$

Similarly $\displaystyle\int_{x_0+3h}^{x_0+6h} f(x)\,dx = \frac{3h}{8}\,[\,y_3 + 3y_4 + 3y_5 + y_6]$

$\vdots \qquad\qquad \vdots$

$$\int_{x_0+(n-3)h}^{x_0+nh} f(x)\,dx = \frac{3h}{8}[y_{n-3} + 3y_{n-2} + 3y_{n-1} + y_n]$$

Adding all these integrals, we get

$$\int_{x_0}^{x_0+nh} f(x)\,dx = \frac{3h}{8}[y_0 + 3(y_1 + y_2 + y_4 + y_5 + \ldots\ldots + y_{n-1}) + 2(y_3 + y_0 + \ldots\ldots + y_{n-3}) + y_n]$$

This formula is known as Simpson's three-eight's rule.

SOLVED EXAMPLES

Example 1. *Use Trapezoidal rule to evaluate* $\int_0^1 x^3\,dx$ *considering five sub-intervals.*

Solution. Divide the interval [0, 1] into five parts each of width $h = \frac{1-0}{5} = 0.2$ and compute the value of $y = x^3$ at each points of sub-interval. These values are given below:

x	0	0.2	0.4	0.6	0.8	1
$y = x^3$	0	0.008	0.064	0.216	0.512	1

By Trapezoidal rule, we have

$$\int_0^1 x^3\,dx = \frac{h}{2}[y_0 + 2(y_1 + y_2 + y_3 + y_4) + y_5]$$

$$= \frac{0.2}{2}[0 + 2(0.008 + 0.064 + 0.216 + 0.512) + 1]$$

$$= \frac{0.2}{2}[2 \times 0.8 + 1] = 0.26.$$ **Ans.**

Example 2. *Evaluate* $\int_0^{\pi/2} e^{\sin x}\,dx$ *correct to four decimal places by Simpson's one-third and three-eighth rule, dividing the interval* $\left(0, \frac{\pi}{2}\right)$ *into three equal parts.*

Solution. Divide the interval $\left(0, \frac{\pi}{2}\right)$ into three parts each of width $h = \frac{(\pi/2) - 0}{3} = \frac{\pi}{6}$ and compute the value of $y = e^{\sin x}$ at each point of sub-interval. These values are given below:

x	0	$\frac{\pi}{6}$	$\frac{\pi}{3}$	$\frac{\pi}{2}$
$y = e^{\sin x}$	1	1.64872	2.45960	2.71828

By Simpson's one-third rule, we have $\int_0^{\pi/2} e^{\sin x}\, dx = \frac{h}{3}\,[\,y_0 + 4(\,y_1) + 2(\,y_2) + y_3]$

$$= \frac{\pi/6}{3}\,[1 + 4 \times 1.64872 + 2 \times 2.45960 + 2.71828]$$

$$= \frac{\pi}{18}[15.23236] = 2.6596. \quad \textbf{Ans.}$$

By Simpson's three-eighth rule, we have

$$\int_0^{\pi/2} e^{\sin x}\, dx = \frac{3h}{8}\,[y_0 + 3(y_1 + y_2) + y_3]$$

$$= \frac{3}{8}\frac{\pi}{6}\,[1 + 3\,(1.\,64872 + 2.45960) + 2.71828]$$

$$= \frac{3}{8}\frac{\pi}{6}\,(16.04324) = 3.1513. \quad \textbf{Ans.}$$

Example 3. *Find the value of* $\log_e 2$ *from* $\int_0^1 \frac{x^2}{1+x^3}\, dx$, *using Simpson's one-third rule, by dividing the range into four equal parts.*

Solution. Divide the interval [0, 1] into four equal parts each of width $h = \frac{1-0}{4} = 0.25$ and compute the value of $f(x) = \frac{x^2}{1+x^3}$ at each point of sub-interval. These values are given below:

x	x^2	x^3	$1 + x^3$	$f(x) = x^2/1 + x^3$
0	0	0	1	0
0.25	0.0625	0.015625	1.015625	0.061538
0.50	0.25	0.125	1.125	0.222222
0.75	0.5625	0.421875	1.421875	0.395604
1	1	1	2	0.5

By Simpson's one-third rule, we have

$$\int_0^1 \frac{x^2}{1+x^3}\, dx = \frac{h}{3}\,[y_0 + 4\,(y_1 + y_3) + 2y_2 + y_4]$$

$$= \frac{0.25}{3} [0 + 4(0.061538 + 0.395604) + 2(0.222222) + 0.5]$$

$$= \frac{0.25}{3} [2.773015] = 0.231084. \quad \textbf{Ans.}$$

Again $$\int_0^1 \frac{x^2}{1+x^3} dx = \frac{1}{3}\int_0^1 \frac{3x^2}{1+x^3} dx = \frac{1}{3}\left[\log(1+x^3)\right]_0^1 = \frac{1}{3}\log 2$$

$$= \frac{1}{3} \times (0.693147) = 0.231049$$

Therefore the error $= 0.231084 - 0.231049 = 0.000034.$ **Ans.**

Example 4. *Evaluate* $\int_0^6 \frac{dx}{1+x^2}$ *by using Trapezoidal, Simpson's one-third and three-eighth rule.*

Solution. Divide the interval (0, 6) into six parts each of width $h = 1$ and compute the value of $y = \frac{1}{1+x^2}$ at each point of sub-interval. These values are given below

x	$y = \frac{1}{1+x^2}$
0	1
1	0.5
2	0.2
3	0.1
4	0.0588
5	0.0385
6	0.027

By Trapezoidal rule, we have

$$\int_0^6 \frac{dx}{1+x^2} = \frac{h}{2} [y_0 + 2(y_1 + y_2 + y_3 + y_4 + y_5) + y_6]$$

$$= \frac{1}{2} [1 + 2\,(0.5 + 0.2 + 0.1 + 0.0588 + 0.0385) + 0.027]$$

$$= \frac{1}{2} [2.8216] = 1.4108. \quad \textbf{Ans.}$$

By Simpson's one-third rule, we have

$$\int_0^6 \frac{dx}{1+x^2} = \frac{h}{3}\,[y_0 + 4\,(y_1 + y_3 + y_5) + 2(y_2 + y_4) + y_6]$$

$$= \frac{1}{3}\,[\,1 + 4(0.5 + 0.1 + 0.0385) + 2(0.2 + 0.0588) + 0.027]$$

$$= \frac{1}{3}\,[1.027 + 4 \times 0.6385 + 2 \times 0.2588]$$

$$= \frac{1}{3}\,[4.0986] = 1.\,3662. \quad \textbf{Ans.}$$

By Simpson's three-eight rule, we have

$$\int_0^6 \frac{dx}{1+x^2} = \frac{3h}{8}\,[\,y_0 + 3(y_1 + y_2 + y_4 + y_5) + 2(y_3) + y_6]$$

$$= \frac{3}{8}\,[1 + 3\,(0.5 + 0.2 + 0.0588 + 0.0385) + 2(0.1) + 0.027]$$

$$= \frac{3}{8}(3.6189) = 1.3570. \quad \textbf{Ans.}$$

Example 5. *Calculate the approximate value of* $\int_{-3}^{3} x^4\, dx$ *by using Trapezoidal rule, Simpson's one-third and three eight rule, by dividing the range in six equal parts.*

Solution. Divide the range (– 3, 3) into six equal parts each of width $h = \frac{3-(-3)}{6} = 1$ and compute the value of $y = x^4$ at each point of sub-interval. These values are given as below:

x	– 3	– 2	– 1	0	1	2	3
$y = x^4$	81	16	1	0	1	16	81

By Trapezoidal rule, we have

$$\int_{-3}^{3} x^4\, dx = \frac{h}{2}\,[y_0 + 2(y_1 + y_2 + y_3 + y_4 + y_5) + y_6]$$

$$= \frac{1}{2}\,[81 + 2\,(16 + 1 + 0 + 1 + 16) + 81]$$

$$= \frac{1}{2}\,[\,162 + 68] = 115. \quad \textbf{Ans.}$$

By Simpson's one-third rule, we have

$$\int_{-3}^{3} x^4\,dx = \frac{h}{3}\,[y_0 + 4(y_1 + y_3 + y_5) + 2(y_2 + y_4) + y_6]$$

$$= \frac{1}{3}[81 + 4(16 + 0 + 16) + 2(1 + 1) + 81]$$

$$= \frac{1}{3}[162 + 4 \times 32 + 2 \times 2] = \frac{1}{3}(294) = 98. \quad \textbf{Ans.}$$

By Simpson's three-eighth rule, we have

$$\int_{-3}^{3} x^4\,dx = \frac{3h}{8}[y_0 + 3(y_1 + y_2 + y_4 + y_5) + 2(y_3) + y_6]$$

$$= \frac{3}{8}[81 + 3(16 + 1 + 1 + 16) + 2.0 + 81]$$

$$= \frac{3}{8}[162 + 3 \times 34] = 99$$

But the exact value of

$$\int_{-3}^{3} x^4\,dx = 2\int_{0}^{3} x^4\,dx = 2\left[\frac{x^5}{5}\right]_0^3 = \frac{2}{5}(3)^5$$

$$= \frac{2}{5} \times 243 = 97.2. \quad \textbf{Ans.}$$

Example 6. *Use Simpson's rule dividing the range into ten equal parts, to show that*

$$\int_0^1 \frac{\log(1 + x^3)}{1 + x^2} = 0.1730.$$

Solution. Dividing the interval (0, 1) into ten equal parts of width $h = \frac{1-0}{10} = 0.1$ and compute the value of $f(x) = \frac{\log(1 + x^3)}{1 + x^2}$ at each points of sub-interval. These values are given below

x	x^2	$1 + x^2$	$\log (1 + x^2)$	$\frac{\log(1+x^2)}{1+x^2}$
0	0	1.0	0	0
0.1	0.01	1.01	0.009950	0.009851
0.2	0.04	1.04	0.039220	0.037712
0.3	0.09	1.09	0.086177	0.079062
0.4	0.16	1.16	0.14842	0.127948
0.5	0.25	1.25	0.223143	0.178514
0.6	0.36	1.36	0.307484	0.226091
0.7	0.49	1.49	0.398776	0.267634
0.8	0.64	1.64	0.494696	0.301644
0.9	0.81	1.81	0.593326	0.327804
1.0	1.0	2.0	0.693147	0.346573

By Simpson's one-third rule, we have

$$\int_0^1 \frac{\log(1+x^2)}{1+x^2}\,dx = \frac{h}{3}[y_0 + 4(y_1 + y_3 + y_5 + y_7 + y_9) + 2(y_2 + y_4 + y_6 + y_8) + y_{10}]$$

$$= \frac{0.1}{3}[0 + 4(0.009851 + 0.079062 + 0.178514 + 0.267634 + 0.327804)$$
$$+ 2(0.037712 + 0.127948 + 0.226091 + 0.301644) + 0.346573]$$

$$= \frac{0.1}{3}[5.184839] = 0.17282793. \quad \textbf{Ans.}$$

Example 7. *Evaluate* $\int_4^{5.2} \log_e x\,dx$ *by Simpson's one-third and three-eighth rule.*

Solution. Dividing the interval (4, 5.2) into six equal parts of width $h = \frac{5.2 - 4}{6} = 0.2$ and compute the value of $f(x) = \log_e x$ at each point of sub-interval. These values are given below:

x	4	4.2	4.4	4.6	4.8	5.0	5.2
$f(x)$	1.386294	1.435084	1.481604	1.526056	1.568615	1.609437	1.648658

By Simpson's one-third rule, we have

$$\int_4^{5.2} \log_e x\,dx = \frac{h}{3}[y_0 + 4(y_1 + y_3 + y_5) + 2(y_2 + y_4) + y_6]$$

$$= \frac{0.2}{3}[1.386294 + 4(1.435084 + 1.526056 + 1.609437)$$
$$+ 2(1.481604 + 1.568615) + 1.648658]$$

$$= \frac{0.2}{3}\,(27.417698) = 1.827847$$

By Simpson's three-eighth rule, we have

$$\int_4^{5.2} \log_e x\,dx = \frac{3h}{8}\,[y_0 + 3(y_1 + y_2 + y_4 + y_5) + 2(y_3) + y_6]$$

$$= \frac{3(0.2)}{8}\,[1.386294 + 3(1.435084 + 1.481604 + 1.568615 + 1.609437) + 2(1.526056) + 1.648658]$$

$$= \frac{0.6}{8} \times (24.371294) = 1.827847. \quad \textbf{Ans.}$$

Example 8. *Evaluate* $\int_{0.5}^{0.7} x^{1/2} e^{-x}\,dx$.

Solution. Dividing the interval (0.5, 0.7) into four equal parts of width $h = \frac{0.7 - 0.5}{4} = 0.05$ and compute the value of $f(x) = x^{1/2}e^{-x}$ at each point of sub-interval. These values are given below:

x	$x^{1/2}$	e^{-x}	$f(x) = x^{1/2}\,e^{-x}$
0.50	0.707106	0.606530	0.428881
0.55	0.741619	0.576949	0.427876
0.60	0.774596	0.548811	0.425107
0.65	0.806225	0.522045	0.420886
0.70	0.836660	0.496585	0.415473

By Trapezoidal rule, we have

$$\int_{0.5}^{0.7} x^{1/2}\,e^{-x}\,dx = \frac{h}{2}[y_0 + 2(y_1 + y_2 + y_3) + y_4]$$

$$= \frac{0.05}{2}\,[0.428881 + 2\,(0.427876 + 0.425107 + 0.420886) + 0.415473]$$

$$= \frac{0.05}{2}\,(3.392092) = 0.0848023. \quad \textbf{Ans.}$$

By Simpson's $\frac{1}{3}$ rule, we have

$$\int_{0.5}^{0.7} x^{1/2}e^{-x}\,dx = \frac{h}{3}[\,y_0 + 4\,(y_1 + y_3) + 2y_2 + y_4]$$

$$= \frac{0.05}{3}[0.428881 + 4(0.427876 + 0.420886) + 2(0.425107) + 0.415473]$$

$$= \frac{0.05}{3}\,[5.089616] = 0.0848269. \quad \textbf{Ans.}$$

EXERCISE 2.1

1. Evaluate $\int_0^1 \frac{dx}{1+x^2}$ by using Simpson's one-third and three-eighth rule. Hence obtain the approximate value of π in each case.

2. A curve is drawn to pass through the points given by the following table:

x	1	1.5	2	2.5	3	3.5	4
y	2.0	2.4	2.7	2.8	3.0	2.6	2.1

Estimate the area bounded by the curve, the x-axis and the lines $x = 1$, $x = 4$ by Simpson's rule.

3. Evaluate $\int_2^{10} \frac{dx}{1+x}$ by dividing the range into eight equal parts by Simpson's one-third rule.

4. Show that $\int_0^1 \frac{dx}{1+x} = \log 2 = 0.69315$ by using Simpson's one-third rule.

5. Calculate the value of the integral $\int_4^{5.2} \log x \, dx$ by using Trapezoidal rule, Simpson's one-third and three-eighth rule.

6. Use Simpson's one-third rule to find the approximate area of the cross-section of a river 80 meter wide, the depth y at a distance x from one bank being given by the following table:

x	0	10	20	30	40	50	60	70	80
y	0	4	7	9	12	15	14	8	3

7. Evaluate $\int_0^4 e^x \, dx$, by Simpson's rule using the data.

x	0	1	2	3	4
e^x	1	2.72	7.39	20.39	54.60

and compare it with the actual value.

8. Calculate an approximate value of the integral $\int_0^{\pi/2} \sin x \, dx$ by (*i*) Trapezoidal rule (*ii*) Simpson's one-third rule (*iii*) Simpson's three-eighth rule.

9. Use Simpson's one-third rule to prove that $\log_e 7$ is approximately 1.9587 using $\int_1^7 \frac{dx}{x}$.

10. Use Simpson's one-third rule to find the value of $\int_1^5 f(x)\,dx$ given.

x	1	2	3	4	5
$f(x)$	10	50	70	80	100

ANSWERS

1. 0.785397 and 0.785395, $\pi = 3.141588$

2. 15.5667

3. 1.29962

5. 1.827648, 1.827847, 1.827847

6. 710 sq. me.

7. 53.873

8. 0.99795, 1.0006, 1.1003

10. 256.66667.

CHAPTER 3

Ordinary Differential Equations of First Order

INTRODUCTION

In this chapter, we will discuss the important methods of solving ordinary differential equation of first order having numerical coefficients and given boundary or initial conditions

$$\left(i.e., \frac{dy}{dx} = f(x, y) \text{ given } y(x_0) = y_0 \right) \text{ numerically.}$$

These method also useful to solve those types of problem related to first order differential equation which cannot be integrated analytically.

For example, $\dfrac{dy}{dx} = x^2 + y^2 - c^2$

Some important methods are:

1. Euler's method

2. Euler's Modified method

3. Picard's method of successive approximation

4. Runge-Kutta method

5. Milne's series method.

3.1. EULER'S METHOD

This is simplest and oldest method was devised by Euler. It illustrates, the basic idea of those numerical methods which seek to determine the change Δy in y corresponding to a small increase in the arguments x.

Consider the differential equation

$$y' = \frac{dy}{dx} = f(x, y) \qquad \text{...(3.1)}$$

with initial condition $y = y_0$ when $x = x_0$, *i.e.*, $y(x_0) = y_0$

We wish to solve (3.1), for the values of y at $x = x_i$

where $x_i = x_0 + ih$, $i = 1, 2, 3, \ldots\ldots$

Now integrate (3.1), we have

$$y_1 = y_0 + \int_{x_0}^{x} f(x, y)\, dx$$

Let $f(x, y) = f(x_0, y_0)$ where $x_0 \le x \le x_1$ ...(3.2)

Now $$y_1 = y_0 + \int_{x_0}^{x_1} f(x_0, y_0)\, dx$$

$$= y_0 + (x_1 - x_0) f(x_0, y_0)$$

$$= y_0 + h f(x_0, y_0) \qquad [\because \; h = x_1 - x_0]$$

Similarly for $x_1 \le x \le x_2$, we have

$$y_2 = y_1 + h f(x_1, y_1)$$

Proceeding in the same way, we have finally

$$y_{n+1} = y_n + h f(x_n, y_n) \qquad \text{...(3.3)}$$

Thus, starting from x_0 when $y = y_0$ we can construct a table of y for given steps of h in x.

The process of Euler's method is very slow and to obtain desired accuracy with Euler's method, h should be taken small.

3.2. EULER'S MODIFIED METHOD

Instead of approximating $f(x, y)$ by $f(x_0, y_0)$ in (3.2), the integral in (3.3) is approximated by Trapezoidal rule to obtain

$$y_1 = y_0 + \frac{h}{2} [f(x_0, y_0) + f(x_1, y_1)] \qquad \text{...(3.4)}$$

Thus we obtain the formula

$$y_1^{(n+1)} = y_0 + \frac{h}{2} [f(x_0, y_0) + f(x_1, y_1^{(n)})] \quad (n = 0, 1, 2 \ldots) \qquad \text{...(3.5)}$$

This is the nth approximation of y_1. The formula (3.3) uses the initial value $y_1^{(0)}$ from Euler's method.

$$y_1^{(0)} = y_0 + h f(x_0, y_0).$$

3.3. PICARD'S METHOD OF SUCCESSIVE APPROXIMATION

Consider the differential equation

$$y' = \frac{dy}{dx} = f(x, y) \qquad \text{...(3.6)}$$

with the initial condition $y = y_0$ for $x = x_0$ *i.e.*, $y(x_0) = y_0$

Integrating the differential equation (3.6), we have

$$y = y_0 + \int_{x_0}^{x} f(x, y)\, dx \qquad ...(3.7)$$

Equation (3.7) in which the unknown function y appears under the integral sign, is called an integral equation. In this method, the first approximation $y^{(1)}$ is obtained by replacing y by y_0 in $f(x, y)$ in R.H.S. of (3.7) and integrating w.r.t. x, we get

i.e.,
$$y^{(1)} = y_0 + \int_{x_0}^{x} f(x, y_0)\, dx \qquad ...(3.8)$$

The second approximation $y^{(2)}$ is obtained by replacing y by $y^{(1)}$ in $f(x, y)$ in R.H.S. of (3.7) and integrating w.r.t. x, we get

$$y^{(2)} = y_0 + \int_{x_0}^{x} f(x, y^{(1)})\, dx \qquad ...(3.9)$$

Proceeding in the same way we obtain $y^{(3)}$, $y^{(4)}$, ..., y^{n-1} and y^n where

$$y^{(n)} = y_0 + \int_{x_0}^{x} f(x, y^{(n-1)})\, dx \qquad ...(3.10)$$

with
$$y^{(0)} = y_0$$

we repeat the steps till the two value of y becomes same to the desired degree of accuracy.

3.4. RUNGE-KUTTA METHOD

This method is most commonly used method and most suitable when computation of higher derivatives is complicated.

Consider the following differential equation

$$\frac{dy}{dx} = f(x, y) \text{ with } y(x_0) = y_0$$

Runge-Kutta method of order four is given by

$$y_{n+1} = y_n + k \quad \text{for} \quad x = x_0 + h$$

where $x = \dfrac{h}{6}\left[k_1 + \dfrac{4[k_1 + k_2]}{2} + k_3\right] = \dfrac{h}{6}\,[k_1 + 2(k_1 + k_2) + k_3]$

where $k_1 = f(x_0, y_0)$

$$k_2 = f\left(x_0 + \frac{h}{2}, y_0 + k_1\frac{h}{2}\right)$$

$$k_3 = f\left(x_0 + \frac{h}{2}, y_0 + k_2\frac{h}{2}\right)$$

$$k_4 = f(x_0 + h, y_0 + k_3 h)$$

3.5. MILNE'S SERIES METHOD

If we solve the differential equation $\frac{dy}{dx} = f(x, y)$ with $y(x_0) = y_0$ by this method, we first obtain the approximate value of y_{n+1} by predictor formula and then improve this value by means of a corrector formula.

The predicted value of the solution is given by

$$y_{n+1} = y_{n-3} + \frac{4h}{3}(2y'_{n-2} - y'_{n-1} + 2y')$$

The predicted value is substituted in corrector formula

$$y_{n+1}^{(1)} = y_{n-1} + \frac{h}{3}\left(y'_{n-1} + 4y'_n + y'_{n+1}\right)$$

to improve that predicted value.

SOLVED EXAMPLES

Example 1. *Given* $\frac{dy}{dx} = \frac{y-x}{y+x}$ *with the initial condition* $y = 1$ *when* $x = 0$ *find* y *for* $x = 0.1$ *in 4 steps by Euler's method.*

Solution. We have $\frac{dy}{dx} = \frac{y-x}{y+x} = f(x, y)$, $x_0 = 0$, $y_0 = 1$ and $h = \frac{0.1}{4} = 0.025$

We know that $y_{n+1} = y_n + hf(x_n, y_n)$

By putting $n = 1, 2, 3$, we get

$$y_1 = y_0 + hf(x_0, y_0) = y_0 + h\,\frac{y_0 - x_0}{y_0 + x_0}$$

$$= 1 + (0.025) \times \frac{1-0}{1+0} = 1.025$$

$$y_1 = 1.025$$

Again

$$y_2 = y_1 + hf(x_1, y_1) = y_1 + h\,\frac{y_1 - x_1}{y_1 + x_1}$$

$$= 1.025 + (0.025) \times \frac{1.025 - 0.025}{1.025 + 0.025} = 1.025 + 0.025 \times \frac{1}{1.05}$$

$$y_2 = 1.0488$$

Now again

$$y_3 = y_2 + hf(x_2, y_2) = y_2 + h\,\frac{y_2 - x_2}{y_2 + x_2}$$

$$= 1.0488 + (0.025) \times \frac{1.0488 - 0.05}{1.0488 + 0.05}$$

$$= 1.0488 + 0.025 \times \frac{1.0438}{1.0988}$$

$$y_3 = 1.07152$$

Again $$y_4 = y_3 + hf(x_3, y_3) = y_3 + h\,\frac{y_3 - x_3}{y_3 + x_3}$$

$$= 1.07152 + 0.025 \times \frac{1.07152 - 0.075}{1.07152 + 0.075}$$

$$y_4 = 1.09324$$

Hence $y_{(0.1)} = 1.09324$ **Ans.**

Example 2. *Use Euler's method compute y(0.5) for differential equation* $\frac{dy}{dx} = y^2 - x^2$ *with* $y = 1$ *when* $x = 0$.

Solution. We have $\frac{dy}{dx} = y^2 - x^2 = f(x, y),\ x_0 = 0,\ y_0 = 1$

and let $h = \frac{0.5}{5} = 0.1$

We known that $y_{n+1} = y_n + h\,f(x_n, y_n)$

By putting $n = 0, 1, 2, 3, 4$, we get

$$y_1 = y_0 + hf(x_0, y_0) = y_0 + h(y_0^2 - x_0^2)$$

$$= 1 + (0.1)(1^2 - 0) = 1.1$$

$$y_2 = y_1 + hf(x_1, y_1) = y_1 + h(y_1^2 - x_1^2)$$

$$= 1.1 + (0.1)[(1.1)^2 - (0.1)^2] = 1.1 + (0.1)(1.21 - 0.01)$$

$$= 1.220 \qquad [\because\ x_1 = x_0 + h]$$

$$y_3 = y_2 + hf(x_2, y_2) = y_2 + h\,(y_2^2 - x_2^2)$$

$$= 1.22 + (0.1)[(1.22)^2 - (0.2)^2] \qquad [\because\ x_2 = x_1 + h]$$

$$= 1.22 + (0.1)(1.4484) = 1.36484$$

$$y_4 = y_3 + hf(x_3, y_3) = y_3 + h(y_3^2 - x_3^2)$$

$$= 1.36484 + (0.1)[(1.36484)^2 - (0.3)^2]$$

$$= 1.36484 + (0.1)(1.7728) = 1.54212$$

$$y_5 = y_4 + hf(x_4, y_4) = y_4 + h(y_4^2 - x_4^2)$$

$$= 1.54212 + (0.1)[(1.54212)^2 - (0.4)^2] = 1.7639$$

Hence $y(0.5) = 1.7639$ **Ans.**

Example 3. *Using Euler's modified method, solve numerically the equation* $\frac{dy}{dx} = x + \sqrt{|y|}$ *with* $y(0) = 1$ *for* $0 \le x \le 0.6$ *in the steps of 0.2.*

Solution. Here $f(x, y) = x + \sqrt{|y|}$, $x_0 = 0$, $y_0 = 1$ and $h = 0.2$

By Euler's method, we have

$$y_1 = y_0 + hf(x_0, y_0) = 1 + (0.2)(0 + \sqrt{1}) = 1.2$$

$$y_1 = 1.2$$

The value of y_1, thus obtained is improved by Euler's modified method

$$y_1^{(n+1)} = y_0 + \frac{h}{2}[f(x_0, y_0) + f(x_1, y_1^{(n)})]$$

Put $n = 0$, we get

$$y_1^{(1)} = y_0 + \frac{h}{2}[f(x_0, y_0) + f(x_1, y_1^{(0)})]$$

$$= 1 + \frac{0.2}{2}[(0 + \sqrt{1}) + (0.2 + \sqrt{1.2})]$$

$$= 1 + 0.2295 = 1.2295$$

Put $n = 1$, we get

$$y_1^{(2)} = y_0 + \frac{h}{2}[f(x_0, y_0) + f(x_1, y_1^{(1)})]$$

$$= 1 + \frac{0.2}{2}[(0 + \sqrt{1}) + (0.2 + \sqrt{1.2295})]$$

$$= 1 + 0.2309 = 1.2309$$

Put $n = 2$, we get

$$y_1^{(3)} = y_0 + \frac{h}{2}[f(x_0, y_0) + f(x_1, y_1^{(2)})]$$

$$= 1 + \frac{0.2}{2}[(0 + \sqrt{1}) + (0.2 + \sqrt{1.2309})]$$

$$= 1 + 0.2309 = 1.2309 \qquad [\because\ y_1^{(2)} = y_1^{(3)}]$$

Hence we take $y_1 = 1.2309$ at $x = 0.2$

Now, we proceed to compute y at $x = 0.4$

Applying Euler's method, we have

$$y_2 = y_1 + hf(x_1, y_1)$$

$$= y_1 + h(x_1 + \sqrt{|y_1|})$$

$$= 1.2309 + 0.2\,(0.2 + \sqrt{1.2309})$$

$$= 1.2309 + 0.2(1.30945) = 1.49279$$

The value of y_2, thus obtained is improved by Euler's modified method

$$y_2^{(n+1)} = y_1 + \frac{h}{2}[f(x_1, y_1) + f(x_2, y_2^{(n)})]$$

Put $n = 0$, $$y_2^{(1)} = y_1 + \frac{h}{2}[f(x_1, y_1) + f(x_2, y_2^{(0)})]$$

$$= 1.2309 + \frac{0.2}{2}[(0.2 + \sqrt{1.2309}) + (0.4 + \sqrt{1.49279})] = 1.52402$$

Put $n = 1$, we get $$y_2^{(2)} = y_1 + \frac{h}{2}[f(x_1, y_1) + f(x_2, y_2^{(1)})]$$

$$= 1.2309 + \frac{0.2}{2}[(0.2 + \sqrt{1.2309}) + (0.4 + \sqrt{1.52402})] = 1.525297$$

Put $n = 2$, we get $$y_2^{(3)} = y_1 + \frac{h}{2}[f(x_1, y_1) + f(x_2, y_2^{(2)})]$$

$$= 1.2309 + \frac{0.2}{2}[(0.2 + \sqrt{1.2309}) + (0.4 + \sqrt{1.525297})] = 1.52535$$

Put $n = 3$, we get $$y_2^{(4)} = y_1 + \frac{h}{2}[f(x_1, y_1) + f(x_2, y_2^{(3)})]$$

$$= 1.2309 + \frac{0.2}{2}[(0.2 + \sqrt{1.2309}) + (0.4 + \sqrt{1.52535})] = 1.52535$$

$\because$ $$y_2^{(3)} = y_2^{(4)}$$

Hence, we take $y_2 = 1.52535$ at $x = 0.4$

Again, we proceed to compute y at $x = 0.6$

Applying Euler's method, we have

$$y_3 = y_2 + hf(x_2, y_2) = y_2 + h(x_2 + \sqrt{|y_2|})$$
$$= 1.52535 + 0.2(0.4 + \sqrt{1.52535}) = 1.85236$$

The value of y_3, thus obtained is improved by Euler's modified method

$$y_3^{(n+1)} = y_2 + \frac{h}{2}[f(x_2, y_2) + f(x_3, y_3^{(n)})]$$

Put $n = 0$ $$y_3^{(1)} = y_2 + \frac{h}{2}[f(x_2, y_2) + f(x_3, y_3^{(0)})]$$

$$= 1.52535 + \frac{0.2}{2}[(0.4 + \sqrt{1.52535}) + (0.6 + \sqrt{1.85236})] = 1.88496$$

Put $n = 1$ $$y_3^{(2)} = y_2 + \frac{h}{2}[f(x_2, y_2) + f(x_3, y_3^{(1)})]$$

$$= 1.52535 + \frac{0.2}{2}[(0.4 + \sqrt{1.52535}) + (0.6 + \sqrt{1.88496})] = 1.88615$$

Put $n = 2$ $$y_3^{(3)} = y_2 + \frac{h}{2}[f(x_2, y_2) + f(x_3, y_3^{(2)})]$$

$$= 1.52535 + \frac{0.2}{2}[(0.4 + \sqrt{1.52535}) + (0.6 + \sqrt{1.88615})] = 1.88619$$

Put $n = 3$ $$y_3^{(4)} = y_2 + \frac{h}{2}[f(x_2, y_2) + f(x_3, y_3^{(3)})]$$

$$= 1.52535 + \frac{0.2}{2}[(0.4 + \sqrt{1.52535}) + (0.6 + \sqrt{1.88619})] = 1.88619$$

$\because$ $y_3^{(3)} = y_3^{(4)}$

Hence, we take $y_3 = 1.88619$ at $x = 0.6$

$\Rightarrow$ $y(0.2) = 1.2309,\ y(0.4) = 1.52535,\ y(0.6) = 1.88619.$ **Ans.**

Example 4. *Using Euler's modified method, compute y(0.1) correct to six decimal figures where* $\frac{dy}{dx} = x^2 + y$ *with y(0) = 0.94.*

Solution. Here $f(x, y) = x^2 + y,\ x_0 = 0,\ y_0 = 0.94,\ h = 0.1$

By Euler's method, we have

$$y_1 = y_0 + hf(x_0, y_0) = y_0 + h\,[x_0^2 + y_0]$$

$$= 0.94 + (0.1)\,[0 + 0.94] = 1.034$$

The value of y_1, thus obtained is improved by Euler's modified method

$$y_1^{(n+1)} = y_0 + \frac{h}{2}[f(x_0, y_0) + f(x_1, y_1^{(n)})]$$

Put $n = 0$ $$y_1^{(1)} = y_0 + \frac{h}{2}[f(x_0, y_0) + f(x_1, y_1^{(0)})]$$

$$= 0.94 + \frac{0.1}{2}\,[(0 + 0.94) + ((0.1)^2 + 1.034)] = 1.0392$$

Put $n = 1$ $$y_1^{(2)} = y_0 + \frac{h}{2}[f(x_0, y_0) + f(x_1, y_1^{(1)})]$$

$$= 0.94 + \frac{0.1}{2}\,[(0 + 0.94) + ((0.1)^2 + (1.0392))] = 1.03946$$

Put $n = 2$ $$y_1^{(3)} = y_0 + \frac{h}{2}[f(x_0, y_0) + f(x_1, y_1^{(2)})]$$

$$= 0.94 + \frac{0.1}{2}\,[(0 + 0.94) + ((0.1)^2 + 1.03946)] = 1.039473$$

Put $n = 3$ $$y_1^{(4)} = y_0 + \frac{h}{2}[f(x_0, y_0) + f(x_1, y_1^{(3)})]$$

$$= 0.94 + \frac{0.1}{2}\,[(0 + 0.94) + ((0.1)^2 + 1.039473)] = 1.039473$$

Here $y_1^{(3)} = y_1^{(4)}$

Hence $y_1 = 1.039473$

$\Rightarrow$ $y(.1) = 1.039473.$ **Ans.**

Example 5. *Use Picard's method to solve* $\frac{dy}{dx} = x - y$ *for* $x = 0.1$ *and* 0.2 *given that* $y = 1$ *when* $x = 0$.

Solution. We have $f(x, y) = x - y$ and $x_0 = 0$, $y_0 = 1$

Now first approximation

$$y^{(1)} = y_0 + \int_{x_0}^{x} f(x, y_0)\, dx = y_0 + \int_{x_0}^{x} (x - y_0)\, dx$$

$$= 1 + \int_0^x (x-1)\, dx = 1 + \left[\frac{x^2}{2} - x\right]_0^x = \frac{x^2}{2} - x + 1$$

Second approximation

$$y^{(2)} = y_0 + \int_{x_0}^{x} f(x, y^{(1)})\, dx = y_0 + \int_{x_0}^{x} (x - y^{(1)})\, dx$$

$$= 1 + \int_0^x \left[x - \left(\frac{x^2}{2} - x + 1\right)\right] dx = 1 + \int_0^x \left(2x - \frac{x^2}{2} - 1\right) dx$$

$$= 1 + \left[x^2 - \frac{x^3}{6} - x\right]_0^x = 1 + x^2 - \frac{x^3}{6} - x$$

$$= -\frac{x^3}{6} + x^2 - x + 1$$

Third approximation

$$y^{(3)} = y_0 + \int_{x_0}^{x} f(x, y^{(2)})\, dx = y_0 + \int_{x_0}^{x} (x - y^{(2)})\, dx$$

$$= 1 + \int_0^x \left(x + \frac{x^3}{6} - x^2 + x - 1\right) dx = 1 + \left(x^2 + \frac{x^4}{24} - \frac{x^3}{3} - x\right)_0^x$$

$$= \frac{x^4}{24} - \frac{x^3}{3} + x^2 - x + 1$$

Fourth approximation

$$y^{(4)} = y_0 + \int_{x_0}^{x} f(x, y^{(3)})\, dx = y_0 + \int_{x_0}^{x} (x - y^{(3)})\, dx$$

$$= 1 + \int_0^x \left(x - \left(\frac{x^4}{24} - \frac{x^3}{3} + x^2 - x + 1\right)\right) dx$$

$$= 1 + \int_0^x \left(2x - \frac{x^4}{24} + \frac{x^3}{3} - x^2 - 1\right) dx$$

$$= 1 + \left[x^2 - \frac{x^5}{120} + \frac{x^4}{12} - \frac{x^3}{3} - x\right]_0^x = -\frac{x^5}{120} + \frac{x^4}{12} - \frac{x^3}{3} + x^2 - x + 1$$

Fifth approximation

$$y^{(5)} = y_0 + \int_{x_0}^x f(x, y^{(4)})\, dx = y_0 + \int_{x_0}^x (x - y^{(4)})\, dx$$

$$= 1 + \int_0^x \left(x - \left(-\frac{x^5}{120} + \frac{x^4}{12} - \frac{x^3}{3} + x^2 - x + 1\right)\right) dx$$

$$= 1 + \int_0^x \left[2x + \left(\frac{x^5}{120} - \frac{x^4}{12} + \frac{x^3}{3} - x^2 - 1\right)\right] dx$$

$$= 1 + \left(x^2 + \frac{x^6}{720} - \frac{x^5}{60} + \frac{x^4}{12} - \frac{x^3}{3} - x\right)_0^x$$

$$= \frac{x^6}{720} - \frac{x^5}{60} + \frac{x^4}{12} - \frac{x^3}{3} + x^2 - x + 1$$

When $x = 0.1$, we have

$$y_0 = 1,\ y^{(1)} = 0.905,\ y^{(2)} = 0.9098,\ y^{(3)} = 0.90967,\ y^{(4)} = 0.90967$$

i.e., $y = 0.90967$ at $x = 0.1$

When $x = 0.2$, we have

$$y_0 = 1,\ y^{(1)} = 0.82,\ y^{(2)} = 0.83867,\ y^{(3)} = 0.83740,\ y^{(4)} = 0.83746$$

$$y^{(5)} = 0.83746$$

i.e., $y = 0.83746$ at $x = 0.2$. **Ans.**

Example 6. *Apply Picard's method to solve* $\frac{dy}{dx} = x + y^2$ *given that when* $x_0 = 0$, $y_0 = 0$ *up to third order of approximation.*

Solution. We have $f(x, y) = x + y^2$. and $x_0 = 0, y_0 = 0$

Now first approximation

$$y^{(1)} = y_0 + \int_{x_0}^x f(x, y_0)\, dx = y_0 + \int_{x_0}^x (x + y_0^2)\, dx = 0 + \int_0^x (x + 0)\, dx$$

$$= \left[\frac{x^2}{2}\right]_0^x = \frac{x^2}{2}$$

Second approximation

$$y^{(2)} = y_0 + \int_{x_0}^{x} f(x, y^{(1)})\, dx = y_0 + \int_{x_0}^{x} (x + (y^{(1)})^2)\, dx$$

$$y^{(2)} = 0 + \int_0^x \left(x + \left(\frac{x^2}{2}\right)^2\right) dx = \int_0^x \left(x + \frac{x^4}{4}\right) dx = \left(\frac{x^2}{2} + \frac{x^5}{20}\right)_0^x = \frac{x^2}{2} + \frac{x^5}{20}$$

Third approximation

$$y^{(3)} = y_0 + \int_{x_0}^{x} f(x, y^{(2)})\, dx = y_0 + \int_{x_0}^{x} [x + (y^{(2)})^2]\, dx$$

$$= 0 + \int_0^x \left(x + \left(\frac{x^2}{2} + \frac{x^5}{20}\right)^2\right) dx = \int_0^x \left(x + \frac{x^4}{4} + \frac{x^{10}}{400} + \frac{x^7}{20}\right) dx$$

$$= \left(\frac{x^2}{2} + \frac{x^5}{20} + \frac{x^8}{160} + \frac{x^{11}}{4400}\right)_0^x$$

$$= \frac{x^2}{2} + \frac{x^5}{20} + \frac{x^8}{160} + \frac{x^{11}}{4400}. \quad \textbf{Ans.}$$

Example 7. *Use Runge-Kutta method to solve* $\frac{dy}{dx} = xy$ *for* $x = 1.2, 1.4$, *initially* $x = 1$, $y = 2$.

Solution. Here $x_0 = 1$, $y_0 = 2, f(x, y) = xy$

$\Rightarrow \qquad f(x_0, y_0) = 1 \times 2 = 2$

Let $\quad h = 0.2$ then $k_1 = f(x_0, y_0) = 2$

$$k_2 = f\left(x_0 + \frac{h}{2}, y_0 + k_1\frac{h}{2}\right) = \left(x_0 + \frac{h}{2}\right)\left(y_0 + k_1\frac{h}{2}\right) = \left(1 + \frac{0.2}{2}\right)\left(2 + 2 \times \frac{0.2}{2}\right)$$

$$= (1.1)(2.2) = 2.42$$

$$k_3 = f\left(x_0 + \frac{h}{2}, y_0 + k_2\frac{h}{2}\right) = \left(x_0 + \frac{h}{2}\right)\left(y_0 + k_2\frac{h}{2}\right) = \left(1 + \frac{0.2}{2}\right)\left(2 + 2.42 \times \frac{0.2}{2}\right)$$

$$= (1.1)\,(2.242) = 2.4662$$

$$k_4 = f(x_0 + h, y_0 + k_3 h) = (x_0 + h)\,(y_0 + k_3 h) = (1 + 0.2)(2 + 2.4662 \times 0.2)$$

$$= (1.2)(2.49324) = 2.9918$$

$$\therefore \quad k = \frac{h}{6}\,[k_1 + 2(k_2 + k_3) + k_4] = \frac{0.2}{6}\,[2 + 2\,(2.42 + 2.4662) + 2.9918]$$

$$= \frac{0.2}{6}\,[2 + 9.7724 + 2.9918]$$

$$k = 0.49214$$

This show that $\qquad x_1 = x_0 + h = 1 + 0.2 = 1.2$

and $\qquad y_1 = y_0 + k = 2 + 0.49214 = 2.4921$

Therefore $\qquad y(1.2) = 2.4921.$ **Ans.**

Now for the second interval

$$x_1 = 1.2,\ y_1 = 2.4921,\ f(x, y) = xy$$

$\Rightarrow$ $$f(x_1, y_1) = x_1y_1 = 1.2 \times 2.4921 = 2.99052$$

Let $h = 0.2$ then $k_1 = f(x_1, y_1) = 2.99052$

$$k_2 = f\left(x_1 + \frac{h}{2}, y_1 + k_1\frac{h}{2}\right) = \left(x_1 + \frac{h}{2}\right)\left(y_1 + k_1\frac{h}{2}\right)$$

$$= (1.2 + 0.1)\left(2.4921 + 2.9905 \times \frac{0.2}{2}\right)$$

$$= (1.3)\,(2.79105) = 3.6283$$

$$k_3 = f\left(x_1 + \frac{h}{2}, y_1 + k_2\frac{h}{2}\right) = \left(x_1 + \frac{h}{2}\right)\left(y_1 + k_2\frac{h}{2}\right)$$

$$= \left(1.2 + \frac{0.2}{2}\right)\left(2.4921 + 3.6283 \times \frac{0.2}{2}\right)$$

$$= (1.3)\,(2.8548) = 3.71143$$

$$k_4 = f(x_1 + h, y_1 + k_3h) = (x_1 + h)\,(y_1 + k_3h) = (1.2 + 0.2)\,(2.4921 + 3.71128 \times 0.2)$$

$$= (1.4)(3.2343) = 4.5281$$

$\therefore$ $$k = \frac{h}{6}\,[(k_1 + 2)k_2 + k_3) + k_4]$$

$$= \frac{0.2}{6}\,[2.9905 + 2\,(3.6283 + 3.7112) + 4.5281] = 0.73992$$

This show that $x_2 = x_1 + h = 1.2 + 0.2 = 1.4$

$$y_2 = y_1 + k = 2.4921 + 0.73992 = 3.2330$$

Therefore $y(1.4) = 3.2321$. **Ans.**

Example 8. *Solve the equation* $\dfrac{dy}{dx} = -\,2xy^2$ *with initial condition* $y(0) = 1$ *by Runge-Kutta's method for* $x = 0.2$ *and* 0.4 *with* $h = 0.2$.

Solution. Here $x_0 = 0,\ y_0 = 1,\ f(x, y) = -\,2xy^2$

$\Rightarrow$ $$f(x_0, y_0) = -\,2x_0y_0^2 = 0$$

Let $h = 0.2$ then $k_1 = f(x_0, y_0) = 0$

$$k_2 = f\left(x_0 + \frac{h}{2}, y_0 + k_1\frac{h}{2}\right) = -\,2\left(x_0 + \frac{h}{2}\right)\left(y_0 + k_1\frac{h}{2}\right)^2$$

$$= -\,2\left(0 + \frac{0.2}{2}\right)\left(1 + 0 \times \frac{0.2}{2}\right) = -\,2\,(0.1)(1) = -\,0.2$$

$$k_3 = f\left(x_0 + \frac{h}{2}, y_0 + k_2\frac{h}{2}\right) = -\,2\left(x_0 + \frac{h}{2}\right)\left(y_0 + k_2\frac{h}{2}\right)^2$$

$$= -\,2\left(0 + \frac{0.2}{2}\right)\left(1 + (-\,0.2) \times \frac{0.2}{2}\right)^2 = -\,2(0.1)(0.98)^2 = -\,0.1920$$

$$k_4 = f(x_0 + h, y_0 + k_3h) = -2\,(x_0 + h)\,(y_0 + k_3h)^2$$
$$= -2(0 + 0.2)[1 + (-0.1920)(0.2)]^2 = -0.36986$$

$\therefore$
$$k = \frac{h}{6}\,[k_1 + 2(k_2 + k_3) + k_4]$$
$$= \frac{0.2}{6}\,[0 + 2((-0.2) + (-0.1920)) + (-0.36986)]$$
$$= \frac{0.2}{6}\,(-1.15386) = -0.03846$$

This show that $x_1 = x_0 + h = 0 + 0.2 = 0.2$

and $y_1 = y_0 + k = 1 + (-0.03846) = 0.96154$

Therefore $y(0.2) = 0.96154$. **Ans.**

Now for the second interval

$$x_1 = 0.2,\ y_1 = 0.9615$$
$$f(x, y) = -2xy^2$$

$\Rightarrow$ $f(x_1, y_1) = -2x_1y_1^2 = -2 \times 0.2 \times (0.9615)^2 = -0.36979$

Let $h = 0.2$ then $k_1 = f(x_1, y_1) = -0.36979$

$$k_2 = f\left(x_1 + \frac{h}{2}, y_1 + k_1\frac{h}{2}\right) = -2\left(x_1 + \frac{h}{2}\right)\left(y_1 + k_1\frac{h}{2}\right)^2$$
$$= -2\left(0.2 + \frac{0.2}{2}\right)\left(0.9615 + \left(-0.36979 \times \frac{0.2}{2}\right)\right)^2$$
$$= -2\,(0.3)\,(0.85473) = -0.51284$$
$$k_3 = f\left(x_1 + \frac{h}{2}, y_1 + k_2\frac{h}{2}\right) = -2\left(x_1 + \frac{h}{2}\right)\left(y_1 + k_2\frac{h}{2}\right)^2$$
$$= -2\left(0.2 + \frac{0.2}{2}\right)\left(0.9615 + \left(-0.51284 \times \frac{0.2}{2}\right)\right)^2$$
$$= -2\,(0.3)\,(0.82849) = -0.49709$$
$$k_4 = f(x_1 + h, y_1 + k_3h) = -2\,(x_1 + h)(y_1 + k_3h)^2$$
$$= -2\,(0.2 + 0.2)\,(0.9615 + (-0.49709) \times 0.2)^2$$
$$= -2\,(0.4)(0.7431) = -0.59454$$

$\therefore$
$$k = \frac{h}{6}\,[k_1 + 2\,(k_2 + k_3) + k_4]$$
$$= \frac{0.2}{6}\,(-0.36979 + 2\,(-0.51284 - 0.49709) - 0.59454) = -0.099473$$

This show that $x_2 = x_1 + h = 0.2 + 0.2 = 0.4$

$$y_2 = y_1 + k = 0.9615 - 0.099473 = 0.86202$$

Therefore $y(0.4) = 0.86202$. **Ans.**

Example 9. *Compute y(2) if y(x) is the solution of* $\frac{dy}{dx} = \frac{1}{2}(x + y)$ *assuming y(0) = 2, y(0.5) = 2.636, y(1) = 3.595, y(1.5) = 4.968.*

Solution. Here $f(x, y) = \frac{1}{2}(x + y)$ then we have

$x_0 = 0$	$y_0 = 2$	$y_0' = \frac{1}{2}(0+2) = 1$
$x_1 = 0.5$	$y_1 = 2.636$	$y_0' = \frac{1}{2}(0.5 + 2.636) = 1.568$
$x_2 = 1$	$y_2 = 3.595$	$y_2' = \frac{1}{2}(1 + 3.595) = 2.2975$
$x_3 = 1.5$	$y_3 = 4.968$	$y_3' = \frac{1}{2}(1.5 + 4.968) = 3.234$

By the predictor formula, we have

$$y_4 = y_0 + \frac{4h}{3}[2y_1' - y_2' + 2y_3']$$

$$= 2 + \frac{4}{3}(0.5)[2 \times 1.568 - 2.2975 + 2 \times 3.234]$$

$$= 2 + \frac{2}{3}[7.3065) = 6.871$$

Now $$y_4' = \frac{1}{2}(x_4 + y_4) = \frac{1}{2}(2 + 6.871) = 4.4355$$

Now by corrector formula

$$y_4 = y_2 + \frac{h}{3}[y_2' + 4y_3' + y_4']$$

$$= 3.595 + \frac{0.5}{3}[2.2975 + 4 \times 3.234 + 4.4355]$$

$$= 3.595 + \frac{0.5}{3}[19.669] = 6.873166 \simeq 6.8732$$

Corrected $$y_4' = \frac{1}{2}(x_4 + y_4) = \frac{1}{2}(2 + 6.8732) = 4.4366$$

Again using corrector formula, we get

$$y_4 = y_2 + \frac{h}{3}[y_2' + 4y_3' + y_4']$$

$$= 3.595 + \frac{0.5}{3}[2.2975 + 4 \times 3.234 + 4.4366]$$

$$= 3.595 + \frac{0.5}{3}[19.6701] = 6.87335 \simeq 6.8734$$

Hence $y(2) = 6.8734$. **Ans.**

Example 10. *Solve initial value problem* $\frac{dy}{dx} = 1 + xy^2$, $y(0) = 1$, $h = 0.1$ *for* $x = 0.4$ *by using Milne's method when it is given*

x	*0.1*	*0.2*	*0.3*
y	*1.105*	*1.223*	*1.355*

Solution. Here $f(x, y) = 1 + xy^2$ then, we have

$x_0 = 0$	$y_0 = 1$	$y_0' = 1 + 0 \times 1^2 = 1$
$x_1 = 0.1$	$y_1 = 1.105$	$y_1' = 1 + (0.1)(1.105)^2 = 1.1221$
$x_2 = 0.2$	$y_2 = 1.223$	$y_2' = 1 + (0.2)(1.223)^2 = 1.2991$
$x_3 = 0.3$	$y_3 = 1.355$	$y_3' = 1 + (0.3)(1.355)^2 = 1.5508$

By the predictor formula, we have

$$y_4 = y_0 + \frac{4h}{3}[2y_1' - y_2' + 2y_3']$$

$$= 1 + \frac{4 \times (0.1)}{3}[2 \times 1.1221 - 1.2991 + 2 \times 1.5508]$$

$$= 1 + \frac{0.4}{3}[4.0467] = 1.53956 \simeq 1.539$$

Now $y_4' = 1 + x_4 y_4^2 = 1 + (0.4)(1.539)^2 = 1.9474$

Now by corrector formula

$$y_4 = y_2 + \frac{h}{3}[y_2' + 4y_3' + y_4']$$

$$= 1.223 + \frac{0.1}{3}[1.2991 + 4 \times 1.5508 + 1.9474]$$

$$= 1.223 + \frac{0.1}{3}[9.4497] = 1.53799$$

Corrected $y_4' = 1 + x_4 y_4^2 = 1 + (0.4)(1.538)^2 = 1.9461$

Again using corrector formula

$$y_4 = y_2 + \frac{h}{3}[y_2' + 4y_3' + y_4']$$

$$= 1.223 + \frac{0.1}{3}[1.2991 + 4 \times 1.5508 + 1.9474]$$

$$= 1.223 + \frac{0.1}{3}[9.4497] = 1.53799$$

Hence $y(0.4) = 1.53799 \cong 1.538.$ **Ans.**

EXERCISE 3.1

1. Use Picard's method to solve $\frac{dy}{dx} = 1 + xy$ with $y(2) = 0$.
2. Use Picard's method to solve $\frac{dy}{dx} = x + y$ for $x = 0.1$ and $x = 0.2$ with $x_0 = 0$, $y_0 = 1$.
3. Use Picard's method to solve $\frac{dy}{dx} = y - x$ with $y = 2$ when $x = 0$ up to third order of approximation.
4. Find the solution of $\frac{dy}{dx} = 1 + xy$ with $y(0) = 1$ in the interval [0, 0.5] by using Picard's method such that the value of y is correct to three decimal place. (Take $h = 0.1$)
5. Use Picard's method to approximation the value of y when $x = 0.1$ given that $y = 1$ when $x = 0$, $\frac{dy}{dx} = 3x + y^3$.
6. Use Euler's method compute $y(0.04)$ for the differential equation $\frac{dy}{dx} = -y$ with $y = 1$ at $x = 0$.
7. Use Euler's method with $h = 0.1$ to find the solution of the differential equation $\frac{dy}{dx} = x^2 + y^2$ with $y(0) = 0$ in the range $0 \le x \le 0.5$.
8. Given $\frac{dy}{dx} = x + y$, find the value of y in the range $0 \le x \le 1$ with $h = 0.1$ given that $y(0) = 1$ using Euler's method.
9. Find $y(2.2)$ for $\frac{dy}{dx} = -xy^2$ where $y(2) = 1$ by Euler's modified method.
10. Using Euler's modified method, compute $y(2)$ in steps of 0.2 given that $\frac{dy}{dx} = 2 + \sqrt{xy}$ with $y(1) = 1$.
11. Given that $\frac{dy}{dx} = \log_{10}(x + y)$ with the initial condition $y(0) = 1$, find y at $x = 0.2, 0.3, 0.4$ and 0.5 in steps of 0.1.
12. Using Runge-Kutta's method, find the approximate value of $x = 0.1$ and $x = 0.2$ if $\frac{dy}{dx} = x + y^2$ given that at $x = 0$, $y = 1$.

13. Solve $\frac{dy}{dx} = \frac{1}{x+y}$ for $x = 0.5, 1.0, 1.5, 2$ $[h = 0.5]$ by using Runge-Kutta's method with $x_0 = 0$, $y_0 = 1$.

14. Solve the equation $\frac{dy}{dx} = x + y$ with initial condition $y(0) = 1$ by Runge-Kutta's rule from $x = 0$ to $x = 0.4$ with $h = 0.1$.

15. Use Runge-Kutta's method to calculate the approximate value of $x = 0.8$ if $\frac{dy}{dx} = \sqrt{x+y}$ given that $y\,(0.4) = 0.41$ and $h = 0.2$.

16. Solve the equation $\frac{dy}{dx} = x - y^2$ for $x = 0.8$ given that $y(0) = 0$, $y(0.2) = 0.02$, $y(0.4) = 0.0795$, $y(0.6) = 0.1762$ by Milne's method.

17. Use Milne's method to solve $\frac{dy}{dx} = x + y$ with initial condition $y(0) = 1$ from $x = 0.20$ to $x = 0.30$

18. Use Milne's method to solve the equation $\frac{dy}{dx} = 2e^x - y$ at $x = 0.4$ given that $y(0) = 2$, $y(0.1) = 2.01$, $y(0.2) = 2.04$, $y(.3) = 2.09$.

19. Solve the equation $\frac{dy}{dx} = 1 + y^2$ for $x = 0.8$ and 1.0 given that $y(0) = 0$, $y(0.2) = .2027$, $y(0.4) = 0.4228$, $y(0.6) = 0.6841$.

ANSWERS

1. $y^{(3)} = \frac{x^5}{15} - \frac{x^4}{4} + \frac{x^3}{3} - \frac{x^2}{2} + x - \frac{22}{15}$

2. $(y)_{x=0.1} = 1.1103$ and $(y)_{x=0.2} = 1.2427$

3. $y^{(3)} = -\frac{x^4}{24} + \frac{x^3}{6} + \frac{x^2}{2} + 2x + 2$

$(y)_{x=0.4} = 1.505$, $(y)_{x=0.5} = 1.677$

4. $(y)_0 = 1$, $(y)_{0.1} = 1.105$, $(y)_{x=0.2} = 1.203$, $(y)_{x=0.3} = 1.355$

5. $(y)_{x=0.1} = 1.127$

6. -0.6705

7. $y(0.5) = 0.03002$

8. $y(1) = 3.1874$

9. 0.7018

10. $y(2) = 5.0516$

11. $y(0.2) = 1.0083$, $y(0.3) = 1.0184$

$y(0.4) = 1.0322$, $y(0.5) = 1.0$

12. $y(0.1) = 1.1165$, $y(0.2) = 1.2736$

13. $y(0.5) = 1.3571$, $y(1) = 1.5837$

$y(1.5) = 1.7555$, $y(2) = 1.8957$

14. $y(0.1) = 1.1103$, $y(0.2) = 1.2428$

$y(0.3) = 1.3997$, $y(0.4) = 1.5836$

15. $y(0.6) = 0.61035$, $y(0.8) = 0.84899$

16. $(y)_{x=0.8} = 0.3046$

17. $(y)_{x=0.20} = 1.2428$ and $(y)_{x=0.30} = 1.3997$

18. $y(0.4) = 2.1621$

19. $(y)_{0.8} = 1.0294$ and $(y)_1 = 1.5557$.

CHAPTER 4

Difference Equations

INTRODUCTION

In this chapter, we shall discuss basic concepts of difference equation, homogeneous and non-homogeneous linear difference equation with constant coefficient.

4.1. DIFFERENCE EQUATIONS

An equation which contain independent variable, dependent variable and the successive differences of the dependent variable is called the difference equation.

Examples

(1) $y_{n+2} - 6y_{n+1} + 9y_n = 0$

(2) $(E^2 + 6E + 9)y_n = 2^n$ *i.e.*, $y_{n+2} + 6y_{n+1} + y_n = 2^n$

(3) $(\Delta^2 + 3\Delta + 2)y_n = 1$ or $\Delta^2 y_n + 3\Delta y_n + 2y_n = 1$

or $$(y_{n+2} - 2y_{n+1} + y_n) + 3(y_{n+1} - y_n) + 2y_n = 1$$

or $$y_{n+2} + y_{n+1} = 1.$$

4.2. ORDER OF DIFFERENCE EQUATION

The order of the difference equation is the difference between the largest and smallest arguments occurring in the difference equation divided by the unit of argument.

Thus, the order of the difference equation

$$= \frac{\text{Largest argument} - \text{Smallest argument}}{\text{Unit of argument}}$$

The order of the difference equation $y_{n+2} - 7y_n = 5$ is

$$\frac{(n+2) - n}{1} = 2.$$

4.3. DEGREE OF DIFFERENCE EQUATION

The highest degree of y_n's in the difference equation is called the degree of the difference equation.

Example 1. The order and degree of $y_{n+2} + 4y_{n+1} + 4y_n = 2^n$ are $\frac{n+2-n}{1} = 2$ and 1 respectively.

Example 2. The order and degree of $y_{n+3} + 3y_{n+2}^2 + 3y_{n+1} + y_n = 0$ are $\frac{n+3-n}{1} = 3$ and 2 respectively.

4.4. SOLUTION OF DIFFERENCE EQUATION

Any function which satisfies the given difference equation is called the solution of the difference equation.

A solution in which the number of arbitrary constants is equal to the order of the difference equation is called **general solution** of the difference equation.

A solution which is obtained from the general solution by assigning particular values is called particular solution.

4.5. FORMATION OF DIFFERENCE EQUATION

Example 1. *Write the difference equation $\Delta^3 y_n + \Delta^2 y_n + \Delta y_n + y_n = 0$ in the subscript notation.*

Solution. We have
$$\Delta y_n = y_{n+1} - y_n$$
$$\Delta^2 y_n = y_{n+2} - 2y_{n+1} + y_n$$
$$\Delta^3 y_n = y_{n+3} - 3y_{n+2} + 3y_{n+1} - y_n$$

Putting in given difference equation

$$(y_{n+3} - 3y_{n+2} + 3y_{n+1} - y_n) + (y_{n+2} - 2y_{n+1} + y_n) + (y_{n+1} - y_n) + y_n = 0$$

or
$$y_{n+3} - 2y_{n+2} + 2y_{n+1} = 0$$
or
$$E^3 y_n - 2E^2 y_n + 2E y_n = 0$$
or
$$(E^3 - 2E^2 + 2E) y_n = 0.$$

Example 2. *Form the difference equation from the equation $y = Ax^2 - Bx$.*

Solution. We have
$$\Delta y = \Delta(Ax^2 - Bx) = A\Delta x^2 - B\Delta x$$
$$= A\{(x+1)^2 - x^2\} - B\{(x+1) - x\}$$
$$\Delta y = A(2x+1) - B \qquad ...(i)$$
and
$$\Delta^2 y = \Delta\{A(2x+1) - B\} = A\,\Delta\,(2x+1) - B\Delta$$
$$= A\{2(x+1) + 1 - (2x+1)\}$$

$$\Delta^2 y = 2A \qquad ...(ii)$$

By (*ii*) $\quad A = \frac{1}{2}\Delta^2 y$

Put in (*i*), we get $\quad B = \frac{1}{2}\Delta^2 y\,(2x + 1) - \Delta y$

or $\quad B = \frac{1}{2}(2x + 1)\Delta^2 y - \Delta y$

Putting these values in given equation, we get

$$y = \left[\frac{1}{2}\Delta^2 y\right]x^2 - \left[\frac{1}{2}(2x+1)\,\Delta^2 y - \Delta y\right]x$$

$$= \left[\frac{1}{2}x^2 - \frac{1}{2}(2x+1)x\right]\Delta^2 y + x\Delta y$$

$$2y = [-x^2 - x]\,\Delta^2 y + 2x\,\Delta y$$

or $$(x^2 + x)\Delta^2 y - x\,\Delta y + 2y = 0$$

or $$(x^2 + x)(y_{n+2} - 2y_{n+1} + y_n) - 2x\,(y_{n+1} - y_n) + 2y_n = 0$$

or $$(x^2 + x)y_{n+2} - (2x^2 + 4x)y_{n+1} + (x^2 + 3x + 2)y_n = 0.$$ **Ans.**

Example 3. *Form by* $y_n = A2^n + B(-2)^n$ *a difference equation by eliminating the constants A and B.*

Solution. We have

$$y_n = A2^n + B(-2)^n \quad \Rightarrow \quad y_n - A2^n - B(-2)^n$$

$$y_{n+1} = 2A2^n - 2B(-2)^n \quad \Rightarrow \quad y_{n+1} - 2A2^n + 2B(-2)^n$$

$$y_{n+2} = 4A2^n + 4B(-2)^n \quad \Rightarrow \quad y_{n+2} - 4A2^n - 4B(-2)^n$$

Eliminate A and B, we get

$$\begin{vmatrix} y_n & -1 & -1 \\ y_{n+1} & -2 & 2 \\ y_{n+2} & -4 & -4 \end{vmatrix} = 0 \quad \text{or} \quad y_{n+2} - 4y_n = 0.$$ **Ans.**

4.6. LINEAR DIFFERENCE EQUATION

A difference equation in which y_n, y_{n+1}, y_{n+2} ...... occur to the first degree only and are not multiplied together is called the linear difference equation.

A linear difference equations of order κ is

$$a_0\, y_{k+n} + a_1 y_{k+n-1} + a_2 y_{k+n-2} + \ldots\ldots + a_k y_n = R(n) \qquad ...(4.1)$$

If $R(n) = 0$ then the equation (4.1) is called the linear homogeneous difference equation otherwise it is called non-homogeneous difference equation.

4.7. HOMOGENEOUS LINEAR DIFFERENCE EQUATION WITH CONSTANT COEFFICIENT

An equation of the form

$$(a_0E^k + a_1E^{k-1} + \ldots\ldots + a_nI)y_n = 0 \qquad \ldots(4.2)$$

or

$$\phi(E)y_n = 0$$

where $a_0, a_1, \ldots\ldots a_n$ are all constants, is known as homogeneous linear difference equation with constant coefficient

Let $y_n = m^n$ be the solution of the difference equation (4.2), then we have

$$a_0m^{n+k} + a_1m^{n+k-1} + \ldots\ldots + a_nm^n = 0$$

$$\Rightarrow \qquad (a_0m^k + a_1m^{k-1} + \ldots\ldots + a_n)m^n = 0 \qquad \ldots(4.3)$$

which show that m^n is the solution of (4.3) if m satisfies

$$a_0m^k + a_1m^{k-1} + \ldots\ldots + a_n = 0 \qquad \ldots(4.4)$$

Equation (4.4) is known as auxiliary equation. This equation has k roots which we take m_1, $m_2,\ldots,m_k$. Here some cases arise.

Case I. When roots are all real and distinct

If $m_1, m_2 \ldots\ldots m_k$ are real and distinct roots of auxiliary equation (4.4), then the solution is

$$y_n = c_1(m_1)^n + c_2(m_2)^n \ldots\ldots + c_k(m_k)^n.$$

Case II. When some of the roots are equal

If two roots be equal *i.e.*, $m_1 = m_2$, then the solution is $y_n = (c_1 + c_2n)(m_1)^n + c_3(m_3)^n + \ldots\ldots + c_k(m_k)^n$.

If three roots be equal *i.e.*, $m_1 = m_2 = m_3$, then the solution is

$$y_n = (c_1 + c_2n + c_3n^2)(m_1)^n + c_4(m_4)^n + \ldots\ldots + c_k(m_k)^n$$

If k roots be equal *i.e.*, $m_1 = m_2 = m_3 \ldots\ldots = m_k$, then the solution is

$$y_n = [c_1 + c_2n + c_3n^2 + \ldots\ldots + c_k n^{k-1}](m_1)^n.$$

Case III. When the roots are complex number

We know that if complex roots occur then they must be conjugate complex number

i.e., if $(\alpha + i\beta)$ is the root then $(\alpha - i\beta)$ is also a root where α and β are real. Then the solution is

$$y_n = c_1(\alpha + i\beta)^n + c_2(\alpha - i\beta)^n$$

which can be written as

$$y_n = \gamma^n (A_1 \cos n\theta + A_2 \sin n\theta)$$

where $\gamma = \sqrt{\alpha^2 + \beta^2}$ and $A_1 = c_1 + c_2, A_2 = i(c_1 - c_2)$

$$\theta = \tan^{-1}(\beta/\alpha).$$

Case IV. When some of the complex roots are equal

Let the root $(\alpha \pm i\,\beta)$ becomes twice then the solution.

$$y_n = \gamma^n [(A_1 + A_2 n)\cos n\theta + (A_3 + A_4 n)\sin n\theta]$$

where $\gamma = \sqrt{\alpha^2 + \beta^2}$ and $\theta = \tan^{-1}(\beta/\alpha)$.

4.8. NON-HOMOGENEOUS LINEAR DIFFERENCE EQUATION WITH CONSTANT COEFFICIENT

The equation of the form

$$(a_0 E^k + a_1 E^{k-1} + \ldots\ldots + a_n I) y_n = R(n) \qquad \ldots(4.5)$$

or

$$\phi(E) y_n = R(n)$$

where $a_0, a_1, a_2, \ldots.. a_n$ are all constants is known as non-homogeneous linear difference equation with constant coefficient.

The general solution of (4.5) consists of two parts, complementary function and particular integral Complementary function is the general solution of the homogeneous equation *i.e.*, left hand side of (4.5) and particular integral $= \dfrac{1}{\phi(E)} R(n)$.

Rules for Obtaining the Particular Integral

The particular integral (P.I.) $= \dfrac{1}{\phi(E)} R(n)$.

Case I. When $R(n) = a^n$

$$\therefore \quad \text{P.I.} = \frac{1}{\phi(E)} a^n = \frac{1}{\phi(a)} a^n \text{ provided } \phi(a) \neq 0$$

If $\phi(a) = 0$, then for the equation

$(E - a)^k y_n = a^n$; k is positive integer.

$$\text{P.I.} = \frac{1}{(E-a)^k} a^n = \frac{n(n-1)(n-2) \ldots\ldots (n-(k-1))}{k!} a^{n-k}$$

Particular case: When $k = 1$

$$\text{P.I.} = \frac{1}{E-a} a^n = n a^{n-1}$$

When $k = 2$

$$\text{P.I.} = \frac{1}{(E-a)^2} a^n = n(n-1) a^{n-2}.$$

Case II. When $R(n) = \sin an$ or $\cos an$

$$\therefore \quad \text{P.I.} = \frac{1}{\phi(E)} \sin an = \frac{1}{\phi(E)} \frac{e^{ian} - e^{-ian}}{2i}$$

$$= \frac{1}{2i}\left\{\frac{1}{\phi(E)}(e^{ia})^n - \frac{1}{\phi(E)}(e^{-ia})^n\right\}$$

$$= \frac{1}{2i}\left\{\frac{1}{\phi(e^{ia})}e^{ian} - \frac{1}{\phi(e^{-ia})}e^{-ian}\right\}$$

Similarly $$\text{P.I.} = \frac{1}{\phi(E)}\cos an = \frac{1}{2}\left\{\frac{1}{\phi(e^{ia})}e^{ian} + \frac{1}{\phi(e^{-ia})}e^{-ian}\right\}$$

Provided $\phi\,(e^{ia}) \neq 0$ and $\phi\,(e^{-ia}) \neq 0$.

Case III. When $R(n) = n^k$

$\therefore$ $$\text{P.I.} = \frac{1}{\phi(E)}n^k = \frac{1}{\phi(1+\Delta)}n^k \qquad [\because\ E = 1 + \Delta]$$

$$= [\phi\,(1 + \Delta)]^{-1}\,n^k.$$

First, we expand $[\phi(1 + \Delta)]^{-1}$ in ascending power of Δ by the Binomial theorem as far as the term in Δ^k, then express n^k in the factorial notation and distribute it each term of the expansion.

Case IV. When $R(n) = a^n\,F(n)$, where $F(n)$ being a polynomial in n.

$\therefore$ $$\text{P.I.} = \frac{1}{\phi(E)}a^n\,F(n) = a^n\,\frac{1}{\phi(aE)}F(n).$$

SOLVED EXAMPLES

Example 1. *Solve the difference equation* $y_{n+2} - 2y_{n+1} - 8y_n = 0$.

Solution. The given equation can be written as

$$(E^2 - 2E - 8)\,y_n = 0$$

The auxiliary equation is $m^2 - 2m - 8 = 0$

The roots are $m = -2, 4$

Hence, the solution is $y_n = c_1\,(-2)^n + c_2 4^n$. **Ans.**

Example 2. *Solve the difference equation* $y_{n+3} - 2y_{n+2} - 5y_{n+1} + 6y_n = 0$.

Solution. The given difference equation can be written as

$$(E^3 - 2E^2 - 5E + 6)y_n = 0$$

The auxiliary equation is $m^3 - 2m^2 - 5m + 6 = 0$

The roots are $m = 1, -2, 3$

Hence, the solution is $y_n = c_1(1)^n + c_2(-2)^n + c_3 3^n$. **Ans.**

Example 3. *Solve* $y_{n+3} + y_{n+2} - 8y_{n+1} - 12y_n = 0$.

Solution. The given difference equation can be written as

$$(E^3 + E^2 - 8E - 12)y_n = 0$$

The auxiliary equation is $m^3 + m^2 - 8m - 12 = 0$

The roots are $m = -2, -2, 3$

Hence, the solution is $y_n = (c_1 + c_2 n)(-2)^n + c_3 3^n$. **Ans.**

Example 4. *Solve the difference equation* $9y_{n+2} - 6y_{n+1} + y_n = 0$.

Solution. The given difference equation can be written as

$$(9E^2 - 6E + 1)y_n = 0$$

The auxiliary equation is $9m^2 - 6m + 1 = 0$

The roots are $m = \frac{1}{3}, \frac{1}{3}$

Hence, the solution is $y_n = (c_1 + c_2 n)\left(\frac{1}{3}\right)^n$.

Example 5. *Solve* $y_{n+2} + 16y_n = 0$.

Solution. The given difference equation can be written as

$$(E^2 + 16)y_n = 0$$

The auxiliary equation is $m^2 + 16 = 0$

The roots are $m = \pm 4i$

Hence, the solution is

$$y_n = c_1(4i)^n + c_2(-4i)^n$$

$$y_n = 4^n \{c_1(i)^n + c_2(-i)^n\}$$

$$= 4^n \left\{c_1\left(\cos\frac{\pi}{2} + i\sin\frac{\pi}{2}\right)^n + c_2\left(\cos\frac{\pi}{2} - i\sin\frac{\pi}{2}\right)^n\right\}$$

$$= 4^n \left\{c_1\left(\cos\frac{n\pi}{2} + i\sin\frac{n\pi}{2}\right) + c_2\left(\cos\frac{n\pi}{2} - i\sin\frac{n\pi}{2}\right)\right\}$$

$$= 4^n \left\{(c_1 + c_2)\cos\frac{n\pi}{2} + i(c_1 - c_2)\sin\frac{n\pi}{2}\right\}$$

$$= 4^n \left[A\cos\frac{n\pi}{2} + B\sin\frac{n\pi}{2}\right].$$

Example 6. *Solve* $y_{n+2} - 4y_{n+1} + 13y_n = 0$.

Solution. The given difference equation can be written as

$$(E^2 - 4E + 13)y_n = 0$$

The auxiliary equation is $m^2 - 4m + 13 = 0$

The roots are $m = \dfrac{4 \pm \sqrt{16 - 4 \times 13}}{2} = \dfrac{4 \pm 6i}{2} = 2 \pm 3i$

Let $2 + 3i = r(\cos\theta + i\sin\theta) \Rightarrow r = (13)^{1/2}$ and $\theta = \tan^{-1}\left(\frac{3}{2}\right)$

Therefore solution is

$$y_n = (13)^{n/2}\{c_1 \cos n\theta + c_2 \sin n\theta\} \text{ where } \theta = \tan^{-1}\left(\frac{3}{2}\right).$$

Example 7. *Solve* $y_{n+2} - y_{n+1} + y_n = 0$ *given that* $y_0 = 1,\ y_1 = \dfrac{1+\sqrt{3}}{2}$.

Solution. The given difference equation can be written as $(E^2 - E + 1)\, y_n = 0$

The auxiliary equation is $m^2 - m + 1 = 0$

The roots are $$m = \frac{1 \pm \sqrt{1-4}}{2} = \frac{1 \pm \sqrt{3}i}{2}$$

Hence the solution is $$y_n = c_1 \left(\frac{1 + i\sqrt{3}}{2}\right)^n + c_2 \left(\frac{1 - i\sqrt{3}}{2}\right)^n$$

$$= c_1 \left(\cos\frac{\pi}{3} + i \sin\frac{\pi}{3}\right)^n + c_2 \left(\cos\frac{\pi}{3} - i \sin\frac{\pi}{3}\right)^n$$

$$= c_1 \left(\cos\frac{n\pi}{3} + i \sin\frac{n\pi}{3}\right) + c_2 \left(\cos\frac{n\pi}{3} - i \sin\frac{n\pi}{3}\right)$$

$$= (c_1 + c_2) \cos\frac{n\pi}{3} + i\,(c_1 - c_2) \sin\frac{n\pi}{3}$$

$$y_n = A \cos\frac{n\pi}{3} + B \sin\frac{n\pi}{3} \qquad \ldots(i)$$

But $y_0 = 1$ and $y_1 = \dfrac{1+\sqrt{3}}{2}$

$\therefore$ $$1 = A \cos 0 + B \sin 0 \Rightarrow A = 1$$

and $$\frac{1+\sqrt{3}}{2} = A \cos\frac{\pi}{3} + B \sin\frac{\pi}{3}$$

$$\frac{1+\sqrt{3}}{2} = 1 . \frac{1}{2} + B . \frac{\sqrt{3}}{2} \Rightarrow B = 1$$

Put $A = 1$ and $B = 1$ in (*i*), we get

$$y_n = \cos\frac{n\pi}{3} + \sin\frac{n\pi}{3}.$$

Example 8. *Solve the difference equation* $y_{n+2} - 2y_{n+1} + 5y_n = 2 . 3^n - 4 . 7^n$.

Solution. The given difference equation can be written as

$$(E^2 - 2E + 5)y_n = 2 . 3^n - 4 . 7^n$$

The auxiliary equation is $m^2 - 2m + 5 = 0$

The roots are $$m = \frac{2 \pm \sqrt{4 - 4 \times 5}}{2 \cdot 1} = 1 \pm 2i$$

Let $$1 + 2i = r(\cos\theta + i \sin\theta) \Rightarrow r = \sqrt{5} \text{ and } \theta = \tan^{-1} 2$$

Therefore complementary solution is

$$= (5)^{n/2} (c_1 \cos n\theta + c_2 \sin n\theta) \text{ where } \theta = \tan^{-1} 2$$

The particular solution is given by

$$\text{P.I.} = \frac{1}{E^2 - 2E + 5}\, 2\,.\,3^n - 4\,.\,7^n$$

$$= 2\,\frac{1}{E^2 - 2E + 5}\, 3^n - 4\,\frac{1}{E^2 - 2E + 5}\, 7^n$$

$$= 2\,\frac{1}{8}\, 3^n - 4\,\frac{1}{40}\, 7^n = \frac{1}{4}\, 3^n - \frac{1}{10}\, 7^n$$

Hence solution is $y_n = \text{C.F.} + \text{P.I.}$

$$= (5)^{n/2}\,\{c_1 \cos n\theta + c_2 \sin n\theta\} + \frac{1}{4}\, 3^n - \frac{1}{10}\, 7^n.$$ **Ans.**

Example 9. *Solve the difference equation* $(E^2 - 3E - 4)y_n = 3^n$.

Solution. The auxiliary equation of given difference equation is $m^2 - 3m - 4 = 0$

The roots are $m = -1, +4$.

Therefore complementary solution is $c_1\,(-1)^n + c_2\, 4^n$.

The particular solution is given by

$$\text{P.I.} = \frac{1}{E^2 - 3E - 4}\, 3^n = \frac{1}{9 - 9 - 4}\, 3^n = -\frac{1}{4}\, 3^n$$

Hence solution is $y_n = \text{C.F.} + \text{P.I.} = c_1(-1)^n + c_2 4^n - \frac{1}{4}\, 3^n$. **Ans.**

Example 10. *Solve the difference equation*

$$y(n+2) - 3y(n+1) + 2y(n) = 6\,.2^n$$

or

$$y_{n+2} - 3y_{n+1} + 2y_n = 6\,.2^n.$$

Solution. The given difference equation can be written as

$$(E^2 - 3E + 2)y_n = 6\,.2^n$$

The auxiliary equation is $m^2 - 3m + 2 = 0$

The roots are $m = 1, 2$.

Therefore the complementary solution is $c_1(1)^n + c_2(2)^n$.

The particular solution is given by

$$\text{P.I.} = \frac{1}{E^2 - 3E + 2}\, 6\,.\,2^n \qquad [\because\ F(2) = 0]$$

$$= \frac{1}{(E-1)(E-2)}\, 6\,.\,2^n = \frac{1}{(2-1)(E-2)}\, 6\,.\,2^n = 6\,\frac{1}{(E-2)}\, 2^n$$

$$= 6.\, n\, 2^{n-1} = 3n2^n.$$

Hence solution is $y_n = \text{C.F.} + \text{P.I.}$

$$y_n = c_1 + c_2 2^n + 3n2^n.$$ **Ans.**

Aliter :

$$\text{P.I.} = \frac{1}{E^2 - 3E + 2} \, 6 \,.\, 2^n$$

$$= 6 \frac{1}{E^2 - 3E + 2} \, 2^n \,.\, 1 \qquad [\because \quad F(2) = 0]$$

$$= 6 \,.\, 2^n \frac{1}{(2E)^2 - 3(2E) + 2} \,.1 = 6 \,.\, 2^n \frac{1}{4E^2 - 6E + 2} \,.1$$

$$= 6 \,.\, 2^n \frac{1}{4(1+\Delta)^2 - 6(1+\Delta) + 2} \,.1$$

$$= 6 \,.\, 2^n \frac{1}{4(1 + 2\Delta + \Delta^2) - 6(1+\Delta) + 2} \, 1 = 6.2^n \frac{1}{4\Delta^2 + 2\Delta} \,.1$$

$$= 3 \,.\, 2^n \frac{1}{\Delta(1+2\Delta)} \,.1 = 3 \,.\, 2^n \frac{1}{\Delta} \, (1 + 2\Delta)^{-1} \,.1$$

$$= 3 \,.\, 2^n \frac{1}{\Delta} \, (1 - 2\Delta + 4\Delta^2 \,......) \,.\, 1 = 3 \,.\, 2^n \frac{1}{\Delta} \, (1 - 0 + 0...)$$

$$\text{P.I.} = 3 \,.\, 2^n \, n.$$

Example 11. *Solve* $y_{n+2} + a^2 y_n = \cos an.$

Solution. The given difference equation can be written as

$$(E^2 + a^2) y_n = \cos an$$

The auxiliary equation is $m^2 + a^2 = 0$

The roots are $m = \pm\, ia.$

Therefore the complementary solution is

$c_1(ia)^n + c_2(-\,ia)^n = a^n \, \{c_1(i)^n + c_2(-\,i)^n)$

$$= a^n \left\{ c_1 \left(\cos \frac{\pi}{2} + i \sin \frac{\pi}{2} \right)^n + c_2 \left(\cos \frac{\pi}{2} - i \sin \frac{\pi}{2} \right)^n \right\}$$

$$= a^n \left\{ c_1 \left(\cos \frac{n\pi}{2} + i \sin \frac{n\pi}{2} \right) + c_2 \left(\cos \frac{n\pi}{2} - i \sin \frac{n\pi}{2} \right) \right\}$$

$$= a^n \left\{ (c_1 + c_2) \cos \frac{n\pi}{2} + i \, (c_1 - c_2) \sin \frac{n\pi}{2} \right\} = a^n \left\{ A \cos \frac{n\pi}{2} + B \sin \frac{n\pi}{2} \right\}.$$

The particular solution is given by

$$\text{P.I.} = \frac{1}{E^2 + a^2} \cos an = \frac{1}{(E^2 + a^2)} \frac{e^{ian} + e^{-ian}}{2}$$

$$= \frac{1}{2} \left\{ \frac{1}{E^2 + a^2} (e^{ia})^n + \frac{1}{(E^2 + a^2)} (e^{-ia})^n \right\}$$

$$= \frac{1}{2} \left\{ \frac{1}{e^{2ia} + a^2} e^{ian} + \frac{1}{e^{-2ia} + a^2} e^{-ian} \right\}$$

$$= \frac{1}{2}\left\{\frac{e^{ian}(e^{-2ia}+a^2)+e^{-ian}(e^{2ia}+a^2)}{(e^{2ia}+a^2)(e^{-2ia}+a^2)}\right\}$$

$$= \frac{1}{2}\left\{\frac{e^{ia(n-2)}+e^{-ia(n-2)}+a^2(e^{ian}+e^{-ian})}{1+a^2(e^{2ia}+e^{-2ia})+a^4}\right\}$$

$$= \frac{1}{2}\frac{2\cos a(n-2)+a^2 2\cos an}{1+2a^2\cos 2a+a^4} = \frac{\cos a(n-2)+a^2\cos an}{1+2a^2\cos 2a+a^4}$$

Hence solution $\quad y_n = \text{C.F.} + \text{P.I.}$

$$y_n = a^n\left\{A\cos\frac{n\pi}{2}+B\sin\frac{n\pi}{2}\right\}+\frac{\cos a(n-2)+a^2\cos an}{1+2a^2\cos 2a+a^4}$$

Aliter: $\quad \text{P.I.} = \frac{1}{E^2+a^2}\cos an$

$$= \text{Real part of } \frac{1}{E^2+a^2}(\cos an + i\sin an)$$

$$= \text{Real part of } \frac{1}{E^2+a^2}e^{ian}$$

$$= \text{Real part of } \frac{1}{E^2+a^2}(e^{ia})^n$$

$$= \text{Real part of } \frac{1}{e^{2ia}+a^2}e^{ian}$$

$$= \text{Real part of } \frac{e^{-2ia}+a^2}{(e^{2ia}+a^2)(e^{-2ia}+a^2)}e^{ian}$$

$$= \text{Real part of } \frac{e^{ia(n-2)}+a^2e^{ian}}{1+a^2(e^{2ia}+e^{-2ia})+a^4}$$

$$= \text{Real part of } \frac{\cos a(n-2)+i\sin a(n-2)+a^2(\cos an+i\sin an)}{1+2a^2\cos 2a+a^4}$$

$$= \text{Real part of } \frac{\cos a(n-2)+a^2\cos an+i(\sin a(n-2)+a^2\sin an)}{1+2a^2\cos 2a+a^4}$$

$$= \frac{\cos a(n-2)+a^2\cos an}{1+2a^2\cos 2a+a^4}.$$

Example 12. *Solve* $y_{n+2} - 2\cos\alpha\, y_{n+1} + y_n = \cos\alpha n$.

Solution. The given difference equation can be written as

$$(E^2 - 2\cos\alpha\, E + 1)y_n = \cos\alpha n$$

The auxiliary equation is $m^2 - 2\cos\alpha\, m + 1 = 0$

The roots are $$m = \frac{2\cos\alpha \pm \sqrt{4\cos^2\alpha - 4}}{2.1} = (\cos\alpha \pm i\sin\alpha)$$

Therefore complementary solution is

$$c_1(\cos\alpha + i\sin\alpha)^n + c_2(\cos\alpha - i\sin\alpha)^n$$
$$= c_1(\cos n\alpha + i\sin n\alpha) + c_2(\cos n\alpha - i\sin n\alpha)$$
$$= (c_1 + c_2)\cos n\alpha + i(c_1 - c_2)\sin n\alpha$$
$$= A\cos n\alpha + B\sin n\alpha$$

The particular solution is given by

$$\text{P.I.} = \frac{1}{E^2 - 2\cos\alpha E + 1}\cos n\alpha$$
$$= \frac{1}{E^2 - 2\frac{(e^{i\alpha} + e^{-i\alpha})}{2}E + 1}\,\frac{e^{in\alpha} + e^{-in\alpha}}{2}$$
$$= \frac{1}{E^2 - (e^{i\alpha} + e^{-i\alpha})E + 1}\,\frac{e^{in\alpha} + e^{-in\alpha}}{2}$$
$$= \frac{1}{2}\,\frac{1}{(E - e^{i\alpha})(E - e^{-i\alpha})}\{(e^{i\alpha})^n + (e^{-i\alpha})^n\}$$
$$= \frac{1}{2}\left\{\frac{1}{(E - e^{i\alpha})(E - e^{-i\alpha})}(e^{i\alpha})^n + \frac{1}{(E - e^{i\alpha})(E - e^{-i\alpha})}(e^{-i\alpha})^n\right\}$$
$$= \frac{1}{2}\left\{\frac{1}{(E - e^{i\alpha})(e^{i\alpha} - e^{-i\alpha})}(e^{i\alpha})^n + \frac{1}{(e^{-i\alpha} - e^{i\alpha})(E - e^{-i\alpha})}(e^{-i\alpha})^n\right\}$$
$$= \frac{1}{2}\left\{\frac{1}{(E - e^{i\alpha})\,2i\sin\alpha}(e^{i\alpha})^n + \frac{1}{(-2i\sin\alpha)(E - e^{-i\alpha})}(e^{-i\alpha})^n\right\}$$
$$= \frac{1}{4i\sin\alpha}\left\{\frac{1}{E - e^{i\alpha}}(e^{i\alpha})^n - \frac{1}{E - e^{-i\alpha}}(e^{-i\alpha})^n\right\}$$
$$= \frac{1}{4i\sin\alpha}\left\{n(e^{i\alpha})^{n-1} - n(e^{-i\alpha})^{n-1}\right\} = \frac{n}{4i\sin\alpha}(e^{i(n-1)\alpha} - e^{-i(n-1)\alpha})$$
$$= \frac{n}{4i\sin\alpha}2i\sin(n-1)\alpha = \frac{n\sin(n-1)\alpha}{2\sin\alpha}$$

Hence solution $$y_n = \text{C.F.} + \text{P.I.}$$

$$y_n = A\cos n\alpha + B\sin n\alpha + \frac{n\sin(n-1)\alpha}{2\sin\alpha}.$$ **Ans.**

Example 13. *Solve* $8y_{n+2} - 6y_{n+1} + y_n = 5 \sin\left(\frac{n\pi}{2}\right)$.

Solution. The given difference equation can be written as

$$(8E^2 - 6E + 1)y_n = 5 \sin\left(\frac{n\pi}{2}\right)$$

The auxiliary equation is $8m^2 - 6m + 1 = 0$

The roots are $m = \frac{1}{4}, \frac{1}{2}$

Therefore complementary solution is $c_1\left(\frac{1}{4}\right)^n + c_2\left(\frac{1}{2}\right)^n$

The particular solution is given by

$$\text{P.I.} = \frac{1}{8E^2 - 6E + 1} 5 \sin\left(\frac{n\pi}{2}\right)$$

$$= \frac{1}{8E^2 - 6E + 1} 5 \left(\frac{e^{i\left(\frac{n\pi}{2}\right)} - e^{-i\left(\frac{n\pi}{2}\right)}}{2i}\right)$$

$$= \frac{5}{2i} \frac{1}{8E^2 - 6E + 1}\left(\left(e^{\frac{i\pi}{2}}\right)^n - \left(e^{\frac{-i\pi}{2}}\right)^n\right) = \frac{5}{2i} \frac{1}{8E^2 - 6E + 1}(i^n - (-i)^n)$$

$$= \frac{5}{2i}\left[\frac{1}{8E^2 - 6E + 1} i^n - \frac{1}{8E^2 - 6E + 1}(-i)^n\right]$$

$$= \frac{5}{2i}\left[\frac{1}{8i^2 - 6i + 1} i^n - \frac{1}{8(-i)^2 - 6(-i) + 1}(-i)^n\right]$$

$$= \frac{5}{2i}\left[-\frac{1}{7 + 6i} i^n + \frac{1}{7 - 6i}(-i)^n\right]$$

$$= \frac{5}{2i}\left[\frac{-(7 - 6i)\left(\cos\frac{\pi}{2} + i \sin\frac{\pi}{2}\right)^n + (7 + 6i)\left(\cos\frac{\pi}{2} - i \sin\frac{\pi}{2}\right)^n}{(7 + 6i)(7 - 6i)}\right]$$

$$= \frac{5}{2i}\left[\frac{-(7 - 6i)\left(\cos\frac{n\pi}{2} + i \sin\frac{n\pi}{2}\right) + (7 + 6i)\left(\cos\frac{n\pi}{2} - i \sin\frac{n\pi}{2}\right)}{49 + 36}\right]$$

$$= \frac{1}{34i}\left[-7\cos\frac{n\pi}{2} - 7i\sin\frac{n\pi}{2} + 6i\cos\frac{n\pi}{2} - 6\sin\frac{n\pi}{2} + 7\cos\frac{n\pi}{2}\right.$$

$$\left. - 7i\sin\frac{n\pi}{2} + 6i\cos\frac{n\pi}{2} + 6\sin\frac{n\pi}{2}\right]$$

$$= -\frac{i}{34}\left\{-14i\sin\frac{n\pi}{2} + 12i\cos\frac{n\pi}{2}\right\} = \frac{1}{17}\left\{6\cos\frac{n\pi}{2} - 7\sin\frac{n\pi}{2}\right\}$$

Hence solution is

$$y_n = c_1\left(\frac{1}{4}\right)^n + c_2\left(\frac{1}{2}\right)^n + \frac{1}{17}\left\{6\cos\frac{n\pi}{2} - 7\sin\frac{n\pi}{2}\right\}. \quad \textbf{Ans.}$$

Example 14. *Solve the difference equation*

$$y_{n+2} - 6y_{n+1} + 8y_n = 3n^2 + 2.$$

Solution. The given difference equation can be written as

$$(E^2 - 6E + 8)y_n = 3n^2 + 2$$

The auxiliary equation is $m^2 - 6m + 8 = 0$

The roots are $m = 2, 4$

Therefore the complementary solution is $c_1 2^n + c_2 4^n$

The particular solution is given by

$$\text{P.I.} = \frac{1}{E^2 - 6E + 8}(3n^2 + 2) = \frac{1}{(1+\Delta)^2 - 6(1+\Delta) + 8}(3n^2 + 2)$$

$$= \frac{1}{1 + 2\Delta + \Delta^2 - 6 - 6\Delta + 8}(3n^2 + 2)$$

$$= \frac{1}{\Delta^2 - 4\Delta + 3}(3n^2 + 2) = \frac{1}{3}\frac{1}{\left(1 - \frac{4\Delta - \Delta^2}{3}\right)}(3n^2 + 2)$$

$$= \frac{1}{3}\left(1 - \frac{4\Delta - \Delta^2}{3}\right)^{-1}(3n^2 + 2)$$

$$= \frac{1}{3}\left(1 + \frac{4\Delta - \Delta^2}{3} + \frac{(4\Delta - \Delta^2)^2}{9} + \ldots\ldots\right)(3n^{(2)} + 3n^{(1)} + 2)$$

$$= \frac{1}{3}(3n^{(2)} + 3n^{(1)} + 2) + \frac{4(6n+3) - 6}{3} + \frac{16 \times 6}{9}$$

$$= \frac{1}{3}\,3n(n-1) + 3n + 2 + 8n + 2 + \frac{48}{3} = n^2 + \frac{8}{3}n + \frac{44}{9}$$

Hence solution $y_n = \text{C.F.} + \text{P.I.} = c_1 2^n + c_2 4^n + n^2 + \dfrac{8n}{3} + \dfrac{44}{9}$.

Example 15. *Solve the difference equation*

$$y_{n+2} - 4y_n = n^2 + n - 1.$$

Solution. The given difference equation can be written as

$$(E^2 - 4)y_n = n^2 + n - 1$$

The auxiliary equation is $m^2 - 4 = 0$

The roots are $m = 2, -2$

Therefore the complementary solution is $c_1(2)^n + c_2(-2)^n$

The particular solution is given by

$$\frac{1}{E^2 - 4}(n^2 + n - 1) = \frac{1}{(1+\Delta)^2 - 4}(n^2 + n - 1)$$

$$= \frac{1}{\Delta^2 + 2\Delta - 3}(n^2 + n - 1)$$

$$= \frac{1}{-3\left(1 - \dfrac{2\Delta + \Delta^2}{3}\right)}(n^2 + n - 1)$$

$$= \frac{-1}{3}\left(1 - \frac{2\Delta + \Delta^2}{3}\right)^{-1}(n^2 + n - 1)$$

$$= -\frac{1}{3}\left(1 + \frac{2\Delta + \Delta^2}{3} + \left(\frac{2\Delta + \Delta^2}{3}\right)^2 + \ldots\ldots\right)(n^{(2)} + 2n^{(1)} - 1)$$

$$[\because \; n^2 + n - 1 = n^{(2)} + 2n^{(1)} - 1]$$

$$= \frac{-1}{3}\left(n^{(2)} + 2n^{(1)} - 1 + \frac{2(2n^{(1)} + 2) + 2}{3} + \frac{4.2}{9}\right)$$

$$= -\frac{1}{3}\left(n(n-1) + 2n - 1 + \frac{4}{3}n + \frac{8}{9}\right)$$

$$= -\frac{n^2}{3} - \frac{7n}{9} - \frac{17}{27}$$

Hence solution $y_n = \text{C.F.} + \text{P.I.}$

$$y_n = c_1 2^n + c_2(-2)^n - \frac{n^2}{3} - \frac{7n}{9} - \frac{17}{27}.$$

EXERCISE 4.1

Solve the following homogeneous difference equations

1. $(E^2 - 2E - 8)y_n = 0$

2. $2y_{n+2} - 5y_{n+1} + 2y_n = 0$

3. $f(n+2) + f(n+1) - 56f(n) = 0$

4. $(\Delta - 2)^2 (\Delta - 5)y_n = 0$

5. $(\Delta^2 - 5\Delta + 4)y_n = 0$

6. $y_{n+3} + y_{n+2} - y_{n+1} - y_n = 0$

7. $4y_{n+2} + 25y_n = 0$

8. $y_{n+4} - 4y_{n+3} + 6y_{n+2} - 4y_{n+1} + y_n = 0$

9. $y_{n+2} + 2y_{n+1} + 4y_n = 0$

10. $y_{n+2} - 2y_{n+1} + 5y_n = 0$

11. Solve the difference equation $y_{n+2} - 4y_{n+1} + 4y_n = 0$ and find the particular solution satisfying the initial conditions $y_0 = 1$ and $y_1 = 3$

12. Solve $2y_{n+2} - 5y_{n+1} + 2y_n = 0$ with $y_0 = 0$, $y_1 = 1$

Solve the following non-homogeneous difference equations

13. $y_{n+2} - 4y_{n+1} + 3y_n = 4^n$.

14. $y_{n+2} - 4y_n = 5 . 3n$.

15. $y_{n+2} + y_n = 5 . 2^n$ with $y_0 = 1$, $y_1 = 0$

16. $f(n+2) - 3f(n+1) + 2f(n) = 4^n$

17. $y_{n+2} - 4y_{n+1} + 4y_n = 2^n$

18. $y_{n+2} - 7y_{n+1} + 12y_n = \cos n$

19. $y_{n+2} - 16y_n = \cos(n/2)$

20. $y_{n+2} + y_n = 4\cos 2n$

21. $y_{n+2} + a^2 y_n = \sin an$

22. $y_{n+2} + y_{n+1} + y_n = n^2 + n + 1$

ANSWERS

1. $y_n = c_1(+2)^n + c_2(-4)^n$

2. $y_n = c_1\left(\frac{1}{2}\right)^n + c_2(2)^n$

3. $f(n) = c_1(7)^n + c_2(-8)^n$

4. $y_n = (c_1 + c_2 n)\, 3^n + c_3 6^n$

5. $y_n = c_1 2^n + c_2 5^n$

6. $y_n = (c_1 + c_2 n)(-1)^n + c_3$

7. $y_n = \left(\frac{5}{2}\right)^n \left[c_1 \cos\frac{n\pi}{2} + c_2 \sin\frac{n\pi}{2}\right]$

8. $y_n = (c_1 + c_2 n + c_3 n^2 + c_4 n^3)(1)^n$

9. $y_n = 2^n\, c_1 \cos\left(-\frac{\pi}{3}\right) n + c_2 \sin\left(-\frac{\pi}{3}\right) n$

10. $y_n = 5^n (c_1 \cos n\theta + c_2 \sin n\theta)$ where $\theta = \tan^{-1} 2$

11. $y_n = (c_1 + c_2 n)\, 2^n$ and $y_n = (n+2)2^{n-1}$

12. $y_n = c_1\left(\frac{1}{2}\right)^n + c_2 2^n$ and $y_n = -\frac{2}{3}\left(\frac{1}{2}\right)^n + \frac{2}{3}(2)^n$

13. $y_n = c_1 + c_2 3^n + \frac{1}{3}\, 4^n$

14. $y_n = c_1 2^n + c_2(-2)^n + 3^n$

15. $y_n = \cos\frac{\pi n}{2} + 2^n$

16. $f(n) = c_1 + c_2 2^n + \frac{1}{6}\, 4^n$

17. $y_n = (c_1 + c_2 n)2^n + n(n-1)\, 2^{n-3}$

18. $y_n = c_1 3^n + c_2 4^n + \frac{1}{170}(18 \cos n - 77 \sin n)$

19. $y_n = c_1 4^n + c_2 (-4)^n + \dfrac{\cos\left(\frac{n}{2} - 1\right) - 16 \cos \frac{n}{2}}{(257 - 32 \cos 1)}$

20. $y_n = c_1 \cos \frac{n\pi}{2} + c_2 \sin \frac{n\pi}{2} + \dfrac{2[\cos(2n-4) + \cos 2n]}{1 + \cos 4}$

21. $a^n \left(c_1 \cos \frac{n\pi}{2} + c_2 \sin \frac{n\pi}{2}\right) + \dfrac{\sin a(n-2) + a^2 \sin an}{1 + 2a^2 \cos 2a + a^4}$

22. $y_n = c_1 \cos\left(\frac{2n\pi}{3}\right) + c_2 \sin\left(\frac{2n\pi}{3}\right) + \frac{1}{3}\left(n^2 - n + \frac{1}{3}\right)$

UNIT III

SPECIAL FUNCTION

Exponential, Logarithmic, trigonometric, hyperbolic, etc., are the elementary functions. While Bessel's functions, Legendre's polynomial, Laguerre polynomial, Hermite polynomial. Chebyshev polynomial, Beta function, Gamma function, Error function, etc., are the special functions.

In this unit we shall discuss only two types of special functions, Bessel's and Legendre's.

Chapter first deals with the Bessel's function of first and second kind, orthogonal property of Bessel's function recurrence relations and generating functions.

Chapter second deals with the Legendre's function of first kind, orthogonal property of Legendre's function recurrence relations, and generating function.

CHAPTER 1

Bessel's Functions

1.1. BESSEL'S EQUATION

The differential equation of the form $x^2\frac{d^2y}{dx^2} + x\frac{dy}{dx} + (x^2 - n^2)\, y = 0$ is called the Bessel's differential equation. The solution of Bessel's equation is called Bessel's function. It is also known as cylindrical and spherical function.

Bessel's functions are appear in

(1) the elasticity **(2)** the fluid motion **(3)** planetary motion

(4) the oscillatory motion of a hanging chain **(5)** dynamical astronomy

(6) electrical fields **(7)** potential theory

(8) Euler's theory of circular membrane.

1.2. SOLUTION OF THE BESSEL'S FUNCTIONS

To find the solution of Bessel's differential equation

$$x^2\frac{d^2y}{dx^2} + x\frac{dy}{dx} + (x^2 - n^2)\, y = 0 \qquad ...(1.1)$$

Let $y = \sum_{r=0}^{\infty} a_r x^{m+r}$ be the series solution of (1.1)

so that $$\frac{dy}{dx} = \sum_{r=0}^{\infty} (m + r)\, a_r\, x^{m+r-1} \qquad ...(1.2)$$

and $$\frac{d^2y}{dx^2} = \sum_{r=0}^{\infty} (m + r)(m + r - 1)\, a_r\, x^{m+r-2} \qquad ...(1.3)$$

Substituting these values in (1.1), we get

$$x^2 \sum_{r=0}^{\infty} (m+r)(m+r-1)\, a_r\, x^{m+r-2} + x \sum_{r=0}^{\infty} (m+r)\, a_r\, x^{m+r-1} + (x^2 - n^2) \sum_{r=0}^{\infty} a_r\, x^{m+r} = 0$$

or
$$\sum_{r=0}^{\infty} a_r \left[(m+r)(m+r-1)\, x^{m+r} + (m+r)\, x^{m+r} + x^{m+r+2} - n^2 x^{m+r}\right] = 0$$

or
$$\sum_{r=0}^{\infty} a_r \left[\{(m+r)(m+r-1) + (m+r) - n^2\}\, x^{m+r} - x^{m+r+2}\right] = 0$$

or
$$\sum_{r=0}^{\infty} a_r \left[\{(m+r)^2 - n^2\}\, x^{m+r} - x^{m+r+2}\right] = 0 \qquad \text{...(1.4)}$$

Since the equation (1.4) is an identity, therefore the coefficients of various powers of x must be zero.

$\therefore$ Equating to zero the coefficients of the lowest power of x (*i.e.* x^m) in (1.4), we get

$$a_0\,[m^2 - n^2] = 0$$

But $a_0 \neq 0$

$\therefore$ $m^2 - n^2 = 0$ or $m = \pm n$ ...(1.5)

Now equating to zero the coefficients of the next lowest degree term of x (*i.e.* x^{m+1}) in (1.4), we have

$$a_1[(m+1)^2 - n^2] = 0$$

But $(m+1)^2 - n^2 \neq 0$ for $m = \pm n$ given by (1.5)

$\therefore$ $a_1 = 0$

Again equating to zero the coefficient of the general term (*i.e.* x^{k+r+2}) in (1.4), we get

$$a_{r+2}\,[(m+r+2)^2 - n^2] + a_r = 0$$

or
$$a_{r+2} = -\frac{1}{(m+r+2)^2 - n^2}\, a_r$$

or
$$a_{r+2} = -\frac{1}{(m+r+2+n)(m+r+2-n)}\, a_r \qquad \text{...(1.6)}$$

Putting $r = 1$ in (1.6), we get

$$a_3 = -\frac{1}{(m+n+3)(m-n+3)}\, a_1 = 0 \quad (\because \; a_1 = 0)$$

Similarly putting $r = 3, 5, 7, \ldots\ldots$, in (1.6), we get each $a_1 = a_3 = a_5 \ldots\ldots = 0$

Now two cases arise here

Case I. When $m = n$, from (1.6), we have

$$a_{r+2} = -\frac{1}{(2n+r+2)(r+2)}\, a_r$$

Putting $r = 0, 2, 4, \ldots\ldots$ etc.

$$a_2 = -\frac{1}{(2n+2)(2)}\, a_0 = -\frac{1}{2^2\, 1!\,(n+1)}\, a_0$$

$$a_4 = -\frac{1}{(2n+4)(4)}\, a_2 = -\frac{1}{2^2 \cdot 2(n+2)}\, a_2 = \frac{1}{2^4\, 2!\,(n+1)(n+2)}\, a_0$$

$$\therefore \quad y = \sum_{r=0}^{\infty} a_r\, x^{m+r} = \sum_{r=0}^{\infty} a_r x^{n+r}$$

$$= a_0 x^n + a_1 x^{n+1} + a_2 x^{n+2} + a_3 x^{n+3} + a_4 x^{n+4} \ldots\ldots$$

$$= a_0 x^n - \frac{1}{2^2\, 1!\,(n+1)}\, a_0\, x^{n+2} + \frac{1}{2^4\, 2!\,(n+1)(n+2)}\, a_0 x^{n+4} \ldots\ldots$$

$$= a_0 \left\{ x^n - \frac{x^{n+2}}{2^2\, 1!\,(n+1)} + \frac{x^{n+4}}{2^4\, 2!\,(n+1)(n+2)} \ldots\ldots \right.$$

$$= a_0 x^n \left[1 - \frac{x^2}{2^2\, 1!\,(n+1)} + \frac{x^4}{2^4\, 2!\,(n+1)(n+2)} \ldots\ldots \right]$$

If $\quad a_0 = \dfrac{1}{2^n\, \Gamma(n+1)}$

The above solution is called Bessel's function of the first kind of order n and denoted by $J_n(x)$

$$\therefore \quad J_n(x) = \frac{x^n}{2^n\, \Gamma(n+1)} \left[1 - \frac{x^2}{2^2\, 1!\,(n+1)} + \frac{x^4}{2^4\, 2!\,(n+1)(n+2)} \ldots\ldots \right]$$

$$= \frac{x^n}{2^n\, \Gamma(n+1)} \sum_{r=0}^{\infty} (-1)^r \frac{x^{2r}}{2^{2r}\, r!\,(n+1)(n+2)\ldots\ldots(n+r)}$$

$$J_n(x) = \sum_{r=0}^{\infty} (-1)^r \left(\frac{x}{2}\right)^{n+2r} \frac{1}{r!\,\Gamma(n+r+1)} \qquad \ldots(1.7)$$

Case II. When $m = -n$

The series solution is obtained by replacing n by $-n$ in the (1.7)

$$\therefore \quad J_{-n}(x) = \sum_{r=0}^{\infty} (-1)^r \left(\frac{x}{2}\right)^{n+2r} \frac{1}{r!\,\Gamma(n+r+1)}$$

1.3. GENERAL SOLUTION OF BESSEL'S EQUATION

The solution of the Bessel's differential equation of the type $y(x) = A\, J_n(x) + BJ_{-n}(x)$ where A and B are arbitrary constants is called general solution.

1.4. INTEGRATION OF BESSEL'S EQUATIONS IN SERIES FOR N = 0

For $n = 0$ the Bessel's differential equation is

$$\frac{d^2y}{dx^2} + \frac{1}{x}\frac{dy}{dx} + y = 0 \qquad ...(1.8)$$

Let us assume that the solution of (1) be

$$y = \sum_{r=0}^{\infty} a_r x^{m+r}$$

$$\therefore \qquad \frac{dy}{dx} = \sum_{r=0}^{\infty} a_r\,(m+r)\,x^{m+r-1}$$

and
$$\frac{d^2y}{dx^2} = \sum_{r=0}^{\infty} a_r\,(m+r)\,(m+r-1)\,x^{m+r-2}$$

Substituting these values in (1.8), we get

$$\sum_{r=0}^{\infty} a_r\,(m+r)\,(m+r-1)\,x^{m+r-2} + \frac{1}{x}\sum_{r=0}^{\infty} a_r\,(m+r)\,x^{m+r-1} + \sum_{r=0}^{\infty} a_r\,x^{m+r} = 0$$

or
$$\sum_{r=0}^{\infty} a_r\left[(m+r)(m+r-1)\,x^{m+r-2} + (m+r)\,x^{m+r-2} + x^{m+r}\right] = 0$$

or
$$\sum_{r=0}^{\infty} a_r\left[(m+r)^2\,x^{m+r-2} + x^{m+r}\right] = 0 \qquad ...(1.9)$$

which is an identity. Equating to zero the coefficient of the lowest power of x(*i.e.* x^{m-2}), we have

$$a_0 m^2 = 0$$

But $a_0 \neq 0$ therefore $m^2 = 0, \quad \therefore \quad m = 0$ ...(1.10)

Now equating to zero the coefficient of the next lowest power of x give (*i.e.* x^{m-1}), we have

$$a_1(m+1)^2 = 0$$

But $(m + 1) \neq 0$ for $m = 0$ given by (1.10)

Now equating to zero the coefficient of the general term (*i.e.* x^{m+r}), we have

$$a_{r+2}\,(m+r+2)^2 + a_r = 0$$

$$\therefore \qquad a_{r+2} = -\frac{a_r}{(m+r+2)^2}$$

When $m = 0$, we have $a_{r+2} = -\dfrac{a_r}{(r+2)^2}$

Putting $r = 1, 3, 5$ ...... etc., we have $a_1 = a_3 = a_5$ ...... $= 0$

Again putting $r = 0, 2, 4$ etc., we have

$$a_2 = -\frac{a_0}{2^2},\ a_4 = -\frac{a_2}{4^2} = \frac{a_0}{2^2 4^2} \text{ etc.}$$

Since $$y = \sum_{r=0}^{\infty} a_r x^{m+r}$$

When $m = 0$

$$y = \sum_{r=0}^{\infty} a_r x^r = a_0 + a_1 x + a_2 x^2 + a_3 x^3 + a_4 x^4 \ldots\ldots$$

$$= a_0 - \frac{a_0}{2^2} x^2 + \frac{a_0}{2^2 4^2} x^4 \ldots\ldots$$

$$= a_0 \left[1 - \frac{x^2}{2^2} + \frac{x^4}{2^2 4^2} - \frac{x^6}{2^2 4^2 6^2} \ldots\ldots\right]$$

If $a_0 = 1$ then $y = J_0(x)$

$\therefore$ $$\mathbf{J_0(x)} = 1 - \frac{x^2}{2^2} + \frac{x^4}{2^2 4^2} - \frac{x^6}{2^2 4^2 6^2} \ldots\ldots$$

$J_0(x)$ is also known as Bessel's function of order zero.

1.5. GENERATING FUNCTION FOR $J_N(X)$

Prove that $J_n(x)$ is the coefficient of z^n in the expansion of $e^{\frac{x}{2}\left(z - \frac{1}{z}\right)}$

Proof. We have $e^{\frac{x}{2}\left(z - \frac{1}{z}\right)} = e^{\frac{xz}{2}} . e^{-\frac{x}{2z}}$

$$= \left[1 + \frac{xz}{2} + \frac{1}{2!}\left(\frac{xz}{2}\right)^2 + \frac{1}{3!}\left(\frac{xz}{2}\right)^3 + \ldots\ldots + \frac{1}{n!}\left(\frac{xz}{2}\right)^n + \frac{1}{(n+1)!}\left(\frac{xz}{2}\right)^{n+1}\right.$$

$$\left. + \frac{1}{(n+2)!}\left(\frac{xz}{2}\right)^{n+2} \ldots\ldots\right] \times \left[1 - \frac{x}{2z} + \frac{1}{2!}\left(\frac{x}{2z}\right)^2 + \ldots\ldots + \frac{(-1)^n}{n!}\left(\frac{x}{2z}\right)^n\right.$$

$$\left. + \frac{(-1)^{n+1}}{(n+1)!}\left(\frac{x}{2z}\right)^{n+1} + \frac{(-1)^{n+2}}{(n+2)!}\left(\frac{x}{2z}\right)^{n+2} + \ldots\ldots\right]$$

Now coefficient of z^n in this expansion

$$= \frac{1}{n!}\left(\frac{x}{2}\right)^n - \frac{1}{(n+1)!}\left(\frac{x}{2}\right)\left(\frac{x}{2}\right)^{n+1} + \frac{1}{(n+2)!}\frac{1}{2!}\left(\frac{x}{2}\right)^2\left(\frac{x}{2}\right)^{n+2} + \ldots\ldots$$

$$= \frac{1}{n!}\left(\frac{x}{2}\right)^n - \frac{1}{(n+1)!}\left(\frac{x}{2}\right)^{n+2} + \frac{1}{2!(n+2)!}\left(\frac{x}{2}\right)^{n+4} + \ldots\ldots$$

$$= \frac{(-1)^0}{\Gamma(n+1)}\left(\frac{x}{2}\right)^n + \frac{(-1)^1}{1!\Gamma(n+2)}\left(\frac{x}{2}\right)^{n+2} + \frac{(-1)^2}{2!\Gamma(n+3)}\left(\frac{x}{2}\right)^{n+4} + \ldots\ldots$$

$$= \sum_{r=0}^{\infty} (-1)^r \frac{1}{r!\,\Gamma(n+r+1)} \left(\frac{x}{2}\right)^{n+2r} = J_n(x)$$

Similarly, the coefficient of z^{-n} in the above expansion

$$= \frac{(-1)^n}{n!}\left(\frac{x}{2}\right)^n + \frac{(-1)^{n+1}}{(n+1)!}\frac{x}{2}\left(\frac{x}{2}\right)^{n+1} + \frac{(-1)^{n+2}}{(n+2)!}\frac{1}{2!}\left(\frac{x}{2}\right)^2\left(\frac{x}{2}\right)^{n+2} + \ldots\ldots$$

$$= (-1)^n \left[\frac{(-1)^0}{\Gamma(n+1)}\left(\frac{x}{2}\right)^n + \frac{(-1)^1}{\Gamma(n+2)}\left(\frac{x}{2}\right)^{n+2} + \frac{(-1)^2}{2!\,\Gamma(n+3)}\left(\frac{x}{2}\right)^{n+4} \ldots\ldots\right]$$

$$= (-1)^n J_n(x)$$

Hence we obtained $e^{\frac{x}{2}\left(z-\frac{1}{z}\right)} = \sum_{n=-\infty}^{\infty} z^n J_n(x)$.

1.6. RECURRENCE RELATIONS FOR $J_n(X)$

(I) $xJ_n'(x) = nJ_n(x) - xJ_{n+1}(x)$

Proof. We know that

$$J_n(x) = \sum_{r=0}^{\infty} \frac{(-1)^r}{r!\,\Gamma(n+r+1)}\left(\frac{x}{2}\right)^{n+2r}$$

Differentiating w.r.t x, we get

$$J_n'(x) = \sum_{r=0}^{\infty} \frac{(-1)^r (n+2r)}{r!\,\Gamma(n+r+1)}\left(\frac{x}{2}\right)^{n+2r-1} \cdot \frac{1}{2}$$

$$\therefore \quad xJ_n'(x) = \sum_{r=0}^{\infty} \frac{(-1)^r (n+2r)}{r!\,\Gamma(n+r+1)}\left(\frac{x}{2}\right)^{n+2r}$$

$$= \sum_{r=0}^{\infty} \frac{(-1)^r n}{r!\,\Gamma(n+r+1)}\left(\frac{x}{2}\right)^{n+2r} + \sum_{r=0}^{\infty} \frac{(-1)^r 2r}{r!\,\Gamma(n+r+1)}\frac{x}{2}\left(\frac{x}{2}\right)^{n+2r-1}$$

$$= nJ_n(x) + x\sum_{r=0}^{\infty} \frac{(-1)^r}{(r-1)!\,\Gamma(n+r+1)}\left(\frac{x}{2}\right)^{n+2r-1}$$

$$= nJ_n(x) - x\sum_{r=0}^{\infty} \frac{(-1)^{r-1}}{(r-1)!\,\Gamma[(n+1)+(r-1)+1]}\left(\frac{x}{2}\right)^{(n+1)+2(r-1)}$$

$$= nJ_n(x) - x\sum_{s=0}^{\infty} \frac{(-1)^s}{s!\,\Gamma[(n+1)+s+1]}\left(\frac{x}{2}\right)^{(n+1)+2s} \quad [\because \text{ Putting } r-1=s]$$

$$\boxed{xJ_n'(x) = nJ_n(x) - xJ_{n+1}(x).}$$ **Hence proved.**

(II) $xJ_n'(x) = -nJ_n(x) + xJ_{n-1}(x)$

Proof. We know that

$$J_n(x) = \sum_{r=0}^{\infty} \frac{(-1)^r}{r!\,\Gamma(n+r+1)} \left(\frac{x}{2}\right)^{n+2r}$$

Differentiating (I) w.r.t "x", we get

$$J'_n(x) = \sum_{r=0}^{\infty} \frac{(-1)^r (n+2r)}{r!\,\Gamma(n+r+1)} \left(\frac{x}{2}\right)^{n+2r-1} \cdot \frac{1}{2}$$

or

$$xJ_n'(x) = \sum_{r=0}^{\infty} \frac{(-1)^r (n+2r)}{r!\,\Gamma(n+r+1)} \left(\frac{x}{2}\right)^{n+2r} = \sum_{r=0}^{\infty} \frac{(-1)^r (2n+2r-n)}{r!\,\Gamma(n+r+1)} \left(\frac{x}{2}\right)^{n+2r}$$

$$= \sum_{r=0}^{\infty} \frac{(-1)^r\, 2(n+r)}{r!\,\Gamma(n+r+1)} \left(\frac{x}{2}\right)^{n+2r} + \sum_{r=0}^{\infty} \frac{(-1)^r - n}{r!\,\Gamma(n+r+1)} \left(\frac{x}{2}\right)^{n+2r}$$

$$= -n \sum_{r=0}^{\infty} \frac{(-1)^r}{r!\,\Gamma(n+r+1)} \left(\frac{x}{2}\right)^{n+2r} + \sum_{r=0}^{\infty} \frac{(-1)^r\, 2(n+r)}{r!\,\Gamma(n+r).(n+r)} \left(\frac{x}{2}\right)^{n+2r-1} \frac{x}{2}$$

$$= -n\, J_n(x) + x \sum_{r=0}^{\infty} \frac{(-1)^r}{r!\,\Gamma(n+r)} \left(\frac{x}{2}\right)^{n+2r-1r}$$

$$= -nJ_n(x) + x \sum_{r=0}^{\infty} \frac{(-1)^r}{r!\,\Gamma[(n-1)+r+1]} \left(\frac{x}{2}\right)^{(n-1)+2r}$$

$$= -nJ_n(x) + xJ_{n-1}(x)$$

$$\boxed{xJ_n'(x) = -nJ_n(x) + xJ_{n-1}(x).}$$ **Hence proved.**

(III) $2J_n'(x) = J_{n-1}(x) - J_{n+1}(x)$

Proof. By recurrence relation (I) and (II), we have

$$xJ_n'(x) = nJ_n(x) - xJ_{n+1}(x) \qquad ...(i)$$

$$xJ_n'(x) = -nJ_n(x) + xJ_{n-1}(x) \qquad ...(ii)$$

On adding, we get

$$2xJ_n'(x) = x\,[J_{n-1}(x) - J_{n+1}(x)]$$

or

$$\boxed{2J_n'(x) = J_{n-1}(x) - J_{n+1}(x).}$$ **Hence proved.**

(IV) $2nJ_n(x) = x\,[J_{n-1}(x) + J_{n+1}(x)]$

Proof. By recurrence relation (I) and (II), we have

$$xJ_n'(x) = nJ_n(x) - xJ_{n+1}(x) \qquad ...(i)$$

$$xJ_n'(x) = -nJ_n(x) + xJ_{n-1}(x) \qquad ...(ii)$$

On subtracting, we get

$$0 = 2nJ_n(x) - xJ_{n+1}(x) - xJ_{n-1}(x)$$

or $$2nJ_n(x) = x[J_{n+1}(x) + J_{n-1}(x)].$$ **Hence proved.**

(V) $\mathbf{\dfrac{d}{dx}\,[x^{-n}\,J_n(x)] = -\,x^{-n}\,J_{n+1}(x)}$

Proof. We have $\dfrac{d}{dx}\,[x^{-n}\,J_n(x)] = -\,nx^{-n-1}\,J_n(x) + x^{-n}\,J_n'(x)$

$$= x^{-n-1}\,[-\,nJ_n(x) + xJ_n'(x)]$$

$$= x^{-n-1}\,[-\,n\,J_n(x) + nJ_n(x) - xJ_{n+1}(x)]$$

By recurrence relation (I)

$$= x^{-n-1}\,[-\,xJ_{n+1}(x)]$$

$$\frac{d}{dx}\,[x^{-n}\,J_n(x)] = -x^{-n}\,J_{n+1}(x).$$ **Hence proved.**

(VI) $\mathbf{\dfrac{d}{dx}\,[\,x^{n}\,J_n(x)] = x^{n}\,J_{n-1}\,(x)}$

Proof. We have $\dfrac{d}{dx}\,[x^{n}\,J_n(x)] = nx^{n-1}\,J_n(x) + x^{n}J_n'(x) = x^{n-1}\,[nJ_n(x) + xJ_n'(x)]$

$$= x^{n-1}\,[n\,J_n(x) - nJ_n(x) + xJ_{n-1}(x)]$$ [By recurrence relation (II)]

$$\frac{d}{dx}\,[\,x^{n}\,J_n(x)\,] = x^{n}J_{n-1}\,(x).$$ **Hence proved.**

1.7. ORTHOGONAL PROPERTY OF BESSEL'S FUNCTIONS

$$\int_0^1 x\,J_n(\alpha\,x)\,J_n\,(\beta x)\,dx = \begin{cases} 0 & \text{if} \quad \alpha \neq \beta \\ \dfrac{1}{2}\,[J_{n+1}(\alpha)]^2 & \text{if} \quad \alpha = \beta \end{cases}$$

where α, β are the roots of $J_n(x) = 0$.

Proof. We know that

$$x^2\frac{d^2y}{dx^2} + x\,\frac{dy}{dx} + (\alpha^2x^2 - n^2)\,y = 0 \qquad ...(1.11)$$

$$x^2\frac{d^2z}{dx^2} + x\frac{dz}{dx} + (\beta^2x^2 - n^2)\,z = 0 \qquad ...(1.12)$$

Solution of (1.11) and (1.12) equations are $y = J_n(\alpha\,x)$, $z = J_n(\beta x)$ respectively. Multiplying (1.11) by $\dfrac{z}{x}$ and (1.12) by $\dfrac{y}{x}$ and subtracting, we get

$$x\left(z\frac{d^2y}{dx^2} - y\frac{d^2z}{dx^2}\right) + \left(z\frac{dy}{dx} - y\frac{dz}{dx}\right) + (\alpha^2 - \beta^2)\,xyz = 0$$

or $$\frac{d}{dx}\left[x\left(z\frac{dy}{dx}-y\frac{dz}{dx}\right)\right]+(\alpha^2-\beta^2)\,xyz=0 \qquad ...(1.13)$$

Integrating w.r.t. x, both sides of (1.13) from 0 to 1, we get

$$\left[x\left(z\frac{dy}{dx}-y\frac{dz}{dx}\right)\right]_0^1+(\alpha^2-\beta^2)\int_0^1 xyzdz=0$$

or $$(\beta^2-\alpha^2)\int_0^1 xyz\,dx=\left[x\left(z\frac{dy}{dx}-y\frac{dz}{dx}\right)\right]_0^1$$

$$=\left(z\frac{dy}{dx}-y\frac{dz}{dx}\right)_{x=1} \qquad ...(1.14)$$

Putting the values of $y=J_n(\alpha\,x) \Rightarrow \dfrac{dy}{dx}=\alpha J_n'(\alpha\,x)$

and $z=J_n(\beta x) \Rightarrow \dfrac{dz}{dx}=\beta J_n'(\beta x)$ in (1.14), we get

$$(\beta^2-\alpha^2)\int_0^1 x\,J_n(\alpha\,x)\,J_n(\beta x)\,dx=[\alpha J_n'(\alpha\,x)\,J_n(\beta x)-\beta J_n'(\beta x)\,J_n(\alpha\,x)]_{x=1}$$

$$=\alpha J_n'(\alpha)\,J_n(\beta)-\beta J_n'(\beta)\,J_n(\alpha) \qquad ...(1.15)$$

Since α, β are the roots of $J_n(x)=0$ therefore $J_n(\alpha)=0$, $J_n(\beta)=0$

Then by (1.15) $(\beta^2-\alpha^2)\int_0^1 x\,J_n(\alpha x)\,J_n(\beta x)\,dx=0$

or $\int_0^1 x\,J_n(\alpha x)\,J_n(\beta x)\,dx=0$ Hence proved.

By (1.15), we have

$$(\beta^2-\alpha^2)\int_0^1 x\,J_n(\alpha x)\,J_n(\beta x)\,dx=\alpha\,J_n'(\alpha)\,J_n(\beta)-\beta J_n'(\beta)\,J_n(\alpha) \qquad ...(1.16)$$

When $\alpha=\beta$

We also know that $J_n(\alpha)=0$ and β as a variable approaching α then from (1.16), we have

$$\lim_{\beta\to\alpha}\int_0^1 x\,J_n(\alpha x)\,J_n(\beta x)\,dx=\lim_{\beta\to\alpha}\frac{\alpha\,J_n'(\alpha)\,J_n(\beta)}{\beta^2-\alpha^2} \qquad \left|\frac{0}{0}\text{ form}\right|$$

Now apply L'Hospitals Rule

$$\int_0^1 x(J_n(\alpha x))^2\,dx=\lim_{\beta\to\alpha}\frac{\alpha\,J_n'(\alpha)\,J_n'(\beta)}{2\beta}$$

$$=\frac{1}{2}\,(J_n'(\alpha))^2=\frac{1}{\alpha}\,(J_{n+1}(\alpha))^2$$

Hence $$\int_0^1 x\,J_n(\alpha x)\,J_n(\beta x)\,dx=\begin{cases}0 & \text{if } \alpha\neq\beta\\ \frac{1}{2}(J_{n+1}(\alpha))^2 & \text{if } \alpha=\beta\end{cases}$$

SOLVED EXAMPLES

Example 1. *Prove that* $J_{-n}(x) = (-1)^n J_n(x)$

Solution. We have $J_{-n}(x) = \sum_{r=0}^{\infty} (-1)^r \left(\frac{x}{2}\right)^{-n+2r} \frac{1}{r!\,\Gamma(-n+r+1)}$

$$= \sum_{r=0}^{n-1} (-1)^r \left(\frac{x}{2}\right)^{-n+2r} \frac{1}{r!\,\Gamma(-n+r+1)} + \sum_{r=n}^{\infty} (-1)^r \left(\frac{x}{2}\right)^{-n+2r} \frac{1}{r!\,\Gamma(-n+r+1)}$$

$$= 0 + \sum_{r=n}^{\infty} (-1)^r \left(\frac{x}{2}\right)^{-n+2r} \frac{1}{r!\,\Gamma(-n+r+1)}$$

[∵ – ve integer = ∞]

On putting $r = n + k$, we have

$$J_{-n}(x) = \sum_{r=0}^{\infty} (-1)^{n+k} \left(\frac{x}{2}\right)^{n+2k} \frac{1}{(n+k)!\,\Gamma(k+1)}$$

$$= (-1)^n \sum_{x=0}^{\infty} (-1)^k \left(\frac{x}{2}\right)^{n+2k} \frac{1}{k!\,\Gamma(n+k+1)}$$

$J_{-n}(x) = (-1)^n J_n(x)$. **Hence proved.**

Example 2. *Prove that*

(i) $J_{1/2}(x) = \sqrt{\left(\frac{2}{\pi x}\right)} \sin x$ *(ii)* $J_{-1/2}(x) = \sqrt{\left(\frac{2}{\pi x}\right)} \cos x$

Solution. We have

$$J_n(x) = \frac{x^n}{2^n\,\Gamma(n+1)} \left[1 - \frac{x^2}{2^2\,1!(n+1)} + \frac{x^4}{2^4\,2!(n+1)(n+2)} + \ldots\ldots\right] \qquad \ldots(i)$$

(I) Substituting $n = 1/2$ in equation (*i*)

$$J_{1/2}(x) = \frac{x^{1/2}}{2^{1/2}\sqrt{\frac{1}{2}+1}} \left[1 - \frac{x^2}{2^2\,1!\left(\frac{1}{2}+1\right)} + \frac{x^4}{2^4\,2!\left(\frac{1}{2}+1\right)\left(\frac{1}{2}+2\right)} \ldots\ldots\right]$$

$$= \frac{\sqrt{x}}{\sqrt{2}\,\frac{1}{2}\sqrt{\pi}} \left[1 - \frac{x^2}{2.3} + \frac{x^4}{2.3.4.5} \ldots\ldots\right]$$

$$= \sqrt{\frac{2.x}{\pi}} \cdot \frac{1}{x} \left[x - \frac{x^3}{13} + \frac{x^5}{15} \ldots\ldots\right]$$

$$= \sqrt{\left(\frac{2}{\pi x}\right)} \sin x. \quad \textbf{Hence proved.}$$

(II) Now substituting $n = -1/2$ in (*i*) we have

$$J_{-1/2}(x) = \frac{x^{-1/2}}{2^{-1/2}\sqrt{-\frac{1}{2}+1}}\left[1 - \frac{x^2}{2^2\, 1!\left(-\frac{1}{2}+1\right)} + \frac{x^4}{2^4\, 2!\left(-\frac{1}{2}+1\right)\left(-\frac{1}{2}+2\right)} + \ldots\ldots\right]$$

$$= \sqrt{\frac{2}{\pi x}}\left[1 - \frac{x^2}{1.2} + \frac{x^4}{1.2.3.4} + \ldots\ldots\right]$$

$$J_{-1/2}(x) = \sqrt{\frac{2}{\pi x}} \cos x. \quad \textbf{Hence proved.}$$

Example 3. *Prove that (i)* $J_{3/2}(x) = \sqrt{\frac{2}{\pi x}}\left(\frac{1}{x}\sin x - \cos x\right)$.

(ii) $J_{5/2}(x) = \sqrt{\frac{2}{\pi x}}\left[\left(\frac{3-x^2}{x^2}\right)\sin x - \frac{3}{x}\cos x\right]$.

Solution. By recurrence relation (IV), we have

$$2nJ_n(x) = x\,[J_{n-1}(x) + J_{n+1}(x)]$$

or $$J_{n+1}(x) = \frac{2n}{x} J_n(x) - J_{n-1}(x) \quad \ldots(i)$$

(I) Putting $n = 1/2$ in (*i*), we get

$$J_{3/2}(x) = \frac{1}{x} J_{1/2}(x) - J_{-1/2}(x)$$

$$= \frac{1}{x}\sqrt{\frac{2}{\pi x}} \sin x - \sqrt{\frac{2}{\pi x}} \cos x$$

$$= \sqrt{\frac{2}{\pi x}}\left(\frac{1}{x}\sin x - \cos x\right). \quad \textbf{Hence proved.}$$

(II) Again putting $n = 3/2$ in (*i*), we get

$$J_{5/2}(x) = \frac{3}{x} J_{3/2}(x) - J_{1/2}(x)$$

$$= \frac{3}{x}\sqrt{\frac{2}{\pi x}}\left(\frac{1}{x}\sin x - \cos x\right) - \sqrt{\frac{2}{\pi x}} \sin x$$

$$= \sqrt{\frac{2}{\pi x}} \left(\frac{3}{x} \left(-\frac{1}{x} \sin x - \cos x \right) - \sin x \right)$$

$$= \sqrt{\frac{2}{\pi x}} \left(\left(\frac{3 - x^2}{x^2} \right) \sin x - \frac{3}{x} \cos x \right). \quad \textbf{Hence proved.}$$

Example 4. *Prove that* $x^2 J_n''(x) = (n^2 - n - x^2) J_n(x) + x J_{n+1}(x)$.

Solution. We have the Bessel's equation is

$$x^2 \frac{d^2 y}{dx^2} + x \frac{dy}{dx} + (x^2 - n^2) y = 0 \qquad ...(i)$$

Since $J_n(x)$ is the solution of eqn. (*i*) so

$$x^2 J_n''(x) + x J_n'(x) + (x^2 - n^2) J_n(x) = 0$$

By the recurrence relation (I) we have

$$x J_n'(x) = n J_n(x) - x J_{n+1}(x) \qquad ...(ii)$$

On putting the value of $x J_n'(x)$ from (*ii*) in (*i*), we get

$$x^2 J_n''(x) + n J_n(x) - x J_{n+1}(x) + (x^2 - n^2) J_n(x) = 0$$

or $$x^2 J_n''(x) = (n^2 - n - x^2) J_n(x) + x J_{n+1}(x). \quad \textbf{Hence proved.}$$

Example 5. *Prove that* $4J_n''(x) = J_{n-2}(x) - 2J_n(x) + J_{n+2}(x)$.

Solution. By recurrence relation (III) we have

$$2J_n'(x) = J_{n-1}(x) - J_{n+1}(x) \qquad ...(i)$$

Differentiating both sides, we have

$$2J_n''(x) = J_{n-1}'(x) - J_{n+1}'(x)$$

or $$4J_n''(x) = 2J_{n-1}'(x) - 2J_{n+1}'(x) \qquad ...(ii)$$

Now putting $n = n - 1$ and $n + 1$ in (*i*)

$$2 J_{n-1}'(x) = J_{n-2}(x) - J_n(x)$$

$$2 J_{n+1}'(x) = J_n(x) - J_{n+2}(x)$$

Substituting these in (*ii*), we get

$$4J_n''(x) = J_{n-2}(x) - J_n(x) - J_n(x) + J_{n+2}(x)$$

$$4J_n''(x) = J_{n-2}(x) - 2J_n(x) + J_{n+2}(x). \quad \textbf{Hence proved.}$$

Example 6. *Prove that* $\frac{d}{dx}(x J_n J_{n+1}) = x(J_n^2 - J_{n+1}^2)$

Solution. We have $\frac{d}{dx}(x J_n J_{n+1}) = J_n J_{n+1} + x J_n' J_{n+1} + x J_n J_{n+1}' \qquad ...(i)$

By the recurrence formula (I) and (II), we have

$$x J_n'(x) = n J_n(x) - x J_{n+1}(x) \qquad ...(ii)$$

and $$x J_n'(x) = -n J_n(x) + x J_{n-1}(x) \qquad ...(iii)$$

Putting $n = n + 1$, in (*iii*), we get

$$xJ'_{n+1}(x) = -(n+1)\,J_{n+1}(x) + xJ_n(x) \qquad \text{...}(iv)$$

On putting the value of $xJ_n'(x)$ and $xJ'_{n+1}(x)$ in equation (*i*), we get

$$\frac{d}{dx}(xJ_nJ_{n+1}) = J_nJ_{n+1} + [nJ_n(x) - xJ_{n+1}(x)]\,J_{n+1} + J_n[-(n+1)\,J_{n+1}(x) + xJ_n(x)]$$

$$= J_nJ_{n+1} + n\,J_nJ_{n+1} - x\,J_{n+1}^2 - (n+1)\,J_{n+1}\,J_n + xJ_n^2(x)$$

$$\frac{d}{dx}(x\,J_nJ_{n+1}) = x(J_n^2 - J_{n+1}^2).$$ **Hence proved.**

Example 7. *Prove that* $J_{n-1} = \frac{2}{x}\,[\,n\,J_n - (n+2)\,J_{n+2} + (n+4)\,J_{n+4}\,......]$

and hence deduce that $\frac{1}{2}\,xJ_n = (n+1)\,J_{n+1} - (n+3)\,J_{n+3} + (n+5)\,J_{n+5}\,.......$.

Solution. By the recurrence formula IV

$$2nJ_n(x) = x[J_{n-1}(x) + J_{n+1}(x)]$$

or

$$J_{n-1} = \frac{2n}{x}\,J_n(x) - J_{n+1}(x) \qquad \text{...}(i)$$

Putting $n = n + 2$ in (*i*), we get

$$J_{n+1} = \frac{2(n+2)}{x}\,J_{n+2}(x) - J_{n+3}(x)$$

Putting the value of $J_{n+1}(x)$ in (*i*), we get

$$J_{n-1} = \frac{2n}{x}\,J_n(x) - \frac{2(n+2)}{x}\,J_{n+2}(x) + J_{n+3}(x) \qquad \text{...}(ii)$$

Again putting $n = n + 4$ in (*i*), we get

$$J_{n+3} = \frac{2(n+4)}{x}\,J_{n+4}(x) - J_{n+5}(x)$$

Putting the value of J_{n+3} in (*ii*), we get

$$J_{n-1} = \frac{2n}{x}\,J_n(x) - \frac{2(n+2)}{x}\,J_{n+2}(x) + \frac{2(n+4)}{x}\,J_{n+4}(x) - J_{n+5}(x)$$

Proceeding in the same way and take $J_n \to 0$ as $n \to \infty$, we have

$$J_{n-1} = \frac{2}{x}\,nJ_n - \frac{2}{x}\,(n+2)\,J_{n+2} + \frac{2}{x}\,(n+4)\,J_{n+4}\,......$$

$$= \frac{2}{x}\,[\,nJ_n - (n+2)\,J_{n+2} + (n+4)\,J_{n+4}\,......\,]. \qquad \text{...}(iii)$$

Hence proved.

Deduction: On putting $n = n + 1$ in (*iii*), we get

$$\frac{1}{2}x\,J_n = (n+1)\,J_{n+1} - (n+3)\,J_{n+3} + (n+5)\,J_{n+5} \;\ldots\ldots .$$

Example 8. *Prove that* $J_3(x) + 3J_0'(x) + 4J_0'''(x) = 0.$

Solution. We have $\quad 2J_n' = J_{n-1} - J_{n+1}$

Differentiating w.r.t. x, we get

$$2J_n'' = J_{n-1}' - J_{n+1}'$$

or
$$2^2J_n'' = 2J_{n-1}' - 2J_{n+1}' = (J_{n-2} - J_n) - (J_n - J_{n+2}) = J_{n-2} - 2J_n + J_{n+2}$$

Differentiating again w.r.t. x, we get

$$2^2J_n''' = J_{n-2}' - 2J_n' + J_{n+2}'$$

or
$$2^3J_n''' = 2J_{n-2}' - 2.2J_n' + 2J_{n+2}'$$
$$= (J_{n-3} - J_{n-1}) - 2(J_{n-1} - J_{n+1}) + J_{n+1} - J_{n+3}$$
$$= J_{n-3} - 3J_{n-1} + 3J_{n+1} - J_{n+3}$$

Putting $\quad n = 0$, we get

$$2^3J_0''' = J_{-3} - 3J_{-1} + 3J_1 - J_3$$
$$= (-1)^3 J_3 - 3(-1)\,J_1 + 3J_1 - J_3 = -2J_3 + 6J_1$$
$$4J_0''' = -J_3 + 3J_1 = -J_3 + 3(-J_0') \qquad (\because\ J_0' = -J_1)$$
$$J_3 + 3J_0' + 4J_0''' = 0$$

$J_3(x) + 3J_0'(x) + 4J_0'''(x) = 0.$ **Hence proved.**

Example 9. *Prove that* $\int_0^{\pi/2} \sqrt{\pi x}\, J_{1/2}(2x)\, dx = 1.$

Solution. We have $J_{1/2}(x) = \sqrt{\dfrac{2}{\pi x}}\,\sin x$

On replacing x by $2x$, we get

$$J_{1/2}(2x) = \sqrt{\frac{2}{\pi 2x}}\,\sin 2x \quad \text{or} \quad \sqrt{\pi x}\, J_{1/2}(2x) = \sin 2x$$

Now integrating w.r.t. "x" from 0 to $\pi/2$, we get

$$\int_0^{\pi/2} \sqrt{\pi x}\, J_{1/2}(2x)\, dx = \int_0^{\pi/2} \sin 2x\, dx = \left[-\frac{\cos 2x}{2}\right]_0^{\pi/2}$$

$$= -\frac{1}{2}\,[\cos \pi - \cos 0) = 1.$$ **Hence proved.**

Example 10. *Prove that if $n > -1$,* $\int_0^x x^{-n} J_{n+1}(x)\, dx = \dfrac{1}{2^n \sqrt{n+1}} - x^{-n} J_n(x)$

Solution. We know that

$$\frac{d}{dx}(x^{-n} J_n(x)) = -x^{-n} J_{n+1}(x)$$

Integrating w.r.t. x from 0 to x, we get

$$+\int_0^x \frac{d}{dx}(x^{-n} J_n(x)\, dx = -\int_0^x x^{-n} J_{n+1}(x)\, dx$$

or
$$\left[x^{-n} J_n(x)\right]_0^x = -\int_0^x x^{-n} J_{n+1}(x)\, dx$$

or
$$\int_0^x x^{-n} J_{n+1}(x)\, dx = -\left[x^{-n} J_n(x) - \lim_{x \to 0} \frac{J_n(x)}{x^n}\right]$$

$$= -x^{-n} J_n(x) + \frac{1}{2^n \sqrt{n+1}}.$$ **Hence proved.**

$$\left[\because \lim_{x \to 0} \frac{J_n(x)}{x^n} = \frac{1}{2^n \sqrt{n+1}}\right]$$

Example 11. *Form the following using Generating function for $J_n(x)$*

(i) $\cos(x \sin\theta) = J_0 + 2J_2 \cos 2\theta + 2J_4 \cos 4\theta + \ldots\ldots$

(ii) $\sin(x \sin\theta) = 2J_1 \sin\theta + 2J_3 \sin 3\theta + \ldots\ldots$

(iii) $\cos(x \cos\theta) = J_0 - 2J_2 \cos 2\theta + 2J_4 \cos 4\theta + \ldots\ldots$

(iv) $\sin(x \cos\theta) = 2J_1 \cos\theta - 2J_3 \cos 3\theta + \ldots\ldots$

(v) $\cos x = J_0 - 2J_2 + 2J_4 - \ldots\ldots$

(vi) $\sin x = 2J_1 - 2J_3 + 2J_5 - \ldots\ldots$

Solution. We know that $e^{\frac{x}{2}\left(z - \frac{1}{z}\right)} = \sum_{n=-\infty}^{\infty} z^n J_n(x)$

$$= J_0 + \left(z - \frac{1}{z}\right) J_1 + \left(z^2 + \frac{1}{z^2}\right) J_2 + \left(z^3 + \frac{1}{z^3}\right) J_3 \qquad \ldots(i)$$

Putting $z = e^{i\theta} \Rightarrow z^n = e^{in\theta} = \cos n\theta + i \sin n\theta$

and $z^{-n} = e^{-in\theta} = \cos n\theta - i \sin n\theta$

$\therefore$ $\left(z^n + \frac{1}{z^n}\right) = 2 \cos n\theta$ and $\left(z^n - \frac{1}{z^n}\right) = 2i \sin n\theta,\ n = 1, 2, 3, \ldots\ldots$

From (*i*), we have

$$e^{i(x \sin \theta)} = J_0 + (2i \sin \theta) J_1 + (2 \cos 2\theta) J_2 + (2i \sin 3\theta) J_3 + \ldots\ldots$$

or $\cos (x \sin \theta) + i \sin (x \sin \theta) = (J_0 + 2J_2 \cos 2\theta + 2J_4 \cos 4\theta \ldots\ldots)$

$$+ i(2J_1 \sin \theta + 2J_3 \sin 3\theta + \ldots\ldots) \quad \ldots(ii)$$

(*i*) Equating the real parts on both sides of (*ii*), we get

$$\cos (x \sin \theta) = J_0 + 2J_2 \cos 2\theta + 2J_4 \cos 4\theta + \ldots\ldots \quad \ldots(iii)$$

(*ii*) Equating the imaginary parts on both sides of (*ii*), we get

$$\sin (x \sin \theta) = 2J_1 \sin \theta + 2J_3 \sin 3\theta + 2J_5 \sin 5\theta + \ldots\ldots \quad \ldots(iv)$$

(*iii*) Putting $\theta = \pi/2 - \theta$ in (*iii*), we get

$$\cos \left(x \sin \left(\frac{\pi}{2} - \theta \right) \right) = J_0 + 2J_2 \cos 2\left(\frac{\pi}{2} - \theta \right) + 2J_4 \cos 4 \left(\frac{\pi}{2} - \theta \right) + \ldots\ldots$$

or

$$\cos (x \cos \theta) = J_0 - 2J_2 \cos 2\theta + 2J_4 \cos 4\theta - \ldots\ldots \quad \ldots(v)$$

(*iv*) Putting $\theta = \dfrac{\pi}{2} - \theta$ in (*iv*), we get

$$\sin \left(x \sin \left(\frac{\pi}{2} - \theta \right) \right) = 2J_1 \sin \left(\frac{\pi}{2} - \theta \right) + 2J_3 \sin 3 \left(\frac{\pi}{2} - \theta \right) + 2J_5 \sin 5 \left(\frac{\pi}{2} - \theta \right) + \ldots\ldots$$

or

$$\sin (x \cos \theta) = 2J_1 \cos \theta - 2J_3 \cos 3\theta + 2J_5 \cos 5\theta + \ldots\ldots \quad \ldots(vi)$$

(*v*) Putting $\theta = \pi/2$ in (*iii*), we get

$$\cos x = J_0 - 2J_2 + 2J_4 - 2J_6 + \ldots\ldots$$

(*vi*) Putting $\theta = \pi/2$ in (*iv*), we get

$$\sin x = 2J_1 - 2J_3 + 2J_5 - 2J_7 + \ldots\ldots$$

EXERCISE 1.1

1. Prove that $[J_{1/2}(x)]^2 + [J_{-1/2}(x)]^2 = \dfrac{2}{\pi x}$.

2. Prove that (*i*) $J_{-3/2}(x) = -\sqrt{\dfrac{2}{\pi x}} \left(\dfrac{1}{x} \cos x + \sin x \right)$

   (*ii*) $J_{-5/2}(x) = \sqrt{\dfrac{2}{\pi x}} \left(\dfrac{3 - x^2}{x^2} \cos x + \dfrac{3}{x} \sin x \right)$.

3. Prove that $J_n(x)$ is even and odd function for even n and for odd n respectively.

4. Prove that $J_{n+3} + J_{n+5} = \dfrac{2}{x} (n + 4) J_{n+4}$.

5. Prove that $\dfrac{d}{dx}[J_n^2 + J_{n+1}^2] = 2\left(\dfrac{n}{x}J_n^2 - \dfrac{n+1}{x}J_{n+1}^2\right)$.

6. Prove that $\dfrac{d}{dx}[J_0(x)] = -J_1(x)$.

7. Prove that (*i*) $J_2(x) = J_0''(x) - \dfrac{1}{x}J_0'(x)$ (*ii*) $J_2(x) - J_0(x) = 2J_0''(x)$.

8. Prove that $J_n' = \dfrac{2}{x}\left[\dfrac{n}{2}J_n - (n+2)J_{n+2} + (n+4)J_{n+4} \ldots\ldots\right]$.

9. Prove that $J_0^2 + 2(J_1^2 + J_2^2 + J_3^2 + \ldots\ldots) = 1$

Deduce that $|J_0(x)| \le 1, |J_n(x)| \le 2^{-1/2}, n \ge 1$.

10. Prove that $x = 2J_0J_1 + 6J_1J_2 + \ldots\ldots 2(2n+1)J_nJ_{n+1} \ldots\ldots$.

11. Prove that $\lim\limits_{x\to 0} \dfrac{J_n(x)}{x^n} = \dfrac{1}{2^2\sqrt{n+1}}$.

12. Prove that $\int J_{n+1}(x)\,dx = \int J_{n-1}(x)\,dx - 2J_n(x) + A$.

13. Prove that (*i*) $\int_0^x x^n J_{n-1}(x)\,dx = x^n J_{n(x)}$ (*ii*) $\int_0^x x^{n+1} J_n(x)\,dx = x^{n+1} J_{n+1}(x)$.

14. Show that when n is integer

(*i*) $\pi J_n = \int_0^\pi \cos(n\theta - x\sin\theta)\,d\theta$ (*ii*) $\pi J_0 = \int_0^\pi \cos(x\cos\phi)\,d\phi$

(*iii*) $\pi J_0 = \int_0^\pi \cos(x\sin\phi)\,d\phi$.

15. Using generating function prove that

$$J_n(x+y) = \sum_{r=-\infty}^{\infty} J_r(x)\,J_{n-r}(y).$$

16. Prove that $\int_0^\infty e^{-ax} J_0(bx)\,dx = \dfrac{1}{\sqrt{a^2+b^2}}, a > 0$.

Chapter 2

Legendre's Functions

2.1. INTRODUCTION

The differential equation

$$(1-x^2)\frac{d^2y}{dx^2}-2x\frac{dy}{dx}+n(n+1)y=0$$

is known as Legendre's equation. The above equation can also be written as

$$\frac{d}{dx}\left\{(1-x^2)\frac{dy}{dx}\right\}+n(n+1)\,y=0;\text{ where } n \text{ is an integer.}$$

2.2. SOLUTION OF LEGENDRE'S EQUATION

The Legendre's differential equation is

$$(1-x^2)\frac{d^2y}{dx^2}-2x\frac{dy}{dx}+n(n+1)\,y=0 \qquad \text{...(2.1)}$$

Let $y=\sum_{r=0}^{\infty}a_r\,x^{m-r}$, $a_0\neq 0$ be the series solution of (2.1) so that

$$\frac{dy}{dx}=\sum_{r=0}^{\infty}a_r\,(m-r)\,x^{m-r-1} \qquad \text{...(2.2)}$$

and

$$\frac{d^2y}{dx^2}=\sum_{r=0}^{\infty}a_r\,(m-r)\,(m-r-1)x^{m-r-2} \qquad \text{...(2.3)}$$

Substituting these values in (2.1), we get

$$(1-x^2)\sum_{r=0}^{\infty} a_r (m-r)(m-r-1)\, x^{m-r-2} - 2x \sum_{r=0}^{\infty} a_r\, (m-r)\, x^{m-r-1}$$

$$+\, n(n+1) \sum_{r=0}^{\infty} a_r\, (x^{m-r}) = 0$$

or $$\sum_{r=0}^{\infty} a_r\, [(1-x^2)(m-r)(m-r-1)\, x^{m-r-2} - 2x\,(m-r)\, x^{m-r-1} + n(n+1)x^{m-r}] = 0$$

or $$\sum_{r=0}^{\infty} a_r\, [\,(m-r)(m-r-1)\, x^{m-r-2} + \{n\,(n+1) - (m-r)(m-r-1) - 2(m-r)\}x^{m-r}] = 0$$

or $$\sum_{r=0}^{\infty} a_r\, [(m-r)(m-r-1)\, m^{x-r-2} + \{n(n+1) - (m-r)(m-r+1)\}\, x^{m-r}] = 0$$

or $$\sum_{r=0}^{\infty} a_r\, [(m-r)(m-r-1)\, x^{m-r-2} + \{(n^2-(m-r)^2 + n - (m-r)\}\, x^{m-r}] = 0$$

$$\sum_{r=0}^{\infty} a_r\, [(m-r)(m-r-1)x^{m-r-2} + \{n-(m-r)\}\,(n+m-r+1)\, x^{m-r}] = 0 \qquad ...(2.4)$$

Since the equaiton (2.4) is an identity, therefore the coefficients of various powers of x must be zero.

$\therefore$ Equating to zero the coefficient of the highest power of x (*i.e.*, x^m), we have

$$a_0\,(n-m)\,(n+m+1) = 0$$

But $a_0 \neq 0 \quad \therefore \quad m = n \quad$ or $\quad m = -\,(n+1)$ ...(2.5)

Now equating to zero the coefficient of the next lower power of x (*i.e.*, x^{m-1}), we have

$$a_1(n-m+1)\,(n+m) = 0$$

But neither $(n-m+1)$ nor $(n+m)$ is zero for $m = n$ or $-\,(n+1)$ given by (2.5)

$$\therefore \qquad a_1 = 0$$

Again equating to zero the coefficient of the general term (*i.e.*, x^{m-r}), we have

$$a_{r-2}\,(m-r+2)\,(m-r+1) + a_r(n-m+r)\,(n+m-r+1) = 0$$

$$\therefore \qquad a_r = -\,\frac{(m-r+2)\,(m-r+1)}{(n-m+r)\,(n+m-r+1)}\; a_{r-2},\; r \geq 2 \qquad ...(2.6)$$

Putting $r = 3,\quad a_3 = -\,\dfrac{(m-1)\,(m-2)}{(n-m+3)\,(n+m-2)}\; a_1 = 0$

Similarly putting $r = 5, 7, 9$ etc. in (2.6), we get each $a_1 = a_3 = a_5 \ldots\ldots = 0$

Now two cases arise here

Case I. When $m = n$ then from (2.6), we have

$$a_r = -\frac{(n-r+2)(n-r+1)}{r(2n-r+1)} a_{r-2}, r \geq 2$$

Putting $r = 2, 4 \ldots\ldots$ etc.

$$a_2 = -\frac{n(n-1)}{2(2n-1)} a_0$$

$$a_4 = -\frac{(n-2)(n-3)}{4(2n-3)} a_2 = \frac{n(n-1)(n-2)(n-3)}{2.4(2n-1)(2n-3)} a_0 \ldots\ldots \text{ and so on}$$

$$\therefore \quad y = \sum_{r=0}^{\infty} a_r x^{m-r} = \sum_{r=0}^{\infty} a_r x^{n-r}$$

$$= a_0 x^n + a_1 x^{n-1} + a_2 x^{n-2} + a_3 x^{n-3} + a_4 x^{n-4} \ldots\ldots$$

$$= a_0 x^n - \frac{n(n-1)}{2(2n-1)} a_0 x^{n-2} + \frac{n(n-1)(n-2)(n-3)}{2.4(2n-1)(2n-3)} a_0 x^{n-4} \ldots\ldots$$

$$= a_0 \left[x^n - \frac{n(n-1)}{2(2n-1)} x^{n-2} + \frac{n(n-1)(n-2)(n-3)}{2.4(2n-1)(2n-3)} x^{n-4} \ldots\ldots \right] \quad \ldots(2.7)$$

which is solution of (2.1) in a series of descending power of x.

Case II. When $m = -(n+1)$ then from (2.6), we have

$$a_r = -\frac{(n+r-1)(n+r)}{r(2n+r+1)} a_{r-2}$$

Putting $r = 2, 4, \ldots\ldots$ etc.

$$a_2 = -\frac{(n+1)(n+2)}{2(2n+3)} a_0$$

$$a_4 = -\frac{(n+3)(n+4)}{4(2n+5)} a_2 = \frac{(n+1)(n+2)(n+3)(n+4)}{2.4(2n+3)(2n+5)} a_0 \ldots\ldots \text{ and so on.}$$

$$\therefore \quad y = \sum_{r=0}^{\infty} a_r x^{m-r} = \sum_{r=0}^{\infty} a_r x^{-(n+1)-r}$$

$$= a_0 x^{-(n+1)} + a_1 x^{-(n+2)} + a_2 x^{-(n+3)} + a_3 x^{-(x+4)} + a_4 x^{-(n+5)} \ldots\ldots$$

$$= a_0 x^{-(n+1)} - \frac{(n+1)(n+2)}{2(2n+3)} a_0 x^{-(n+3)}$$

$$+ \frac{(n+1)(n+2)(n+3)(n+4)}{2.4(2n+3)(2n+5)} a_0 x^{-(n+5)} \ldots\ldots$$

$$= a_0\left[x^{-(n+1)} - \frac{(n+1)(n+2)}{2(2n+3)}x^{-(n+3)} + \frac{(n+1)(n+2)(n+3)(n+4)}{2.4\,(2n+3)(2n+5)}x^{-(n+5)}\,.......\right] \qquad ...(2.8)$$

which is another solution of (2.1) in a series of descending power of x.

Legendre's Polynomial $P_n(x)$

If n is a positive integer and $a_0 = \dfrac{1.3.5.......(2n-1)}{n!}$ the solution (2.7) is $P_n(x)$ and called Legendre's function of the first kind so that

$$P_n(x) = \frac{1.3.5.......(2n-1)}{n!} = \left[x^n - \frac{n(n-1)}{2(2n-1)}x^{n-2} + \frac{n(n-1)(n-2)(n-3)}{2.4\,(2n-1)(2n-3)}x^{n-4}\,......\right]$$

Legendre's Polynomial $Q_n(x)$

If n is a positive integer and $a_0 = \dfrac{n!}{1.3.5.....(2n+1)}$ the solution (2.8) is $Q_n(x)$ and called Legendre's function of the second kind so that

$$Q_n(x) = \frac{n!}{1.3.5.....(2n+1)}\left[x^{-(n+1)} - \frac{(n+1)(n+2)}{2(2n+3)}x^{-(n+3)} + ...\right].$$

2.3. GENERAL SOLUTION OF LEGENDRE'S EQUATION

The solution of the Legendre's differential equation of the type

$$y(x) = AP_n(x) + BQ_n(x)$$

where A and B are arbitrary constant is called general solution.

2.4. GENERATING FUNCTION OF LEGENDRE'S POLYNOMIAL $P_N(X)$

To show that $P_n(x)$ is the coefficient of h^n in the expansion of $(1-2xh+h^2)^{-1/2}$ in ascending powers of h

We have $(1-2xh+h^2)^{-1/2} = \{1-h(2x-h)\}^{-1/2}$

$$= 1 + \frac{1}{2}h(2x-h) + \frac{1.3}{2.4}h^2(2x-h)^2 + ...$$

$$+ \frac{1.3......(2n-3)}{2.4......(2n-2)}h^{n-1}(2x-h)^{n-1} + \frac{1.3......(2n-1)}{2.4......2n}h^n(2x-h)^n + ...$$

Now the coefficient of h^n in above expansion is

$$= \frac{1.3\ldots\ldots(2n-1)}{2.4\ldots\ldots 2n}(2x)^n - \frac{1.3\ldots\ldots(2n-3)}{2.4\ldots\ldots(2n-2)}\,{}^{n-1}C_1\,(2x)^{n-2}$$

$$+ \frac{1.3\ldots\ldots(2n-5)}{2.4\ldots\ldots(2n-4)}\,{}^{n-2}C_2\,(2x)^{n-4}\ldots\ldots$$

$$= \frac{1.3\ldots\ldots(2n-1)}{2.4\ldots\ldots 2n}2^n\left[x^n - \frac{2n}{2n-1}(n-1)\frac{x^{n-2}}{2^2} + \frac{2n(2n-2)}{(2n-1)(2n-3)}\times\frac{(n-2)(n-3)}{2!}\frac{x^{n-4}}{2^4}\ldots\ldots\right.$$

$$= \frac{1.3\ldots\ldots(2n-1)}{n!}\left[x^n - \frac{2n}{2n-1}(n-1)\frac{x^{n-2}}{2^2} + \frac{(2n)(2n-2)}{(2n-1)(2n-3)}\frac{(n-2)(n-3)}{2!}\frac{x^{n-4}}{2^4}\ldots\ldots\right.$$

$$= \frac{1.3\ldots\ldots(2n-1)}{n!}\left[x^n - \frac{n(n-1)}{2.(2n-1)}x^{n-2} + \frac{n(n-1)(n-2)(n-3)}{2.4(2n-1)(2n-3)}x^{n-4} + \ldots\ldots\right]$$

$= P_n(x)$. **Hence proved.**

Hence, we obtained

$$\sum_{n=0}^{\infty} h^n P_n(x) = (1 - 2xh + h^2)^{-1/2}.$$

2.5. ORTHOGONAL PROPERTIES OF LEGENDRE'S POLYNOMIALS

(*i*) $\int_{-1}^{1} P_n(x)\,P_m(x)\,dx = 0$ if $m \neq n$

(*ii*) $\int_{-1}^{1} (P_n(x))^2\,dx = \frac{2}{2n+1}$

Proof. (*i*) We have Legendre's equation

$$(1 - x^2)\frac{d^2y}{dx^2} - 2x\frac{dy}{dx} + n(n+1)\,y = 0 \qquad \ldots(2.1)$$

This equation can also be written as

$$\frac{d}{dx}\left\{(1-x^2)\frac{dy}{dx}\right\} + n\,(n+1)\,y = 0 \qquad \ldots(2.9)$$

Since $P_n(x)$ is a solution of (2.1) therefore

$$\frac{d}{dx}\left\{(1-x^2)\frac{d\,P_n(x)}{dx}\right\} + n(n+1)\,P_n(x) = 0 \qquad \ldots(2.10)$$

Similarly $\frac{d}{dx}\left\{(1-x^2)\frac{dP_m(x)}{dx}\right\} + m(m+1)\,P_m(x) = 0$ ...(2.11)

Now multiplying (2.10) by $P_m(x)$ and (2.11) by $P_n(x)$ and then subtract, we get

$$P_m(x)\,\frac{d}{dx}\left\{(1-x^2)\frac{dP_n(x)}{dx}\right\} - P_n(x)\,\frac{d}{dx}\left\{(1-x^2)\frac{dP_m(x)}{dx}\right\}$$

$$+\{n(n+1) - m(m+1)\}\,P_n(x)\,P_m(x) = 0 \quad ...(2.12)$$

Integrating (2.12) w.r.t. x from -1 to 1, we get

$$\int_{-1}^{1} P_m(x)\frac{d}{dx}\left\{(1-x^2)\frac{dP_n(x)}{dx}\right\}dx - \int_{-1}^{1} P_n(x)\frac{d}{dx}\left\{(1-x^2)\frac{dP_m(x)}{dx}\,dx\right\}dx$$

$$+\{n(n+1) - m(m+1)\}\int_{-1}^{1} P_n(x)P_m(x)\,dx = 0$$

After integrating, we get

$$\left[P_m(x)(1-x^2)\frac{dP_n(x)}{dx}\right]_{-1}^{1} - \int_{-1}^{1}\frac{dP_m(x)}{dx}\left\{(1-x^2)\frac{dP_n(x)}{dx}\right\}dx - \left[P_n(x)(1-x^2)\frac{dP_m(x)}{dx}\right]_{-1}^{1}$$

$$+\int_{-1}^{1}\frac{dP_n(x)}{dx}\left\{(1-x^2)\frac{dP_m(x)}{dx}\right\}dx + \{n(n+1) - m(m+1)\}\int_{-1}^{1} P_n(x)\,P_m(x)\,dx = 0$$

or $$\{n(n+1) - m(m+1)\}\int_{-1}^{1} P_n(x)\,P_m(x)\,dx = 0$$

Hence $\int_{-1}^{1} P_n(x)\,P_m(x)\,dx = 0$ if $m \neq n$.

(*ii*) We have by generating function

$$\sum_{n=0}^{\infty} h^n P_n(x) = (1 - 2xh + h^2)^{-1/2}$$

Squaring both sides, we get

$$[P_0(x) + hP_1(x) + h^2 P_2(x) + \ldots\ldots h^n P_n(x) + \ldots\ldots]^2 = (1 - 2xh + h^2)^{-1}$$

or $$(P_0(x))^2 + (hP_1(x))^2 + (h^2 P_2(x))^2 + \ldots\ldots (h^n P_n(x))^2 + \ldots\ldots$$

$$+ 2\,[\,P_0(x)\,hP_1(x) + P_0(x)\,h^2P_2(x) + \ldots\ldots P_0(x)h^n P_n(x) \ldots\ldots$$

$$+ hP_1(x)\,h^2P_2(x) + hP_1(x)h^3P_3(x) + \ldots\ldots\,] = (1 - 2xh + h^2)^{-1}$$

or $$\sum_{n=0}^{\infty} h^{2n}(P_n(x))^2 + 2\sum_{\substack{m,n=0\\ m\neq n}}^{\infty} h^{m+n}P_n(x)P_m(x) = \frac{1}{1-2xh+h^2}$$

Integrating both sides w.r.t. x from -1 to 1, we get

$$\int_{-1}^{1}\sum_{n=0}^{\infty} h^{2n}(P_n(x))^2\,dx + 2\int_{-1}^{1}\sum_{\substack{m,n=0\\ m\neq n}}^{\infty} h^{m+n}P_n(x)P_m(x)\,dx = \int_{-1}^{1}\frac{dx}{1-2xh+h^2}$$

or
$$\sum_{n=0}^{\infty} h^{2n}\int_{-1}^{1}(P_n(x))^2\,dx + 2\sum_{\substack{m,n=0\\ m\neq n}}^{\infty} h^{m+n}\int_{-1}^{1}P_n(x)P_m(x)\,dx = \int_{-1}^{1}\frac{dx}{1-2xh+h^2}$$

or
$$\sum_{n=0}^{\infty} h^{2n}\int_{-1}^{1}(P_n(x))^2\,dx = \frac{-1}{2h}\Big[\log(1-2xh+h^2)\Big]_{-1}^{1}$$

$$= -\frac{1}{2h}[\log(1-2h+h^2) - \log(1+2h+h^2)]$$

$$= -\frac{1}{2h}[\log(1-h)^2 - \log(1+h)^2]$$

$$\sum_{n=0}^{\infty} h^{2n}\int_{-1}^{1}(P_n(x))^2\,dx = \frac{1}{2h}\log\frac{(1+h)^2}{(1-h)^2} = \frac{1}{2h}\log\left(\frac{1+h}{1-h}\right)^2$$

$$= \frac{1}{h}\log\frac{1+h}{1-h} = \frac{1}{h}2\left\{h+\frac{h^3}{3}+\frac{h^5}{5}+\ldots\ldots\right\}$$

$$= 2\left\{1+\frac{h^2}{3}+\frac{h^4}{5}\ldots\ldots\frac{h^{2n}}{2n+1}\ldots\ldots\right\}$$

Equating the coefficients of h^{2n}, we have

$$\int_{-1}^{1}(P_n(x))^2\,dx = \frac{2}{2n+1}.$$

2.6. LAPLACE'S FIRST INTEGRAL FOR $P_N(X)$

$$P_n(x) = \frac{1}{\pi}\int_0^{\pi}[x \pm \sqrt{(x^2-1)}\cos\theta]^n\,d\theta,$$ where n is a positive integer.

Proof. We know that

$$\int_0^{\pi}\frac{d\theta}{a \pm b\cos\theta} = \frac{\pi}{\sqrt{a^2-b^2}} \quad \text{where } a^2 > b^2 \qquad \ldots(2.13)$$

Let us taking $a = 1 - hx$ and $b = h\sqrt{(x^2-1)}$

and putting in (2.13), we get

$$\frac{\pi}{\sqrt{(1-hx)^2 - h^2(x^2-1)}} = \int_0^{\pi} \frac{d\theta}{1 - hx \pm h\sqrt{(x^2-1)}\cos\theta}$$

or
$$\frac{\pi}{\sqrt{1 + h^2x^2 - 2hx - h^2x^2 + h^2}} = \int_0^{\pi} \frac{d\theta}{1 - h[x \pm \sqrt{(x^2-1)}\cos\theta]}$$

or
$$\pi\,(1 - 2hx + h^2)^{-1/2} = \int_0^{\pi} \frac{d\theta}{1 - ht} \quad \text{where} \quad t = x \pm \sqrt{(x^2-1)}\cos\theta$$

$$= \int_0^{\pi} (1 - ht)^{-1}\, d\theta$$

$$\pi \sum_{n=0}^{\infty} h^n P_n(x) = \int_0^{\pi} (1 + ht + h^2t^2 + \ldots\ldots h^n t^n + \ldots\ldots)\, d\theta$$

$$= \int_0^{\pi} \sum_{n=0}^{\infty} h^n\, t^n\, d\theta$$

Equating the coefficient of h^n on both sides, we get

$$\pi P_n(x) = \int_0^{\pi} t^n\, d\theta$$

$$P_n(x) = \frac{1}{\pi}\int_0^{\pi} [x \pm \sqrt{(x^2-1)}\cos\theta]^n\, d\theta$$

$$= \frac{1}{\pi}\int_0^{\pi} [x \pm \sqrt{(x^2-1)}\cos\theta]^n\, d\theta\,.$$

2.7. LAPLACE'S SECOND INTEGRAL FOR $P_N(X)$

$$P_n(x) = \frac{1}{\pi}\int_0^{\pi} \frac{d\theta}{\left[x \pm \sqrt{x^2-1}\cos\theta\right]^{n+1}} \quad \text{where } n \text{ is a positive integer}$$

Proof. We know that $\displaystyle\int_0^{\pi} \frac{d\theta}{a \pm b\cos\theta} = \frac{\pi}{\sqrt{a^2 - b^2}}, \quad a^2 > b^2$ ...(2.14)

Let us taking $a = xh - 1$ and $b = h\sqrt{x^2-1}$ on putting in (2.14), we get

$$\frac{\pi}{\sqrt{(xh-1)^2 - h^2(x^2-1)}} = \int_0^{\pi} \frac{d\theta}{xh - 1 \pm h\sqrt{(x^2-1)}\cos\theta}$$

$$\frac{\pi}{\sqrt{x^2h^2 + 1 - 2xh - h^2x^2 + h^2}} = \int_0^{\pi} \frac{d\theta}{-1 + h\left[x \pm \sqrt{(x^2-1)}\cos\theta\right]}$$

$$\pi\,(1-2xh+h^2)^{-1/2} = \int_0^\pi \frac{d\theta}{-1+ht} \qquad \text{where} \quad t = \left[x \pm \sqrt{x^2-1}\cos\theta\right]$$

$$\frac{\pi}{h}\left(1-2x\frac{1}{h}+\frac{1}{h^2}\right)^{-1/2} = \int_0^\pi \frac{1}{ht}\left(1-\frac{1}{ht}\right)^{-1} d\theta$$

$$\pi \sum_{n=0}^{\infty} \frac{1}{h^n} P_n(x) = \int_0^\pi \frac{1}{t}\left(1+\frac{1}{ht}+\frac{1}{h^2t^2}\ldots\ldots\frac{1}{h^nt^n}\ldots\ldots\right)d\theta = \int_0^\pi \sum_{n=0}^{\infty} \frac{1}{h^n t^{n+1}}\,d\theta$$

Equating the coefficient of $\frac{1}{h^n}$ on both sides, we have

$$\pi\,P_n(x) = \int_0^\pi \frac{d\theta}{t^{n+1}} = \int_0^\pi \frac{d\theta}{[x \pm \sqrt{x^2-1}\cos\theta]^{n+1}}$$

$$P_n(x) = \frac{1}{\pi}\int_0^\pi \frac{d\theta}{[x \pm \sqrt{x^2-1}\cos\theta]^{(n+1)}}.$$

2.8. RODRIGUE'S FORMULA

To prove that $$P_n(x) = \frac{1}{2^n\,n!}\frac{d^n}{dx^n}(x^2-1)^n$$

Proof. Let $$y = (x^2-1)^n \qquad \ldots(2.15)$$

Differentiating w.r.t. x, we get

$$y_1 = n(x^2-1)^{n-1}\,2x$$

or $$(x^2-1)\,y_1 = 2nx(x^2-1)^n$$

or $$(x^2-1)y_1 = 2nxy$$

Again differentiating w.r.t. x, we get

$$(x^2-1)\,y_2 + 2xy_1 = 2nxy_1 + 2ny$$

or $$(x^2-1)y_2 + 2(1-n)xy_1 - 2ny = 0$$

Differentiating both sides n times by Leibnitz's theorem, we get

$$(x^2-1)y_{n+2} + n2x\,y_{n+1} + \frac{n(n-1)}{2!}\,2y_n + 2(1-n)\,[x\,y_{n+1} + ny_n] - 2ny_n = 0$$

or $$(x^2-1)y_{n+2} + 2x[n+(1-n)]y_{n+1} + [n\,(n-1) + 2\,(1-n)n - 2n]y_n = 0$$

or $$(x^2-1)y_{n+2} + 2x\,y_{n+1} - n(n+1)\,y_n = 0$$

or $$(1-x^2)\frac{d^{n+2}y}{dx^{n+2}} - 2x\frac{d^{n+1}y}{dx^{n+1}} + n(n+1)\frac{d^ny}{dx^n} = 0 \qquad \ldots(2.16)$$

Let $\dfrac{d^n y}{dx^n} = v$ put in (2.16), we get

$$(1-x^2)\frac{d^2v}{dx^2} - 2x\frac{dv}{dx} + n(n+1)\,v = 0$$

This show that $v = \dfrac{d^n y}{dx^n}$ is a solution of legendre's equation

Therefore $$C\,\frac{d^n y}{dx^n} = P_n(x) \quad \text{...(2.17)}$$

Putting $x = 1$, we get $C\left(\dfrac{d^n y}{dx^n}\right)_{x=1} = P_n(1) = 1$ $\quad [\because\ P_n(1) = 1]$ ...(2.18)

Now $$y = (x^2-1)^n = (x+1)^n (x-1)^n$$

Differentiating w.r.t. x n times by Leibnitz's theorem, we get

$$\frac{d^n y}{dx^n} = (x+1)^n \frac{d^n}{dx^n}(x-1)^n + {}^nC_1\, n(x+1)^{n-1}\frac{d^{n-1}}{dx^{n-1}}(x-1)^n$$
$$+ {}^nC_2\, n(n-1)(x+1)^{n-2}\frac{d^{n-2}}{dx^{n-2}}(x-1)^n$$
$$+ {}^nC_3\, n(n-1)(n-2)(x+1)^{n-3}\frac{d^{n-3}}{dx^{n-3}}(x-1)^n + \ldots\ldots$$
$$+ \ldots\ldots {}^nC_n\left(\frac{d^n}{dx^n}(x+1)^n\right)(x-1)^n$$

$$= (x+1)^n\, n! + {}^nC_1\, n(x+1)^{n-1}\frac{n!}{1!}(x-1)^1 + {}^nC_2\, n(n-1)(x+1)^{n-2}.\ \frac{n!}{2!}(x-1)^{n-3} + \ldots\ldots$$
$$+ .\ n!\,(x-1)^n$$

Put $x = 1$, we get

$$\left(\frac{d^n y}{dx^n}\right)_{x=1} = 2^n\, n!$$

Put in (2.18) $$C = \frac{1}{2^n\, n!}$$

Substituting the value of C in (2.17), we get

$$P_n(x) = \frac{1}{2^n\, n!}\frac{d^n y}{dx^n}$$

$$P_n(x) = \frac{1}{2^n\, n!}\frac{d^n (x^2-1)^n}{dx^n}. \qquad [\because\ y = (x^2-1)^n]$$

2.9. RECURRENCE RELATIONS

(I) $(2n + 1)\, x\, P_n(x) = (n + 1)\, P_{n+1}(x) + nP_{n-1}(x)$

Proof. By the generating function, we have

$$(1 - 2xh + h^2)^{-1/2} = \sum_{n=0}^{\infty} h^n P_n(x) \qquad ...(i)$$

Differentiating (*i*) both sides w.r.t. *h*, we get

$$-\frac{1}{2}(1 - 2xh + h^2)^{-3/2}(-2x + 2h) = \sum_{n=0}^{\infty} nh^{n-1} P_n(x)$$

or $$(x - h)(1 - 2xh + h^2)^{-1/2} = (1 - 2xh + h^2) \sum_{n=0}^{\infty} nh^{n-1} P_n(x)$$

or $$(x - h) \sum_{n=0}^{\infty} h^n P_n(x) = (1 - 2xh + h^2) \sum_{n=0}^{\infty} nh^{n-1} P_n(x)$$

or $$x \sum_{n=0}^{\infty} h^n P_n(x) - \sum_{n=0}^{\infty} h^{n+1} P_n(x)$$

$$= \sum_{n=0}^{\infty} nh^{n-1} P_n(x) - 2x \sum_{n=0}^{\infty} nh^n P_n(x) + \sum_{n=0}^{\infty} nh^{n+1} P_n(x) \qquad ...(ii)$$

Equating the coefficients of h^n on both sides of (*ii*), we get

$$xP_n(x) - P_{n-1}(x) = (n + 1)\, P_{n+1}(x) - n2xP_n(x) + (n - 1)\, P_{n-1}(x)$$

or $$\boxed{(2n + 1)\, x\, P_n(x) = (n + 1)\, P_{n+1}(x) + nP_{n-1}(x).}$$

(II) $nP_n(x) = xP_n'(x) - P_{n-1}'(x)$

Proof. By the generating function, we have

$$(1 - 2xh + h^2)^{-1/2} = \sum_{n=0}^{\infty} h^n P_n(x) \qquad ...(i)$$

Differentiating (*i*) both sides w.r.t. *h*, we get

$$-\frac{1}{2}(1 - 2xh + h^2)^{-3/2}(-2x + 2h) = \sum_{n=0}^{\infty} nh^{n-1} P_n(x)$$

or $$(x - h)(1 - 2xh + h^2)^{-3/2} = \sum_{n=0}^{\infty} nh^{n-1} P_n(x) \qquad ...(ii)$$

Again differentiating (*i*) both sides w.r.t. x, we get

$$-\frac{1}{2}(1-2xh+h^2)^{-3/2}(-2h)=\sum_{n=0}^{\infty} h^n P_n'(x)$$

or $$h(1-2xh+h^2)^{-3/2}=\sum_{n=0}^{\infty} h^n P_n'(x) \qquad ...(iii)$$

Multiplying (*ii*) by h and (*iii*) by $(x-h)$ then subtract, we get

$$0=h\sum_{n=0}^{\infty} nh^{n-1}P_n(x)-(x-h)\sum_{n=0}^{\infty} h^n P_n'(x)$$

or $$\sum_{n=0}^{\infty} nh^n P_n(x)=x\sum_{n=0}^{\infty} h^n P_n'(x)-\sum_{n=0}^{\infty} h^{n+1} P_n'(x). \qquad ...(iv)$$

Equating the coefficient of h^n on both sides of (*iv*), we get

$$\boxed{nP_n(x)=xP_n'(x)-P_{n-1}'(x).}$$

(III) $P'_{n+1}(x)-P'_{n-1}(x)=(2n+1)P_n(x)$

Proof. By recurrence relation I, we have

$$x(2n+1)P_n(x)=(n+1)P_{n+1}(x)+nP_{n-1}(x) \qquad ...(i)$$

Differentiating (*i*) w.r.t. x, we get

$$x(2n+1)P'_n(x)+(2n+1)P_n(x)=(n+1)P'_{n+1}(x)+nP'_{n-1}(x) \qquad ...(ii)$$

By recurrence relation II, we have

$$xP'_n(x)=nP_n(x)+P'_{n-1}(x)$$

Putting $xP'_n(x)$ in (*ii*), we get

$$(2n+1)(nP_n(x)+P'_{n-1}(x))+(2n+1)P_n(x)=(n+1)P'_{n+1}(x)+nP'_{n-1}(x)$$

or $$(2n+1)(n+1)P_n(x)=(n+1)P'_{n+1}(x)+[n-(2n+1)]P'_{n-1}(x)$$

or $$(2n+1)(n+1)P_n(x)=(n+1)P'_{n+1}(x)-(n+1)P'_{n-1}(x)$$

or $$\boxed{(2n+1)P_n(x)=P'_{n+1}(x)-P'_{n-1}(x).}$$

(IV) $(n+1)P_n(x)=P'_{n+1}(x)-xP'_n(x)$

Proof. By recurrence relation (II) and (III), we have.

$$nP_n(x)=xP'_n(x)-P'_{n-1}(x) \qquad ...(i)$$

and $$(2n+1)P_n(x)=P'_{n+1}(x)-P'_{n-1}(x) \qquad ...(ii)$$

Subtract (*i*) from (*ii*), we get

$$(2n+1)P_n(x)-nP_n(x)=P'_{n+1}(x)-P'_{n-1}(x)-xP'_n(x)+P'_{n-1}(x)$$

$$\boxed{(n+1)P_n(x)=P'_{n+1}(x)-xP'_n(x).}$$

(V) $(1-x^2)P'_n(x) = n(P_{n-1} - xP_n)$

Proof. By recurrence relation (II) and (IV), we have

$$nP_n(x) = xP'_n(x) - P'_{n-1}(x) \qquad ...(i)$$

and
$$(n+1)P_n(x) = P'_{n+1}(x) - xP'_n(x) \qquad ...(ii)$$

Putting $(n-1)$ in place of n in (ii), we get

$$n\,P_{n-1}(x) = P'_n(x) - xP'_{n-1}(x) \qquad ...(iii)$$

Now multiplying (i) by x and subtract from (iii), we get

$$nP_{n-1}(x) - nxP_n(x) = P'_n(x) - xP'_{n-1}(x) - x^2P'_n(x) - xP'_{n-1}(x)$$

$$\boxed{(1-x^2)P'_n(x) = n[P_{n-1}(x) - xP_n(x)].}$$

(VI) $(1-x^2)P'_n(x) = (n+1)(xP_n(x) - P_{n+1}(x))$

Proof. By recurrence relation (I) and (V), we have

$$(2n+1)\,x\,P_n(x) = (n+1)\,P_{n+1}(x) + nP_{n-1}(x) \qquad ...(i)$$

and
$$(1-x^2)P'_n(x) = n\,[P_{n-1}(x_1) - xP_n(x)] \qquad ...(ii)$$

By (i)
$$nP_{n-1}(x) = (2n+1)\,x\,P_n(x) - (n+1)\,P_{n+1}(x)$$

Putting in (ii), we get

$$(1-x^2)P'_n(x) = (2n+1)\,x\,P_n(x) - (n+1)\,P_{n+1}(x) - nx\,P_n(x)$$

$$\boxed{(1-x^2)P'_n(x) = (n+1)\,[x\,P_n(x) - P_{n+1}(x)].}$$

SOLVED EXAMPLES

Example 1. *Show that* $P_0(x) = 1,\ P_1(x) = x,\ P_2(x) = \dfrac{3x^2-1}{2},\ P_3(x) = \dfrac{5x^3-3x}{5}$ *and* $P_4(x) = \dfrac{35x^4 - 30x^2 + 3}{8}$.

Solution. By the Rodrigue's formula, we have

$$P_n(x) = \frac{1}{n!2^n}\frac{d^n}{dx^n}(x^2-1)^n \qquad ...(i)$$

Put $n = 0$ in (i), we have

$$P_0(x) = \frac{1}{0!2^0}\frac{d^0}{dx^0}(x^2-1)^0 = 1$$

$$\Rightarrow \qquad P_0(x) = 1$$

Put $n = 1$ in (*i*), we have

$$P_1(x) = \frac{1}{1!2^1}\frac{d^1}{dx^1}(x^2-1)^1 = \frac{1}{2}\,2x = x$$

$$\Rightarrow \qquad P_1(x) = x$$

Put $n = 2$ in (*i*), we have

$$P_2(x) = \frac{1}{2!2^2}\frac{d^2}{dx^2}(x^2-1)^2 = \frac{1}{2.4}\frac{d}{dx}2(x^2-1)\,2x = \frac{1}{2}\frac{d}{dx}(x^3-x)$$

$$= \frac{1}{2}(3x^2-1)$$

$$\Rightarrow \qquad P_2(x) = \frac{1}{2}(3x^2-1)$$

Put $n = 3$ in (*i*), we have

$$P_3(x) = \frac{1}{3!2^3}\frac{d^3}{dx^3}(x^2-1)^3 = \frac{1}{6.8}\frac{d^2}{dx^2}3(x^2-1)^2\,2x = \frac{1}{8}\frac{d^2}{dx^2}(x^2-1)^2 x$$

$$= \frac{1}{8}\frac{d}{dx}(5x^4-6x^2+1) = \frac{1}{8}(20x^3-12x) = \frac{1}{2}(5x^3-3x)$$

Put $n = 4$ in (*i*), we have

$$P_4(x) = \frac{1}{4!2^4}\frac{d^4}{dx^4}(x^2-1)^4 = \frac{1}{24.16}\frac{d^3}{dx^3}4(x^2-1)^3\,2x = \frac{1}{3.16}\frac{d^2}{dx^2}[(x^2-1)^2(7x^2-1)]$$

$$= \frac{1}{3.16}\frac{d}{dx}(x^2-1)(42x^3-18x) = \frac{1}{3.16}(210x^4-180x^2+18) = \frac{1}{8}(35x^4-30x^2+3).$$

Example 2. *Express $3 + 4x - x^2$ in term of Legendre polynomials.*

Solution. We have

$$P_0(x) = 1 \Rightarrow 1 = P_0(x),\ P_1(x) = x \Rightarrow x = P_1(x),$$

$$\Rightarrow \qquad P_2(x) = \frac{3x^2-1}{2} \Rightarrow x_2 = \frac{2P_2(x)+1}{3}$$

On putting these values in the given equation, we have

$$3 + 4x - x^2 = 3P_0(x) + 4P_1(x) - \frac{2P_2(x)+P_0(x)}{3} = \frac{9P_0(x)+12P_1(x)-2P_2(x)-P_0(x)}{3}$$

$$= \frac{1}{3}[8P_0(x) + 12P_1(x) - 2P_2(x)].$$

Example 3. *Show that (i) $P_n(1) = 1$, (ii) $P_n(-x) = (-1)^n P_n(x)$, (iii) $P_n(-1) = (-1)^n$.*

Solution. (*i*) We have $\sum_{n=0}^{\infty} h^n P_n(x) = (1 - 2xh + h^2)^{-1/2}$ ...(*i*)

Putting $x = 1$ in (i) on both sides, we have

$$\sum_{n=0}^{\infty} h^n P_n(1) = (1 - 2h + h^2)^{-1/2} = (1 - h)^{-1} = 1 + h + h^2 + \ldots\ldots + h^n + \ldots\ldots.$$

$$\sum_{n=0}^{\infty} h^n P_n(1) = \sum_{n=0}^{\infty} h^n$$

Equating the coefficients of h^n on both sides, we have

$$P_n(1) = 1 \qquad \ldots(ii)$$

(ii) Putting $-h$ in place of h on both sides of (i), we have

$$\sum_{n=0}^{\infty} (-h)^n P_n = (1 + 2xh + h^2)^{-1/2} \qquad \ldots(iii)$$

Again putting $-x$ in place of x on both sides of (i), we have

$$\sum_{n=0}^{\infty} h^n P_n\,(-x) = (1 + 2xh + h^2)^{-1/2} \qquad \ldots(iv)$$

By (ii) and (iii), we have

$$\sum_{n=0}^{\infty} (-1)^n h^n P_n(x) = \sum_{n=0}^{\infty} h^n P_n\,(-x)$$

Equating the coefficient of h^n on both sides, we have

$$(-1)^n P_n(x) = P_n(-x)$$

or

$$P_n(-x) = (-1)^n P_n(x) \qquad \ldots(v)$$

(iii) Putting $x = 1$ in (v), we have

$$P_n(-1) = (-1)^n P_n(1) = (-1)^n. \qquad \text{Since } P_n(1) = 1 \text{ By } (ii)$$

Example 4. *Prove that $(1 - 2xz + z^2)^{-1/2}$ is a solution of the equation*

$$z\,\frac{\partial^2(zv)}{\partial z^2} + \frac{\partial}{\partial x}\left\{(1 - x^2)\frac{\partial v}{\partial x}\right\} = 0.$$

Solution. Let $\qquad v = (1 - 2xz + z^2)^{-1/2}$

$$v = \sum_{n=0}^{\infty} z^n P_n(x) \qquad \ldots(i)$$

Then

$$zv = \sum_{n=0}^{\infty} z^{n+1} P_n(x)$$

Differentiating partially both sides w.r.t. z two times, we have

$$\frac{\partial^2}{\partial z^2}(zv) = \sum_{n=0}^{\infty} (n+1)nz^{n-1} \; P_n(x)$$

or $$z\frac{\partial^2}{\partial z^2}(zv) = \sum_{n=0}^{\infty} n(n+1)z^{n} \; P_n(x) \qquad ...(ii)$$

Now differentiating partially (i) both sides w.r.t. x, we have

$$\frac{\partial v}{\partial x} = \sum_{n=0}^{\infty} z^n P_n'(x)$$

or $$(1-x^2)\frac{\partial v}{\partial x} = (1-x^2)\sum_{n=0}^{\infty} z^n P_n'(x)$$

Again differentiating partially both sides w.r.t. x, we have

$$\frac{\partial}{\partial x}\left\{(1-x^2)\frac{dv}{dx}\right\} = \frac{\partial}{\partial x}\left\{(1-x^2)\sum_{n=0}^{\infty} z^n \; P_n'(x)\right\}$$

$$= (1-x^2)\sum_{n=0}^{\infty} z^n \; P_n''(x) - 2x\sum_{n=0}^{\infty} z^n \; P_n'(x) \qquad ...(iii)$$

On adding (ii) and (iii), we get

$$z\frac{\partial^2}{\partial z^2}(zv) + \frac{\partial}{\partial x}\left\{(1-x^2)\frac{dv}{dx}\right\} = \sum_{n=0}^{\infty} n(n+1)z^n P_n(x) + (1-x^2)\sum_{n=0}^{\infty} z^n \; P_n''(x) - 2x\sum_{n=0}^{\infty} z^n \; P_n'(x)$$

$$= \sum_{n=0}^{\infty} [(1-x^2)\,P_n''(x) - 2x\,P_n'(x) + n(n+1)\,P_n(x)]\,z^n$$

$$= \sum_{n=0}^{\infty} 0\; z^n = 0 \qquad [\because \; P_n(x) \text{ is the solution of Legendre's equation}]$$

This shows that $v = (1-2xz+z^2)^{-1/2}$ is the solution of the given equation.

Example 5. *Prove that* $\dfrac{1-z^2}{(1-2xz+z^2)^{3/2}} = \sum_{n=0}^{\infty} (2n+1)\,P_n(x)z^n$.

Solution. We have $(1-2xz+z^2)^{-1/2} = \sum_{n=0}^{\infty} z^n \; P_n(x) \qquad ...(i)$

Differentiating on both sides w.r.t. z, we have

$$-\frac{1}{2}(1-2xz+z^2)^{-3/2}(-2x+2z)=\sum_{n=0}^{\infty} nz^{n-1}P_n(x)$$

or
$$(x-z)(1-2xz+z^2)^{-3/2}=\sum_{n=0}^{\infty} nz^{n-1}P_n(x)$$

or
$$2z(x-z)(1-2xz+z^2)^{-3/2}=\sum_{n=0}^{\infty} 2nz^{n}P_n(x) \qquad ...(ii)$$

Adding (i) and (ii)

$$(1-2xz+z^2)^{-1/2}+2z(x-z)(1-2xz+z^2)^{-3/2}=\sum_{n=0}^{\infty} z^n P_n(x)+\sum_{n=0}^{\infty} 2nz^n P_n(x)$$

or
$$\frac{1-2xz+z^2+2z(x-z)}{(1-2xz+z^2)^{3/2}}=\sum_{n=0}^{\infty} z^n(1+2n)P_n(x)$$

or
$$\frac{1-z^2}{(1-2xz+z^2)^{3/2}}=\sum_{n=0}^{\infty}(2n+1)z^n P_n(x).$$

Example 6. *Show that* $P'_{n+1}(x)+P'_n(x)=P_0+3P_1+5P_2+\ldots\ldots+(2n+1)P_n=\sum_{r=0}^{n}(2r+1)P_r(x).$

Solution. By the recurrence relation III, we have

$$(2n+1)P_n=P'_{n+1}-P'_{n-1} \qquad ...(i)$$

Putting $n = 1, 2, 3, 4, 5, \ldots\ldots, n$ in (i), we get

$$\begin{aligned}
3P_1 &= P'_2-P'_0\\
5P_2 &= P'_3-P'_1\\
7P_3 &= P'_4-P'_2\\
&\vdots\\
(2n-3)P_{n-2} &= P'_{n-1}-P'_{n-3}\\
(2n-1)P_{n-1} &= P'_n-P'_{n-2}\\
(2n+1)P_n &= P'_{n+1}-P'_{n-1}
\end{aligned}$$

Adding above results, we get

$$3P_1+5P_2+7P_3+\ldots\ldots(2n+1)P_n=P'_n+P'_{n+1}-P'_0-P'_1 \qquad ...(ii)$$

But $P_0 = 1$, and $P_1 = x$

$\therefore$ $P'_0 = 0$ and $P'_1 = 1 = P_0$

Put in (ii)

$$3P_1+5P_2+7P_3+\ldots\ldots(2n+1)P_n=P'_n+P'_{n+1}-0-P_0$$

Hence $P'_{n+1}(x) + P'_n(x) = P_0 + 3P_1 + 5P_2 + 7P_3 + \ldots\ldots + (2n+1)P_n$

$$= \sum_{r=0}^{n}(2r+1)P_r \quad \text{Proved.}$$

Example 7. *Prove that*

$$P'_{2n+1}(x) = (2n+1)P_{2n}(x) + 2nx\,P_{2n-1}(x) + (2n-1)x^2 P_{2n-2}(x) + \ldots\ldots + 2x^{2n-1}p_1 + x^{2n}.$$

Solution. By the recurrence relation IV, we have

$$P'_n = nP_{n-1} + xP'_{n-1} \qquad \ldots(i)$$

Putting $n = 2, 3, \ldots\ldots (2n-1), 2n, (2n+1)$, we have

$$P'_2 = 2P_1 \ldots\ldots xP'_1$$
$$P'_3 = 3P_2 \ldots\ldots xP'_2$$
$$\vdots$$
$$P'_{2n-1} = (2n-1)P_{2n-2} + xP'_{2n-2}$$
$$P'_{2n} = 2n\,P_{2n-1} + xP'_{2n-1}$$
$$P'_{2n+1} = (2n+1)P_{2n} + xP'_{2n}$$

Multiplying the equation in above $x^{2n-1}, x^{2n-2}, \ldots, x^2, x, 1$ respectively and adding, we get

$$P'_{2n+1} = (2n+1)P_{2n} + 2nx\,P_{2n-1} + (2n-1)x^2 P_{2n-2} + \ldots\ldots + 2x^{2n-1}P_1 + x^{2n} \text{ where } P'_1 = 1.$$

Example 8. *Show that*

$$(i)\ \int P_n(x)\,dx = \frac{1}{2n+1}[P_{n+1}(x) - P_{n-1}(x)] + C$$

$$(ii)\ \int_x^1 P_n(x)\,dx = \frac{1}{2n+1}[P_{n-1}(x) - P_{n+1}(x)].$$

Solution. By the recurrence relation III, we have

$$(2n+1)P_n(x) = P'_{n+1}(x) - P'_{n-1}(x)$$

or
$$P_n(x) = \frac{1}{2n+1}[P'_{n+1}(x) - P'_{n-1}(x)] \qquad \ldots(i)$$

(*i*) Integrating, both sides of (*i*) w.r.t. *x*, we have

$$\int P_n(x)dx = \frac{1}{2n+1}[P_{n+1}(x) - P_{n-1}(x)] + c$$

(*ii*) Integrating both sides of (*i*) w.r.t. *x* from *x* to 1, we get

$$\int_x^1 P_n(x)\,dx = \frac{1}{2n+1}\Big[P_{n+1}(x) - P_{n-1}(x)\Big]_x^1$$

$$= \frac{1}{(2n+1)} [P_{n+1}(1) - P_{n-1}(1) - P_{n+1}(x) + P_{n-1}(x)]$$

$$[\because \; P_{n+1}(x) = 1 = P_{n-1}(x)]$$

$$\therefore \qquad \int_x^1 P_n(x)\, dx = \frac{1}{(2n+1)} [P_{n-1}(x) - P_{n+1}(x)].$$

Example 9. *Prove that* $\int_{-1}^{1} (x^2 - 1) P_{n+1} P_n' \, dx = \frac{2n(n+1)}{(2n+1)(2n+3)}$.

Solution. From recurrence relation (V), we have

$$(x^2 - 1) P_n' = n (x P_n - P_{n-1})$$

or
$$(x^2 - 1) P_{n+1} P_n' = n (x P_n - P_{n-1}) P_{n+1} \qquad ...(i)$$

Integrating both sides of (*i*) w.r.t. x from -1 to 1, we get

$$\int_{-1}^{1} (x^2 - 1) P_{n+1} P_n' \, dx = \int_{-1}^{1} n(xP_n - P_{n-1}) P_{n+1} \, dx$$

$$= n \int_{-1}^{1} xP_n P_{n+1} \, dx - n \int_{-1}^{1} P_{n-1} P_{n+1} \, dx$$

$$= n \int_{-1}^{1} xP_n P_{n+1} \, dx \qquad \left[\because \int_{-1}^{1} P_n(x) P_m(x)\, dx = 0 \text{ if } m \neq n\right]$$

$$= n \int_{-1}^{1} \frac{(n+1) P_{n+1} + nP_{n-1}}{2n+1} P_{n+1} \, dx \qquad \text{(By recurrence formula I)}$$

$$= \frac{n(n+1)}{2n+1} \int_{-1}^{1} P_{n+1}^2 \, dx + \frac{n^2}{2n+1} \int_{-1}^{1} P_{n-1} P_{n+1} \, dx$$

$$= \frac{n(n+1)}{2n+1} \frac{2}{2(n+1)+1} + 0 \qquad \left[\because \int_{-1}^{1} P_n(x) P_m(x)\, dx = \begin{cases} 0 & \text{if } m \neq n \\ \dfrac{2}{2n+1} & \text{if } m = n \end{cases}\right]$$

$$= \frac{2n(n+1)}{(2n+1)(2n+3)}.$$

Example 10. *Prove that* $\int_{-1}^{1} (1 - x^2) P_m' P_n' \, dx = \frac{2m(m+1)}{(2m+1)} \delta_{mn}$ *where* δ_{mn} *is Kronecker delta*

i.e., $\delta_{mn} = \begin{cases} 0 & m \neq n \\ 1 & m = 1 \end{cases}$.

Solution. We have $\int_{-1}^{1} (1 - x^2) P_m' P_n' \, dx$

$$= \left[(1-x^2)\,P'_m P_n\right]_{-1}^{1} - \int_{-1}^{1}\left[P_n \frac{d}{dx}\{(1-x^2)P'_m\}\right]dx$$

$$= -\int_{-1}^{1} P_n \frac{d}{dx}\{(1-x^2)P'_m\}\,dx \qquad ...(i)$$

Since P_m is a solution of the Legendre's equation therefore

$$(1-x^2)\,P''_m - 2xP'_m + m(m+1)\,P_m = 0$$

or
$$\frac{d}{dx}\,\{(1-x^2)P'_m\} = -\,m\,(m+1)\,P_m$$

Put in (*i*), we get

$$\int_{-1}^{1}(1-x^2)\,P'_m\,P'_n\,dx = -\int_{-1}^{1} P_n\{-\,m(m+1)P_m\}\,dx$$

$$= m(m+1)\int_{-1}^{1} P_n P_m\,dx = m\,(m+1)\frac{2}{2m+1}\,\delta_{mn} = \frac{2m(m+1)}{(2m+1)}\,\delta_{mn}$$

where δ_{mn} is Kronecker delta.

EXERCISE 2.1

1. Express $f(x) = x^4 + 2x^3 + 2x^2 - x - 3$ in term of Legendre's polynomials.
2. Prove that $P_n(x)$ is an even or odd function of x according as n is even or odd respectively.
3. Show that

   (*i*) $P_n(0) = 0$, if n is odd (*ii*) $P_n\,(0) = \dfrac{(-1)^{n/2}\,n!}{2^n\,(n/2\,!)^2}$, if n is even.

4. Prove that $\dfrac{1+z}{z\sqrt{1-2xz+z^2}} - \dfrac{1}{z} = \displaystyle\sum_{n=0}^{\infty}\,[P_n(x) + P_{n+1}\,(x)]z^n$.
5. Prove that

   (*i*) $P'_n\,(1) = \dfrac{1}{2}\,n(n+1)$ (*ii*) $P'_n\,(-1) = (-1)^{n-1}\,\dfrac{n(n+1)}{2}$.

6. Prove that $(n+1)\,[P_n\,P'_{n+1} - P_{n+1}\,P'_n] = (n+1)^2\,P_n^{\,2} - (x^2-1)\,P'^{\,2}_n$.
7. Prove that $\displaystyle\int_{-1}^{1} xP_nP_{n-1}\,dx = \frac{2n}{4n^2-1}$.
8. $\displaystyle\int_{-1}^{1}\frac{P_n(x)\,dx}{\sqrt{1-2xh+h^2}} = \frac{2h^n}{2n+1}$, prove it.

9. Prove that $\int_{-1}^{1} x^2 P_{n+1} P_{n-1}\, dx = \dfrac{2n(n+1)}{(2n-1)(2n+1)(2n+3)}$.

10. Prove that

(*i*) $\int_{-1}^{1} (1-x^2)(P_n')^2\, dx = \dfrac{2n(n+1)}{2n+1}$, if $m = n$. (*ii*) $\int_{-1}^{1} (1-x^2)\, P_m'\, P_n'\, dx = 0$, if $m \neq n$.

11. Prove that

(*i*) $\int_{-1}^{1} P_n(x)\, dx = 0, \quad n \neq 0$ (*ii*) $\int_{-1}^{1} P_0(x)\, dx = 2$.

12. Prove that if m is an integer less than n, $\int_{-1}^{1} x^m P_n(x)\, dx = 0$, $\int_{-1}^{1} x^n P_n(x)\, dx = \dfrac{2^{n+1}(n!)^2}{(2n+1)!}$.

13. Evaluate $\int_{-1}^{1} x^4 P_6(x)\, dx$.

14. Prove that $\sum_{n=0}^{\infty} P_n(x) = \dfrac{1}{\sqrt{2-2x}}$.

15. Show that $P_{2n}(0) = (-1)^n \dfrac{1.3.5 \ldots\ldots (2n-1)}{2}$.

16. Prove that $[P_n(x)] < 1$ when $-1 < x < 1$.

17. Let $P_n(x)$ be the Legendre polynomial of degree n, show that for any function $f(x)$ for which the nth derivative is continuous.

ANSWER

13. 0.

UNIT IV

STATISTICS AND PROBABILITY

Probability theory is used in many situation which involve an element of uncertainty. It is used to make important decision on business, to determine premiums on insurance policies, to determine demand in inventory control. Apart from it is used in Engineering. Science, Social Science, Genetics etc. Probability theory is also used in medical sciences.

In this unit, we shall discuss some basic idea of probability, addition and multiplication theorems of probability, Baye's theorem with simple applications. Expected value, Theoretical probability distributions—Binomial, Poisson and normal distributions and applications of probability in many area of human activities. Lines of regression, correlation and rank-correlation are also discussed.

Chapter 1 deals with basic idea of probability, theorems of probability and Baye's theorem with their applications.

Chapter 2 deals with random variable, probability distribution: Binomial, Poisson and Normal distributions.

In chapter 3, the lines of regression, concept of simple correlation and rank-correlation will be discussed.

In the 16 Century an italian mathematician Jerome Cardon (1501–1576) wrote the first book on the subject of probability theory "Book on Games of Chance'. Besides Cardon, Pascal (1623–1662). Fermat (1601–1665), J. Bernoulli (1654–1705). De Moiure (1667–1754), Chebychev (1821–1894), A.A. Markov (1856–1922) and A.N.Kolmogorov (1903 ......) gives outstanding contributions in probability theory.

Probability theory is designed to estimate the degree of uncertainty regarding the happening of a given phenomenon. The word probable itself indicates such a situation. Its dictionary meaning is "likely though not certain to occur". Hence, when a coin is tossed a tail is likely to occur but may not occur. When a die is thrown, it may or may not show the number 6.

CHAPTER 1

Theory of Probability

INTRODUCTION

Before we discuss formal definition of probability, first we shall define certain terminologies and notations, which are used in defining probability.

1.1. TERMINOLOGY AND NOTATIONS

1. **Random experiment:** Any experiment whose outcomes cannot be predicted or determined in advance is a random experiment.

OR

A random experiment is an experiment whose outcomes or result is not unique and therefore cannot be predicted with certainty.

Tossing of a coin, head or tail may occur, throwing a die, 1, 2, 3, 4, 5 or 6 may appear and drawing a card from a well shuffled pack of cards are examples of random experiments.

2. **Sample space and sample point:** The set of all possible outcomes of a random experiment is called the sample space and each element of a sample space is called a sample point.

If a die is thrown, it will land with anyone of its 6 faces pointing upward, resulting in anyone of the numbers 1, 2, 3, 4, 5, 6 appearing on the top face. Hence, the number of each face is a possible result. We write

$$S = \{1, 2, 3, 4, 5, 6\}$$

The set S is a sample space of throwing a die and 1, 2, 3, 4, 5, 6 are called sample point.

When two dice is thrown, then the sample space consists of the following 36 points:

$$S = \begin{bmatrix} (1,1), & (1,2), & (1,3), & (1,4), & (1,5), & (1,6) \\ (2,1), & (2,2), & (2,3), & (2,4), & (2,5), & (2,6) \\ (3,1), & (3,2), & (3,3), & (3,4), & (3,5), & (3,6) \\ (4,1), & (4,2), & (4,3), & (4,4), & (4,5), & (4,6) \\ (5,1), & (5,2), & (5,3), & (5,4), & (5,5), & (5,6) \\ (6,1), & (6,2), & (6,3), & (6,4), & (6,5), & (6,6) \end{bmatrix}$$

Note. The set of all possible outcomes of a single performance of a random experiment is exhaustive events or sample space.

3. **Event:** Any subset A of a sample space S is called an event.

Consider the experiment of tossing of a coin, we have

$$S = \{H, T\}$$

Here are 4 subsets of S, ϕ, $\{H\}$, $\{T\}$, $\{H, T\}$. Each subset of S is an event.

4. **Simple event:** An event is said to be simple event, if it has only one sample point. In tossing of a coin, the events $\{H\}$ and $\{T\}$ are simple events.

5. **Compound event:** An event is said to be compound event, if it has more than one sample point. When two coins are tossed once then sample space $(S) = \{HH, HT, TH, TT\}$. Each event of S is compound event.

6. **Complement of an event:** The complement of an event A with respect to the sample space S is the set of all elements of S which are not in A.

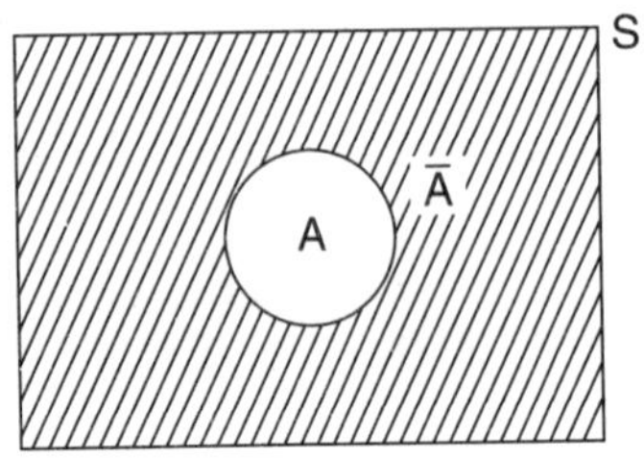

$$\bar{A} = S - A$$

7. **Favourable events:**The number of cases favourable to an event in a trial is the number of outcomes which entail the happening of the event. For example, in throwing of two dice, the number of cases favourable to getting the sum 6 is (1, 5), (2, 4), (3, 3), (3, 3), (4, 2), (5, 1), *i.e.* 6.

8. **Equally likely events:** Events are said to be equally likely, if there is no reason to expect any one in preference to any other. For example, In throwing a die, all the six faces are equally likely to come.

9. **Mutually exclusive events:** Two events A and B are said to be mutually exclusive events iff $A \cap B = \phi$.

If a die is thrown, we will get either an 'even number' or an 'odd number'. The event A "getting an even number" and event B "getting an odd number". We say that A and B are mutually exclusive events because both the events cannot occur at the same time. We have

$$A = \{2, 4, 6\} \quad \text{and} \quad B = \{1, 3, 5\}$$

Then, $\quad A \cap B = \phi$

In tossing of a coin, we will get either a head or tail. The event A "getting a head" and event B "getting a tail". We say that A and B are mutually exclusive events because both the events cannot occur at the same time.

We have $\quad A = \{H\} \quad \text{and} \quad B = \{T\}$

Then, $\quad A \cap B = \phi$

Note. 1. Simple events of a sample space are always mutually exclusive.

2. The events, which ensure the required happening is called favourable events.

10. **Odd in favour of an event and odd against an event:** Let there are m outcomes favourable to a certain event and n outcomes are not favourable to the event in a sample space, then

$$\text{odd in favour of the event} = \frac{m}{n}$$

and $$\text{odd against the event} = \frac{n}{m}.$$

11. **Permutation:** A permutation is an arrangement of objects in a definite order.

The number of permutations of n objects used r at a time, denoted by ${}^nP_r$ is

$${}^nP_r = \frac{n!}{(n-r)!}$$

12. **Combination:** A combination is a selection of objects without regard to order.

The number of combinations of n distinct objects selected r at a time, denoted by ${}^nC_r$ is

$${}^nC_r = \frac{n!}{r!(n-r)!}$$

1.2. DEFINITIONS

The chance of happening of an event when expressed quantitatively is called probability. However, we give three definitions of probability:

(1) Classical or Mathematical definition of probability.

(2) Statistical or Empirical definition of probability.

(3) Axiomatic definition of probability.

1. Classical or Mathematical definition of probability: If an event A can happen m ways out of possible n mutually exclusive and equally likely outcomes of a random experiment then probability of A is

$$P(A) = \frac{\text{Number of favourable cases}}{\text{Total numbers of possible cases}} = \frac{m}{n}$$

The probability of not happening of A is

$$P(\bar{A}) = 1 - \frac{m}{n} = \frac{n-m}{n}$$

Thus, $$P(A) + P(\bar{A}) = 1,\ 0 \le P(A) \le 1 \quad \text{and} \quad 0 \le P(\bar{A}) \le 1.$$

Note. 1. If $P(A) = 1$, then A is called certain event.

2. If $P(A) = 0$, then A is called impossible event.

3. This definition is fail when outcomes are not equally likely and number of outcomes is infinite.

4. The probability of an event is a number between 0 and 1. If occurrence is certain, its probability is 1. If the event cannot occur, its probability is 0.

2. Statistical or Empirical definition of probability: If trial be repeated for a large number of times, say n, under the same conditions, and a certain event A happen on m times then the probability of the event A is

$$P(A) = \lim_{n \to \infty} \frac{m}{n}$$

where, the limit is unique and finite.

3. Axiomatic definition of probability: Let S be a finite sample space and A be any event in S, then probability of A is defined by following three conditions.

(*i*) For every event A in S, $0 \le P(A) \le 1$

(*ii*) $P(S) = 1$.

(*iii*) If A and B are mutually exclusive events in S, then

$$P(A \cup B) = P(A) + P(B).$$

1.3. ELEMENTARY THEOREMS ON PROBABILITY

Theorem 1: *The probability of the impossible event is zero i.e.* $P(\phi) = 0$.

Proof: Since, impossible event contains no sample point, therefore, the certain event S and the impossible event ϕ are mutually exclusive

$$\Rightarrow \qquad S \cup \phi = S$$

$$\Rightarrow \qquad P(S \cup \phi) = P(S)$$

$$\Rightarrow \qquad P(S) + P(\phi) = P(S)$$

$$\Rightarrow \qquad P(\phi) = 0.$$

Theorem 2: *The probability of the complementary event* $\bar{A}$ *of* A *is given by*

$$P(\bar{A}) = 1 - P(A).$$

Proof: Since, A and $\bar{A}$ are mutually disjoint events, so that

$$A \cup \bar{A} = S$$

$$\Rightarrow \qquad P(A \cup \bar{A}) = P(S)$$

$$\Rightarrow \qquad P(A) + P(\bar{A}) = P(S) = 1 \qquad [\because \ P(S) = 1]$$

$$\Rightarrow \qquad P(\bar{A}) = 1 - P(A)$$

Note. 1. If ϕ is the impossible event and S is the sample space then $\bar{S} = \phi$

$$\Rightarrow \qquad P(\bar{S}) = P(\phi)$$

and $$P(\bar{S}) = 1 - P(S)$$
$$= 1 - 1 = 0$$

i.e., $$P(\phi) = 0$$

Theorem 3: *For any two events A and B.*

(*i*) $P(A \cap \bar{B}) = P(A) - P(A \cap B)$ (*ii*) $P(\bar{A} \cap B) = P(B) - P(A \cap B)$

Proof: (*i*) From the Venn diagram, we have

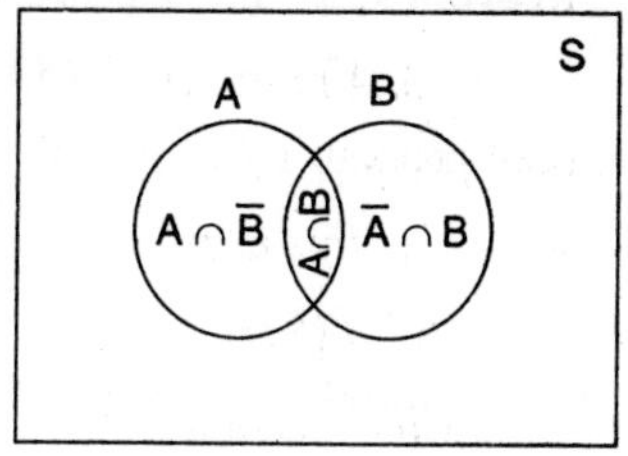

$$A = (A \cap B) \cup (A \cap \bar{B})$$

where, $A \cap B$ and $A \cap \bar{B}$ are disjoint events.

$$\Rightarrow \quad P(A) = P(A \cap B) + P(A \cap \bar{B})$$

$$\Rightarrow \quad P(A \cap \bar{B}) = P(A) - P(A \cap B).$$

(*ii*) From the Venn diagram, we have

$$B = (A \cap B) \cup (\bar{A} \cap B)$$

where, $A \cap B$ and $\bar{A} \cap B$ are disjoint events.

$$\Rightarrow \quad P(B) = P(A \cap B) + P(\bar{A} \cap B)$$

$$\Rightarrow \quad P(\bar{A} \cap B) = P(B) - P(A \cap B)$$

Theorem 4: *If $B \subset A$, then*

(*i*) $P(A \cap \bar{B}) = P(A) - P(B)$ (*ii*) $P(B) \le P(A)$

(*iii*) $P(A \cap B) \le P(A)$ *and* $P(A \cap B) \le P(B)$.

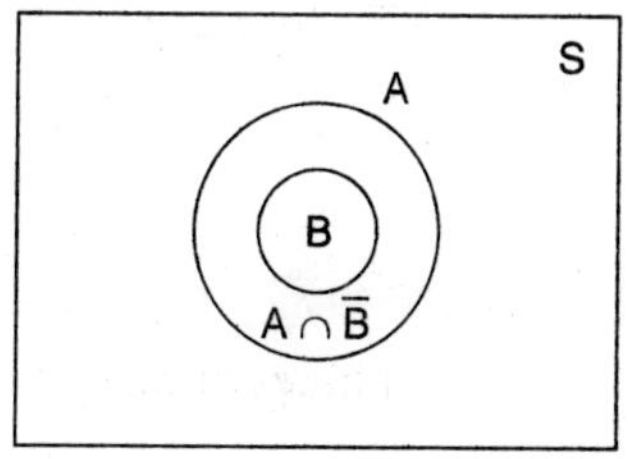

Proof: (*i*) If $B \subset A$, then B and $A \cap \bar{B}$ are mutually exclusive events, so that

$$A = B \cup (A \cap \bar{B})$$

$$\Rightarrow \quad P(A) = P(B) + P(A \cap \bar{B})$$

$$\Rightarrow \quad P(A \cap \bar{B}) = P(A) - P(B) \quad \text{...(1)}$$

(*ii*) If $B \subset A$, then by (1), we have

$$P(A \cap \bar{B}) = P(A) - P(B)$$

$$\Rightarrow \quad P(A \cap \bar{B}) \ge 0$$

$$\Rightarrow \quad P(A) - P(B) \ge 0$$

$$\Rightarrow \quad P(A) \ge P(B)$$

or $$P(B) \le P(A) \quad \text{...(2)}$$

(*iii*) If $B \subset A$, then by (2), we have

$$P(B) \le P(A)$$

Since, $A \cap B \subset A$ and $A \cap B \subset B$

$\Rightarrow$ $P(A \cap B) \le P(A)$ and $P(A \cap B) \le P(B)$.

1.4. ADDITION THEOREM OF PROBABILITY

Theorem 1: *If A and B are two events, then*

$$P(A \cup B) = P(A) + P(B) - P(A \cap B)$$

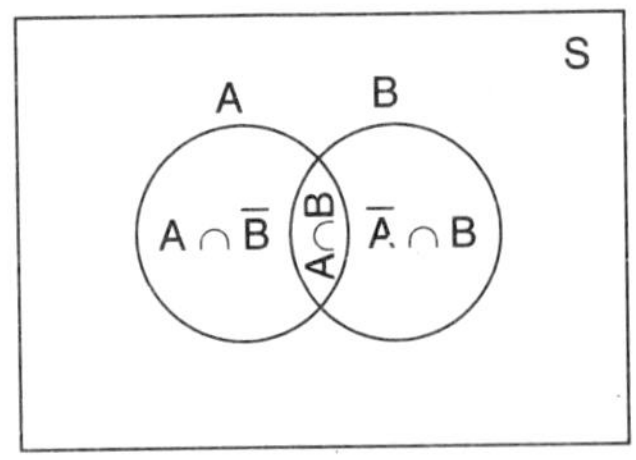

Proof: From the Venn diagram, we have

$$A \cup B = A \cup (\bar{A} \cap B)$$

where, A and $\bar{A} \cap B$ are mutually disjoint events.

$$\Rightarrow \quad P(A \cup B) = P(A) + P(\bar{A} \cap B)$$

$$= P(A) + P(B) - P(A \cap B) \qquad [\because \ P(\bar{A} \cap B) = P(B) - P(A \cap B)] \ ...(1)$$

Note. 1. If A and B are disjoint events, then

$$P(A \cap B) = P(\phi) = 0$$

Then by (1), we have

$$P(A \cup B) = P(A) + P(B)$$

2. $P(A \cup B)$ is also written as $P(A + B)$ and $P(A \cap B)$ is also written as $P(AB)$.

Theorem 2: *If A, B and C are any three events, then*

$$P(A \cup B \cup C) = P(A) + P(B) + P(C) - P(A \cap B) - P(B \cap C) - P(C \cap A) + P(A \cap B \cap C).$$

Proof: Using the above theorem 1 for two events, we have

$$\begin{aligned} P(A \cup B \cup C) &= P[A \cup (B \cup C)] \\ &= P(A) + P(B \cup C) - P[A \cap (B \cup C)] \\ &= P(A) + [P(B) + P(C) - P(B \cap C)] - P[(A \cap B) \cup (A \cap C)] \\ &= P(A) + P(B) + P(C) - P(B \cap C) - [P(A \cap B) + P(A \cap C) - P(A \cap B \cap C)] \\ &= P(A) + P(B) + P(C) - P(A \cap B) - P(B \cap C) - P(C \cap A) + P(A \cap B \cap C) \end{aligned}$$

Theorem 3: *If $A_1, A_2, A_3, \ldots\ldots, A_n$ are n mutually exclusive events, then the probability of the happening of one of them is*

$$P(A_1 \cup A_2 \cup A_3 \cup \ldots\ldots \cup A_n) = P(A_1) + P(A_2) + P(A_3) + \ldots\ldots + P(A_n).$$

Proof: Let N be the total number of mutually exclusive, exhaustive and equally likely cases of which m_1 are favourable to A_1, m_2 are favourable to A_2, m_3 are favourable to A_3 and so on.

Probability of happening of event $A_1 = P(A_1) = \dfrac{m_1}{N}$.

Probability of happening of event $A_2 = P(A_2) = \frac{m_2}{N}$.

....

....

....

Probability of happening of event $A_n = P(A_n) = \frac{m_n}{N}$

$\therefore$ Probability of happening of one of the events $A_1, A_2, A_3, ..., A_n$ is

$$P(A_1 \cup A_2 \cup A_3 \cup ...... \cup A_n) = \frac{m_1}{N} + \frac{m_2}{N} + \frac{m_3}{N} + ... + \frac{m_n}{N}$$

$$= P(A_1) + P(A_2) + P(A_3) + ... + P(A_n).$$

SOLVED EXAMPLES

Example 1: *What is the chance that a leap year, selected at random, will contain 53 Sundays?*

Solution: We know that a leap year consists of 366 days and contains 52 complete weeks and 2 days over.

Combinations of these two days are as follows:

(1) Monday, Tuesday

(2) Tuesday, Wednesday

(3) Wednesday, Thursday

(4) Thursday, Friday

(5) Friday, Saturday

(6) Saturday, Sunday

(7) Sunday, Monday

Of these seven $P(S) = 7$ are likely cases only and last two are favourable *i.e.* $n(A) = 2$.

The required probability $= \frac{n(A)}{n(S)} = \frac{2}{7}$.

Example 2: *From a pack of 52 cards, one is drawn at random. Find the probability of getting a king.*

Solution: From a pack of 52 cards 1 card can be drawn in 52 ways, *i.e.* $n(S) = 52$.

Number of ways in which a card can be king = 4

i.e. $n(A) = 4$

The required probability $= \frac{n(A)}{n(S)} = \frac{4}{52} = \frac{1}{13}$.

Example 3: *What is the probability of throwing a number greater than 3 with an ordinary die whose faces are numbered from 1 to 6.*

Solution: There are 6 possible ways in which the die can fall, and all of these there (4, 5, 6) are favourable to the event required

i.e. $$n(S) = 6 \quad \text{and} \quad n(A) = 3$$

$$\Rightarrow \qquad P(>3) = \frac{n(A)}{n(S)} = \frac{3}{6} = \frac{1}{2}.$$

Example 4: *A coin is tossed. If it shows head, we draw a ball from a bag consisting of 3 blue and 4 white balls; if it shows tail, we throw a die. Describe the sample space of the experiment.*

Solution: Let us denote the blue balls by B_1, B_2, B_3 and white balls by W_1, W_2, W_3, W_4.

The sample space of the experiment is

$$S = \{HB_1, HB_2, HB_3, HW_1, HW_2, HW_3, HW_4, T_1, T_2, T_3, T_4, T_5, T_6\}$$

Example 5: *From a pack of 52 cards there are drawn at random. Find the chance that they are a king, a queen and a knave.*

Solution: From a pack of 52 cards a draw of 3 can be made in ${}^{52}C_3$ ways.

Thus, $$n(S) = {}^{52}C_3$$

In a pack of 52 cards are 4 kings, 4 queens and 4 knaves. A king can be drawn in ${}^4C_1$ ways, a queen in ${}^4C_1$ ways and a knave in ${}^4C_1$ ways. Each of these ways be withdrawn in ${}^4C_1 \times {}^4C_1 \times {}^4C_1$ ways.

Thus, $$n(A) = {}^4C_1 \times {}^4C_1 \times {}^4C_1$$

The required probability $$P(A) = \frac{n(A)}{n(S)} = \frac{{}^4C_1 \times {}^4C_1 \times {}^4C_1}{{}^{52}C_3}$$

$$= \frac{4 \times 4 \times 4 \times 1 \times 2 \times 3}{52 \times 51 \times 50} = \frac{16}{5525}.$$

Example 6: *In a given race the odds in favour of four horses, A, B, C, D are 1 : 3, 1 : 4, 1 : 5, 1 : 6 respectively. Assuming that a dead heat is impossible, find the chance that one of them wins the race.*

Solution: Let p_1, p_2, p_3 and p_4 be the probabilities of winning of the horses A, B, C and D respectively, then

$$P(A) = p_1 = \frac{1}{4},\ P(B) = p_2 = \frac{1}{5}$$

$$P(C) = p_3 = \frac{1}{6},\ P(D) = p_4 = \frac{1}{7}.$$

Since, these are mutually exclusive events the chance that one of them wins

$$= P(A) + P(B) + P(C) + P(D)$$

$$= p_1 + p_2 + p_3 + p_4$$

$$= \frac{1}{4} + \frac{1}{5} + \frac{1}{6} + \frac{1}{7} = \frac{319}{420}.$$

Example 7: *If from a lottery of 30 tickets, marked 1, 2, 3, ... four tickets be drawn, what is the chance that marked 1 and 2 are among them.*

Solution: We have, $n(S) = {}^{30}C_4$ (four tickets can be selected)

When 2 tickets are to be included always, remaining two can be selected in ${}^{28}C_2$ ways, *i.e.* $n(A) = {}^{28}C_2$

$$\text{The required chance } P(A) = \frac{n(A)}{n(S)} = \frac{{}^{28}C_2}{{}^{30}C_2} = \frac{2}{145}.$$

Example 8: *A manufacturer supplies cheap quarter horse power motors in a hot of 25. A buyer before taking a lot, tests a random sample of 5 motors and accepts the lot if they are all good, otherwise he rejects the lot. Find the probability that*

(*i*) *he will accept a lot containing 5 defective motors.*

(*ii*) *he will reject a lot containing only one defective motor.*

Solution: The buyer can choose a random sample of 5 motors out of 25 in ${}^{25}C_5$ ways.

(*i*) The buyer will accept a lot, if in his sample all the motors are good. That means his sample consists of 5 motors from 20 nondefective motors. So the number of ways of selecting the sample for acceptance of the lot, will be ${}^{20}C_5$

$$\text{The required probability} = \frac{{}^{20}C_5}{{}^{25}C_5} = \frac{\dfrac{20!}{5!15!}}{\dfrac{25!}{5!20!}}$$

$$= \frac{20 \times 19 \times 18 \times 17 \times 16}{25 \times 24 \times 23 \times 22 \times 21} = \frac{2584}{8855} = 0.292$$

(*ii*) When he is rejecting a lot containing one defective motor, so his sample will contain the only defective motor and 4 others which are chosen from the 24 non-defective motors. So number of ways of selecting this sample is

$$= {}^{24}C_4 \times {}^{1}C_1$$

$$\text{The required probability} = \frac{{}^{24}C_4}{{}^{25}C_5} = \frac{\dfrac{24!}{4!20!}}{\dfrac{25!}{5!20!}} = \frac{5}{25} = \frac{1}{5} = 0.2$$

EXERCISE 1.1

1. What is the chance of throwing a 3 with an ordinary die?
2. What is the chance of that a non-leap year should have fifty three Sundays?
3. An integer is chosen at random from the first two hundred digits. What is the probability that the integer chosen is divisible by 6 or 8.

4. In a class of 66 students 13 are boys and the rest are girls. Find the probability that a student selected will be a girl.
5. A bag contains 7 white and 12 black balls. Find the probability of drawing a white ball.
6. A card is drawn from an ordinary pack and a gambler bets that it is a spade or an ace. What are the odds against his winning this bet?
7. From a set of 17 cards, number 1, 2, 3, ... 16, 17, one is drawn at random. Show that the chance that its number is divisible by 3 or 5 or 7 is $\frac{9}{17}$.
8. A bag contains 9 discs of which 4 are red, 3 are blue and 2 are yellow. A disc drawn at random from the bag. Calculate the probability that it will be (*i*) red (*ii*) yellow (*iii*) blue (*iv*) not blue.
9. In the odds in favour of an event are 4 to 5, find the probability that it will occur.
10. In a single throw of two dice, find the probability of getting a total of 9 or 11.
11. Find the probability of drawing either an ace or a spade or both from a pack of cards.
12. Probability that a boy will pass an examination is $\frac{2}{5}$ and that for a girl it is $\frac{2}{5}$. What is the probability that at least one of them passes examination?

ANSWERS

1. $\frac{1}{6}$	**2.** $\frac{1}{7}$	**3.** $\frac{1}{4}$	**4.** $\frac{53}{66}$
5. $\frac{7}{19}$	**6.** 9 to 4	**8.** $\frac{4}{9}, \frac{2}{9}, \frac{1}{3}, \frac{2}{3}$	**9.** $\frac{4}{9}$
10. $\frac{1}{6}$	**11.** $\frac{4}{13}$	**12.** $\frac{19}{25}$.	

1.5. INDEPENDENT EVENTS

Two events are said to be independent events, if the occurrence of one does not effect the occurrence of the other.

OR

Two events A and B are said to be independent events iff $P(A \cap B) = P(A) \cdot P(B)$.

When a die is thrown, let A be the event "number appearing is a multiple of 3" and B be the event "number appearing is even". We have

$$A = \{3, 6\}, B = \{2, 4, 6\}$$

and

$$P(A) = \frac{2}{6} = \frac{1}{3}$$

$$P(B) = \frac{3}{6} = \frac{1}{2}$$

$$A \cap B = \{6\}$$

$$P(A \cap B) = \frac{1}{6}$$

Using, $$P(A \cap B) = P(A) \,.\, P(B)$$

$$\frac{1}{6} = \frac{1}{3} \times \frac{1}{2}$$

Hence, A and B are independent events.

1.6. CONDITIONAL PROBABILITY

Let A and B be events such that $P(B) \neq 0$. The conditional probability of A given B, denoted by $P(A/B)$, is defined by

$$P(A/B) = \frac{P(A \cap B)}{P(B)}$$

where, $P(A/B)$ is the probability of occurrence of A given that B has already happened.

Theorem: *If the two events A and B are independent events, then*

$$P(A/B) = P(A) \quad \text{and} \quad P(B/A) = P(B)$$

Proof: We know that if A and B are given to be independent events, then

$$P(A \cap B) = P(A) \,.\, P(B)$$

we have, $$P(A/B) = \frac{P(A \cap B)}{P(B)} = \frac{P(A) \,.\, P(B)}{P(B)} = P(A)$$

and $$P(B/A) = \frac{P(A \cap B)}{P(A)} = \frac{P(A) \,.\, P(B)}{P(A)} = P(B).$$

1.7. MULTIPLICATIVE THEORY OF PROBABILITY OR THEOREM OF COMPOUND PROBABILITY

If there are two events, the respective probability of which are known then the probability that both will happen simultaneously is the product of the probability of one and the conditional probability of the other, given that the first has occurred, *i.e.*

$$P(A \cap B) = P(A)\ P(B/A)$$

or $$P(B \cap A) = P(B) \,.\, P(A/B)$$

Proof: Let out of n outcomes, m_1 be the number of cases favourable to the happening of A.

$\therefore$ Probability of happening of the event $A = P(A) = \frac{m_1}{n}$.

Let m_2 be the number of cases favourable to the happening of B.

$\therefore$ Conditional probability of B, given that A has happened $= P(B/A) = \frac{m_2}{m_1}$

Thus, the number of cases favourable to the happening of both A and B are m_2 out of n.

$$\therefore \quad P(A \cap B) = \frac{m_2}{n} = \frac{m_1}{n} \times \frac{m_2}{m_1} = P(A) \,.\, P(B/A).$$

Note. 1. If A and B are independent events, then

$$P(B/A) = P(B)$$

$$\therefore \quad P(A \cap B) = P(A) \,.\, P(B)$$

2. $P(A \cap B)$ is also written as $P(A$ and $B)$ or $P(AB)$.

3. $P(A \cup B)$ is also written as $P(A$ or $B)$.

4. Term 'independent' is defined in terms of "probability of events" whereas mutually exclusive is defined in terms of "events". Moreover, mutually exclusive events never have an outcome common, but independent events do have an outcome common, provided each event is non-empty. Clearly 'independent' and 'mutually exclusive' do not have the same meaning.

5. If two events A and B are independent, then

(*i*) A and $\overline{B}$ are independent event.

(*ii*) $\overline{A}$ and B are independent event.

(*iii*) $\overline{A}$ and $\overline{B}$ are independent event.

SOLVED EXAMPLES

Example 1: *A die is thrown twice and the sum of the numbers appearing is observed to be 6. What is the conditional probability that the number 4 has appeared at least once?*

Solution: Let A = number 4 appears at least once

B = the sum of the numbers appearing is 6

then,

$$A = \{(4, 1), (4, 2), (4, 3), (4, 4), (4, 5), (4, 6), (1, 4), (2, 4), (3, 4), (5, 4), (6, 4)\}$$

and

$$B = \{(1, 5), (2, 4), (3, 3), (4, 2), (5, 1)\}$$

The required probability $P(A/B) = \dfrac{P(A \cap B)}{P(B)} = \dfrac{2}{5}$

Example 2: *A die is rolled twice and the sum of the numbers appearing on them is observed to be 7. What is the conditional probability that the number 2 has appeared at least once?*

Solution: Let A = getting the number 2 at least once

B = getting 7 as the sum of the numbers on two dice.

Then, $A = \{(2, 1), (2, 2), (2, 3), (2, 4), (2, 5), (2, 6), (1, 2), (3, 2), (4, 2), (5, 2), (6, 2)\}$

and $B = \{(2, 5), (3, 4), (4, 3), (5, 2), (6, 1), (1, 6)\}$

The required probability $P(A/B) = \dfrac{P(A \cap B)}{P(B)} = \dfrac{2}{6} = \dfrac{1}{3}.$

Example 3: *A family has two children. What is the conditional probability that both are boys given that at least one of them is a boy?*

Solution: Let the sample space S is given by

$$S = \{(b, b), (b, g), (g, b), (g, g)\}$$

Let A denote the event that both children are boys, B the event that at least one of them is a boy and all outcomes are equally likely, then the desired probability is

$$P(A/B) = \frac{P(A \cap B)}{P(B)} = \frac{\frac{1}{4}}{\frac{3}{4}} = \frac{1}{3}.$$

Example 4: *Ajeet can either take a course in computers or in maths. If Ajeet takes the computer course, then he will receive an A grade with probability* $\frac{1}{2}$*; if he takes the maths course then he will receive an A grade with probability* $\frac{1}{3}$*. Ajeet decides to base his decision on the flip of a fair coin. What is the probability that Ajeet will get an A in maths?*

Solution: Let C be the event that Ajeet takes maths and A denote the event that he receives on A in whatever course he takes, then the desired probability is

$$P(A \cap C) = P(C)\, P(A/C)$$

$$= \frac{1}{2} \times \frac{1}{3} = \frac{1}{6}$$

Example 5: *Ashu and Ankit appear in an interview for two vacancies in the same post. The probability of Ashu's selection is* $\frac{1}{7}$ *and that of Ankit's selection is* $\frac{1}{5}$*. What is the probability that*

(i) both of them will be selected? *(ii) none of them will be selected?*

(iii) only one of them is selected?

Solution: Let A denote the Ashu is selected and B denote the Ankit is selected then A and B are independent.

1. P(both of them will be selected)

$$= P(A \cap B) = P(A)\, P(B) = \frac{1}{7} \times \frac{1}{5} = \frac{1}{35}$$

2. P(none of them will be selected)

$$= P(\text{not } A \text{ and not } B)$$
$$= P(\text{not } A) \,.\, P(\text{not } B)$$
$$= [1 - P(A)]\,[1 - P(B)]$$
$$= \left(1 - \frac{1}{7}\right)\left(1 - \frac{1}{5}\right) = \frac{6}{7} \times \frac{4}{5} = \frac{24}{35}.$$

3. P(only one of them will be selected)

$$= P(\text{`}A \text{ and not } B\text{' or `}B \text{ and not } A\text{'})$$
$$= P(A \text{ and not } B) + P(B \text{ and not } A)$$
$$= P(A)\, P\,(\text{not } B) + P(B)\, P(\text{not } A)$$
$$= P(A)\,[1 - P(B)] + P(B)[1 - P(A)]$$
$$= \frac{1}{7} \times \left(1 - \frac{1}{5}\right) + \frac{1}{5} \times \left(1 - \frac{1}{7}\right) = \frac{1}{7} \times \frac{4}{5} + \frac{1}{5} \times \frac{6}{7} = \frac{10}{35} = \frac{2}{7}.$$

Example 6: *The probability that a teacher will give an unannounced test during any class meeting is $\frac{1}{5}$. If a student is absent twice, what is the probability that he will miss at least one test?*

Solution: Let T_1 be the event of I test held on his first day of absence and T_2 be the event of II test held on second day of his absence. Since, T_1 and T_2 are independent, then the required probability

P(probability that he will miss at least one test)

$$P(T_1 \cup T_2) = P(T_1) + P(T_2) - P(T_1 \cap T_2)$$
$$= P(T_1) + P(T_2) - P(\text{T}_1)\, P(T_2)$$
$$= \frac{1}{5} + \frac{1}{5} - \frac{1}{5} \times \frac{1}{5} = \frac{2}{5} - \frac{1}{25} = \frac{9}{25}.$$

Example 7: *Three groups of children contain 3 girl and 1 boy; 2 girls and 2 boys; 1 girl and 3 boys respectively. One child is selected at random from each group. Find the chance that there selected comprise 1 girl and 2 boys.*

Solution: Let A, B and C be the three groups. Given that

	A	B	C
Boys	1	2	3
Girls	3	2	1

Let B_1, B_2, B_3 be the events of selecting a boy from A, B and C group respectively and let G_1, G_2, G_3 be the events of selecting a girl from A, B and C group respectively. Then B_1, B_2, B_3, G_1, G_2, G_3 are independent events such that

$$P(B_1) = \frac{1}{4},\ P(B_2) = \frac{2}{4} = \frac{1}{2},\ P(B_3) = \frac{3}{4},$$

$$P(G_1) = \frac{3}{4},\ P(G_2) = \frac{2}{4} = \frac{1}{2},\ P(G_3) = \frac{1}{4}.$$

Now one girl and 2 boys can be chosen in the following three mutually exclusive ways.

Groups	*A*	*B*	*C*
(*I*)	Girl	Boy	Boy
(*II*)	Boy	Girl	Boy
(*III*)	Boy	Boy	Girl

The required probability = $P(I) + P(II) + P(III)$

$$= P(G_1 \cap B_2 \cap B_3) + P(B_1 \cap G_2 \cap B_3) + P(B_1 \cap B_2 \cap G_3)$$

$$= P(G_1)\, P(B_2)\, P(B_3) + P(B_1)\, P(G_2)\, P(B_3) + P(B_1)\, P(B_2)\, P(G_3)$$

$$= \frac{3}{4} \times \frac{1}{2} \times \frac{3}{4} + \frac{1}{4} \times \frac{1}{2} \times \frac{3}{4} + \frac{1}{4} \times \frac{1}{2} \times \frac{1}{4}$$

$$= \frac{13}{32}.$$

Example 8: *A and B are two independent witness in a case. The probability that A will speak the truth is x and the probability that B will speak the truth is y. A and B agree in a certain statement. Show that the probability that this statement is true:* $\dfrac{xy}{1-x-y+2xy}$.

Solution: Let E_1 be the event that A and B agree in a statement and E_2 be the event that their statement is correct.

Then, $\quad P(E_1) = xy + (1-x)(1-y)$

$$= 1 - x - y + 2xy$$

and $\quad P(E_1 \cup E_2) = xy$

We know that

$$P(E_1 \cup E_2) = P(E_1)\, P(E_2/E_1) = P(E_2)\, P(E_1/E_2)$$

$$\therefore \quad P(E_2/E_1) = \frac{P(E_1 \cup E_2)}{P(E_1)}$$

$$= \frac{xy}{1-x-y+2xy}.$$

EXERCISE 1.2

1. If $P(A) = \frac{3}{5}$ and $P(B) = \frac{1}{3}$, find

   (*i*) $P(A \cup B)$, if A and B are mutually exclusive events.

   (*ii*) $P(A \cap B)$, if A and B are independent events.

2. In the two dice experiment, if A is the event of getting the sum of numbers on dice as 11 and B is the event of getting a number other than 5 on the first die, find $P(A \cap B)$. Are A and B independent events?

3. If A and B are independent events, then show that

   (*i*) $\bar{A}, B$ (*ii*) $A, \bar{B}$ and

   (*iii*) $\bar{A}, \bar{B}$ are also independent events.

4. A problem of statistics is given to three students A, B and C whose chances of solving it are 1/2, 3/4 and 1/4 respectively. What is the probability that the problem will be solved?

5. A bag contains 3 red and 5 black balls and a second bag contains 6 red and 4 black balls. A ball is drawn from each bag. Find the probability that

   (*i*) both are red, (*ii*) both are black,

   (*iii*) one is red and one is black.

6. A bag contains 19 tickets, numbered from 1 to 19. A ticket is drawn and then another ticket is drawn without replacement. Find the probability that both the tickets will show an even number.

7. A bag contains 5 white, 7 red and 8 black balls. If four balls are drawn one by one without replacement, find the probability of getting all white balls.

8. An article manufactured by a company consists of two parts A and B. In the process of manufacture of part A, 9 out of 100 are likely to be defective. Similarly, 5 out of 100 are likely to be defective in the manufacture of part B. Find the probability (*i*) that the assembled part will not be defective (*ii*) that the assembled part will be defective.

9. A bag contains 4 white and 6 black balls. If two balls are drawn in succession, what is the probability that one is white and other is black?

10. A bag contain 10 white and 15 black balls. Two balls are drawn in succession. What is the probability that first is white and second is black?

ANSWERS

1. (*i*) $\frac{14}{25}$ (*ii*) $\frac{1}{5}$ **2.** $\frac{1}{36}$, Not. **4.** $\frac{29}{32}$.

5. (*i*) $\frac{9}{40}$ (*ii*) $\frac{1}{4}$ (*iii*) $\frac{21}{40}$ **6.** $\frac{4}{19}$ **7.** $\frac{1}{969}$.

8. (*i*) 0.8645 (*ii*) 0.1355 **9.** $\frac{8}{15}$ **10.** $\frac{1}{4}$.

1.8. THEOREM OF TOTAL PROBABILITY

Theorem: *Let $B_1, B_2, B_3, \ldots\ldots, B_n$ be subsets of a sample space S such that*

(*i*) Each B_i is a proper subset of S *i.e.*,

$$B_i \subset S, i = 1, 2, \ldots\ldots n \text{ and } B_i \neq S.$$

(*ii*) $B_1 \cup B_2 \cup \ldots\ldots \cup B_n = S$ and

$$B_i \cap B_j = \phi\ , i, j = 1, 2, \ldots\ldots n, i \neq j.$$

Then, for any event A of S

$$P(A) = \sum_{i=1}^{n} P(B_i \cap A) = \sum_{i=1}^{n} P(B_i)\, P(A/B_i) \text{ with } P(B_i) \neq 0, i = 1, 2, \ldots\ldots n.$$

Proof: It is given that

$$S = B_1 \cup B_2 \cup \ldots\ldots \cup B_n = \bigcup_{i=1}^{n} B_i$$

and $\quad B_i \cap B_j = \phi$ for any i and j

i.e., their union is S and B_i's are mutually disjoint sets.

Therefore,

$$\begin{aligned} A &= A \cap S = A \cap \left(\bigcup_{i=1}^{n} B_i\right) \\ &= A \cap (B_1 \cup B_2 \cup \ldots\ldots \cup B_n) \\ &= (A \cap B_1) \cup (A \cap B_2) \cup \ldots\ldots \cup (A \cap B_n) \end{aligned}$$

The sets $A \cap B_1, A \cap B_2, \ldots\ldots, A \cap B_n$ are all mutually disjoint sets. We have

$$\begin{aligned} P(A) &= P[(A \cap B_1) \cup (A \cap B_2) \cup \ldots\ldots \cup P(A \cap B_n)] \\ &= P(A \cap B_1) + P(A \cap B_2) + \ldots\ldots + P(A \cap B_n) \\ &= P(B_1 \cap A) + P(B_2 \cap A) + \ldots\ldots + P(B_n \cap A) \\ &= \sum_{i=1}^{n} P(B_i \cap A) \\ &= \sum_{i=1}^{n} P(B_i)\, P(A/B_i) \qquad \text{(Using by multiple theorem)} \end{aligned}$$

1.9. BAYE'S THEOREM

Let $B_1, B_2, B_3, \ldots\ldots, B_n$ are mutually exclusive events with $P(B_i) \neq 0$ (i = 1, 2, ... *n*). For any arbitrary event A in sample space S with $P(A) \neq 0$ and for $1 \leq k \leq n$.

$$P(B_k/A) = \frac{P(B_k)\, P(A/B_k)}{\sum_{i=1}^{n} P(B_i)\, P(A/B_i)}$$

Proof: By the definition of conditional probability

$$P(B_k/A) = \frac{P(B_k \cap A)}{P(A)} \quad ...(1)$$

By the theorem of total probability

$$P(A) = \sum_{i=1}^{n} P(B_i \cap A) = \sum_{i=1}^{n} P(B_i)\, P(A/B_i) \quad ...(2)$$

Also from the theory of multiplication

$$P(B_k \cap A) = P(B_k)\, P\,(A/B_k) \quad ...(3)$$

Putting the values of $P(A)$ and $P(B_k \cap A)$ from (2) and (3) in equation (1), we get

$$P(B_k/A) = \frac{P(B_k)\, P(A/B_k)}{\sum_{i=1}^{n} P(B_i)\, P(A/B_i)}.$$

SOLVED EXAMPLES

Example 1: *In 1888 there will be three candidates for the position of Director—Dr. Singhal, Dr. Mehra and Dr. Chatterji—whose chances of getting the appointment are in the proportion 4 : 2 : 3 respectively. The probability that Dr. Singhal, if selected, will abolish co-education in the college is 0.3. The probability of Dr. Mehra and Dr. Chatterji doing the same are respectively 0.5 and 0.8. What is the probability that co-education will be abolished in the college?*

Solution: Let B_1, B_2 and B_3 be the probability of being appointed of Dr. Singhal, Dr. Mehra and Dr. Chatterji respectively. Let the probability of co-education being abolished be A. Then, by the theorem of total probability, we have

$$P(A) = P(B_1)\, P(A/B_1) + P(B_2)\, P(A/B_2) + P(B_3)\, P(A/B_3)$$

$$= \frac{4}{9} \times (0.3) + \frac{2}{9} \times (0.5) + \frac{3}{9} \times (0.8)$$

$$= \frac{2}{15} + \frac{1}{9} + \frac{4}{15} = \frac{23}{45}.$$

Example 2: *An urn I contains 3 white and 4 red balls and an urn II contains 5 white and 6 red balls. One ball is drawn at random from one of the urn and is found to be white, find the probability that it was drawn from urn I.*

Solution: Let B_1: the ball is drawn from urn *I*

B_2: the ball is drawn from urn *II*

A: the ball is white

To find $P(B_1/A)$.

By Baye's theorem

$$P(B_1/A) = \frac{P(B_1)\,P(A/B_1)}{P(B_1)\,P(A/B_1) + P(B_2)\,P(A/B_2)} \qquad ...(1)$$

Since, two urn are equally likely to be selected,

$$P(B_1) = P(B_2) = \frac{1}{2}$$

$$P(A/B_1) = \text{Probability of a white ball is drawn from urn } I = \frac{3}{7}$$

$$P(A/B_2) = \text{Probability of a white ball is drawn from urn } II = \frac{5}{11}$$

By (1), we have

$$P(B_1/A) = \frac{\frac{1}{2}\times\frac{3}{7}}{\frac{1}{2}\times\frac{3}{7}+\frac{1}{2}\times\frac{5}{11}} = \frac{33}{68}.$$

Example 3: *The probability that a person can hit a target is* $\frac{3}{5}$ *and the probability that another person can hit the same target is* $\frac{2}{5}$ *. But the first person can fire 8 shoots in the time the second person fires 10 shoots. They fire together, what is the probability that the second person shoots the target.*

Solution: Let A denote the event of shooting the target and B_1 denote the event that the first person shoots the target and B_2 denote the event that the second person shoots the target. Therefore, $P(B_2/A)$ is the probability that the second person shoots the targets.

Now, we have $\quad P(A/B_1) = \frac{3}{5},\; P(A/B_2) = \frac{2}{5}$

It is given that the ratio of the shoots of the first person to those of the second person in the same time is $\frac{8}{10}$ *i.e.* $\frac{4}{5}$.

Thus, we have $\quad P(B_1) = \frac{4}{5}\,P(B_2)$

By Baye's theorem, we have

$$P(B_2/A) = \frac{P(B_2)\,P(A/B_2)}{P(B_1)\,P(A/B_1) + P(B_2)\,P(A/B_2)}$$

$$= \frac{P(B_2).\frac{2}{5}}{\frac{4}{5}\,P(B_2).\frac{3}{5} + P(B_2).\frac{2}{5}} = \frac{\frac{2}{5}}{\frac{3}{5}\times\frac{4}{5}+\frac{2}{5}} = \frac{5}{11}$$

Example 4: *The contents of urns I, II and III are as follows:*

1 White, 2 black and 3 red balls,

2 White, 1 black and 1 red balls,

and 4 white, 5 black and 3 red balls.

One urn is chosen at random and two balls drawn. They happen to be white and red. What is the probability that they come from urns I, II or III?

Solution: Let B_1: Urn *I* is chosen;

B_2: Urn *II* is chosen;

B_3: Urn *III* is chosen

and A: the two balls are white and red.

To find $P(B_1/A)$, $P(B_2/A)$ and $P(B_3/A)$.

Now, $P(B_1) = P(B_2) = P(B_3) = \frac{1}{3}$. Since, three urns are equally likely to be selected.

$P(A/B_1)$ = Probability of a white and a red ball are drawn from urn I

$$= \frac{{}^1C_1 \times {}^3C_1}{{}^6C_2} = \frac{1}{5}$$

$P(A/B_2)$ = Probability of a white and a red ball are drawn from urn II

$$= \frac{{}^2C_1 \times {}^1C_1}{{}^4C_2} = \frac{1}{3}$$

$P(A/B_3)$ = Probability of a white and a red ball are drawn from urn III

$$= \frac{{}^4C_1 \times {}^3C_1}{{}^{12}C_2} = \frac{2}{11}$$

By Baye's theorem, we have

$$P(B_1/A) = \frac{P(B_1)\,P(A/B_1)}{P(B_1)\,P(A/B_1) + P(B_2)\,P(A/B_2) + P(B_3)P(A/B_3)}$$

$$= \frac{\frac{1}{3} \times \frac{1}{5}}{\frac{1}{3} \times \frac{1}{5} + \frac{1}{3} \times \frac{1}{3} + \frac{1}{3} \times \frac{2}{11}} = \frac{33}{118}$$

Similarly, $P(B_2/A) = \frac{55}{118}$ and $P(B_3/A) = \frac{15}{59}$.

Example 5: *A bag contains 3 black and 4 red balls. Two balls are drawn at random one at a time without replacement. What is the probability that the first ball selected is black if the second ball is known to be red.*

Solution: Let B_1 be the event of the first ball being black (*bb*, *br*)

Let B_2 be the event of the first ball being red (*rb*, *rr*)

Let A be the event of second ball being red (br, rr)

To find $P(B_1/A)$.

We have $$A = (A \cap B_1) \cap (A \cap B_2) \quad ...(1)$$

$\therefore$ By theorem of multiplication or compound probability

$$P(A \cap B_1) = P(B_1)\, P(A/B_1)$$
$$= \frac{3}{7} \times \frac{4}{6} = \frac{2}{7}$$
$$P(A \cap B_2) = P(B_2)\, P(A/B_2)$$
$$= \frac{4}{7} \times \frac{3}{6} = \frac{2}{7}.$$

By (1), we have $$P(A) = P(A \cap B_1) + P(A \cap B_2)$$
$$= \frac{2}{7} + \frac{2}{7} = \frac{4}{7}$$

The required probability

$$P(B_1/A) = \frac{P(A \cap B_1)}{P(A)} = \frac{\frac{2}{7}}{\frac{4}{7}} = \frac{1}{2}$$

Example 6: *In a bolt factory, machines M_1, M_2 and M_3 manufacture respectively 25%, 35% and 40% of the total of their output 5, 4 and 2 percent are defective bolts. A bolt is drawn at random from the product and is found to be defective. What is the probability that it was manufactured by machine M_1, M_2 and M_3?*

Solution: Let A, B, C denote the events that the bolt was manufactured by machines M_1, M_2 and M_3 respectively and let D denote the event of its being defective. Then

$$P(A) = 0.25,\ P(B) = 0.35,\ P(C) = 0.40$$

The probability that a defective bolt is drawn from those manufactured by M_1 is

$$P(D/A) = \frac{5}{100} = 0.05$$

Similarly, $$P(D/B) = \frac{4}{100} = 0.04,\ P(D/C) = \frac{2}{100} = 0.02$$

By Baye's theorem, we have

$$P(A/D) = \frac{P(A)\, P(D/A)}{P(A)\, P(D/A) + P(B)\, P(D/B) + P(C)\, P(D/C)}$$
$$= \frac{0.25 \times 0.05}{0.25 \times 0.05 + 0.35 \times 0.04 + 0.40 \times 0.02}$$

$$= \frac{0.0125}{0.0345} = \frac{125}{345} = \frac{25}{69}$$

Similarly, $P(B/D) = \dfrac{140}{345}$ and $P(C/D) = \dfrac{80}{345}$.

Example 7: *In a certain college 25% of boys and 10% of girls are studying mathematics. The girls constitute 60% of the student body (a) what is the probability that mathematics is being studied? (b) If a student is selected at random and is found to be studying mathematics, find the probability that student is a girl? (c) a boy?*

Solution: Given that probability of a boy $P(B)$

$$= \frac{40}{100} = \frac{2}{5}$$

and Probability of a girl $P(G)$

$$= \frac{60}{100} = \frac{3}{5}$$

Probability that maths is studied given that the student is a boy

$$= P(M/B) = \frac{25}{100} = \frac{1}{4}$$

Similarly, $P(M/G) = \dfrac{10}{100} = \dfrac{1}{10}$

(*a*) Probability that math is studied

$$P(M) = P(G)\, P(M/G) + P(B)\, P(M/B) \qquad \text{(by theorem on total probability)}$$

$$= \frac{3}{5} \cdot \frac{1}{10} + \frac{2}{5} \cdot \frac{1}{4} = \frac{4}{25}.$$

(*b*) By Baye's theorem, we have

Probability that a math student is a girl

$$P(G/M) = \frac{P(G)\, P(M/G)}{P(M)}$$

$$= \frac{\frac{3}{5} \cdot \frac{1}{10}}{\frac{4}{25}} = \frac{3}{8}$$

(*c*) Probability that a maths student is a boy

$$P(B/M) = \frac{P(B)\, P(M/B)}{P(M)}$$

$$= \frac{\frac{2}{5} \cdot \frac{1}{4}}{\frac{4}{25}} = \frac{5}{8}.$$

EXERCISE 1.3

1. A bag X contains 2 white and 3 red balls and a bag Y contains 4 white and 5 red balls. One ball is drawn at random from one of the bags and is found to be red. Find the probability that it was drawn from bag Y.
2. Three urns contain 6 red, 4 black; 4 red 6 black; 5 red, 5 black balls respectively. One of the urns is selected at random and a ball is drawn from it. If the ball drawn is red, find the probability that it drawn from the first urn.
3. A businessman goes to hotels H_1, H_2, H_3, 20%, 50%, 30% of the time, respectively. It is known that 5%, 4%, 8% of the rooms in H_1, H_2, H_3 hotels have faulty plumbing (*i*) Determine the probability that the businessman goes to hotel with faulty plumbing (*ii*) what is the probability that business's room having faulty plumbing is assigned to hotel H_3?
4. There are three boxes containing respectively 1 white, 2 red, 3 black balls; 2 white, 3 red, 1 black balls; 3 white, 1 red, 2 black balls. A box is chosen at random and from it two balls are drawn at random. The two balls are one red and one white. What is the probability that they come from the (*i*) first box, (*ii*) second box (*iii*) third box.
5. An urn contains 10 white, 9 black, 8 red and 3 blue balls. Balls are drawn one by one at random from the urn until 2 blue balls are obtained at the 11th draw. Find the probability of drawing 2 blue balls upto 11th draw.
6. Suppose the supply of transistors is produced by three systems S_1, S_2 and S_3. Further, suppose that S_1 produces 20%, S_2 produces 30% and S_3 produces 50% of the supply and that the defective (D) rates for three systems S_1, S_2, S_3 are respectively 0.01, 0.02 and 0.05. If a transistor is randomly selected from the supply and is found to be defective, find (*i*) $P(S_1/D)$ (*ii*) $P(S_2/D)$ (*iii*) $P(S_3/D)$.
7. The chance that a doctor will diagnose a disease correctly is 70%. The chance that a patient dies by his treatment after correct diagonals is 35% and the chance of death by wrong diagnosis is 80%. If a patient dies after taking his treatment, what is the chance that the disease was diagnosed correctly.
8. Companies B_1, B_2, B_3 produce 30%, 45% and 25% of the cars respectively. It is known that 2%, 3% and 2% of the cars produced from B_1, B_2 and B_3 are defective.

   (*i*) What is the probability that a car purchased is defective?

   (*ii*) If a car purchased is found to be defective, what is the probability that this car is produced by company B_3?

ANSWERS

1. $\frac{25}{52}$ **2.** $\frac{2}{5}$ **3.** (*i*) 0.054 (*ii*) $\frac{4}{9}$

4. (*i*) $\frac{2}{11}$ (*ii*) $\frac{6}{11}$ (*iii*) $\frac{3}{11}$ **5.** $\frac{19}{406}$

6. (*i*) 0.061 (*ii*) 0.182 (*iii*) 0.757 **7.** $\frac{49}{97}$

8. (*i*) 0.0245 (*ii*) $\frac{10}{49}$.

1.10. BINOMIAL THEOREM

Let the probability of the happening of an event in one trial is p and $q = 1 - p$ be the probability that the event fails in one trial. Let us find out the probability of exactly r successes in n trails.

The chance that an event happens at least r times in n trials is given

$$P = p^n + {}^nC_1 p^{n-1} q + {}^nC_2 p^{n-2} q^2 + \ldots + {}^nC_r p^r q^{n-r}$$

Hence, the probability that the event will happen exactly r times in n trails is the $(n + 1)^{\text{th}}$ term in the expansion of $(q + p)^n$.

1.11. MULTINOMIAL THEOREM

If a dice has f faces marked with 1, 2, 3, ... f, the probability of throwing a total p with n dice is given by

$$= \frac{\text{Coefficient of } x^p \text{ in the expansion of } (x^1 + x^2 + x^3 + \ldots\ldots + x^f)^n}{f^n}$$

$$= \frac{\text{Coeff. of } x^p \text{ in } \dfrac{(1 - x^f)^n}{x^n (1 - x)^n}}{f^n}$$

$$= \frac{\text{Coeff. of } x^{p-n} \text{ in } (1 - x^f)^n (1 - x)^{-n}}{f^n}$$

Note. If n dice are different with faces $f_1, f_2, f_3, \ldots\ldots, f_n$, then the required chance is

$$= \frac{\text{Coeff. of } x^p \text{ in } (x^1 + x^2 + \ldots\ldots + x^{f_1})(x^1 + x^2 + \ldots\ldots + x^{f_2})\ldots\ldots(x^1 + x^2 + \ldots\ldots + x^{f_n})}{f_1 f_2 f_3 \ldots\ldots f_n}$$

1.12. RANDOM VARIABLE

A variable, which takes a definite set of values with a definite probability associated with each value is called a random variable.

Hence, if a variable x takes the values $x_1, x_2, x_3, \ldots\ldots, x_n$ with respective probability $p_1, p_2, p_3, \ldots\ldots, p_n$ so that $\sum_{i=1}^{n} p_i = 1$. Then a discrete probability distribution is defined. The function $P(x)$, which has respective probabilities $p_1, p_2, p_3, \ldots\ldots, p_n$ for $x_1, x_2, x_3, \ldots\ldots, x_n$ is known as the probability function of x. The variable x in such a case is called the random variable.

1.13. EXPECTED VALUE

When X is a discrete variable, which may take n mutually exclusive values x_i ($i = 1, 2, 3, ..., n$) and no other, with respective probabilities p_i ($i = 1, 2, 3, ..., n$) the expectation of X is

$$E(X) = p_1x_1 + p_2x_2 + p_3x_3 + ... + p_nx_n = \sum_{i=1}^{n} p_i x_i$$

when X is a continuous variable, the expectation of x is given by

$$E(X) = \int_{-\infty}^{\infty} x\,\phi(x)\,dx$$

where, $\phi(x)$ is the probability density defining the function

Note 1. If X and Y are random variables, then

$$E(X + Y) = E(X) + E(Y)$$

2. If X and Y are independent random variables, then

$$E(XY) = E(X)\,E(Y)$$

3. The expected value of a random variable X, denoted by $E(X)$ or μ, is defined as

$$E(X) \text{ or } \mu = \begin{cases} \sum_i x_i\, f(x_i) & \text{(if } X \text{ is discrete)} \\ \int_{-\infty}^{\infty} x\, f(x)\, dx & \text{(if } X \text{ is continuous)} \end{cases}$$

4. Mean $(\overline{X}) = E(X) = \mu_1'$

5. $\mu_2' = E(X^2)$

6. Variance $\sigma^2 = E(X^2) - [E(X)]^2 = \mu_2' - \mu_1'^2$

7. $\mu_r' = E(X^r)$

8. $E(aX + b) = aE(X) + b$

9. $E\{a\,\phi(x)\} = a\,E\{\phi(x)\}$

10. $V(aX + b) = a^2V(X)$, where $V(X)$ is the variance of X.

SOLVED EXAMPLES

Example 1: *In a single throws, with a pair of dice, what is the chance of throwing doublets or not.*

Solution: When throw a pair of dice, we get only 6 doublets namely (1, 1); (2, 2); (3, 3); (4, 4); (5, 5); (6, 6).

The chance of throwing a doublet in one throw $= \frac{6}{6^2} = \frac{1}{6}$

And the chance of not throwing a doublet $= 1 - \frac{1}{6} = \frac{5}{6}$

Example 2: *A dice is thrown five times. Find the probability of 3 coming up (i) exactly 3 times (ii) at least 3 times.*

Solution: The probability of 3 coming up $= \frac{1}{6}$.

Let $p = 1/6$ and $q = 5/6$.

The probability of 3 successes in 5 trails is ${}^5C_3 \left(\frac{1}{6}\right)^3 \left(\frac{5}{6}\right)^2$.

Probability of at least 3 successes = Probability of 3 or 4 or 5 successes

$$= {}^5C_3 \left(\frac{1}{6}\right)^3 \left(\frac{5}{6}\right)^2 + {}^5C_4 \left(\frac{1}{6}\right)^4 \left(\frac{5}{6}\right) + {}^5C_5 \left(\frac{1}{6}\right)^5.$$

Example 3: *Four dice are thrown, what is the probability that the sum of the number appearing on the dice is 18?*

Solution: The probability that the sum of the number appearing on the dice is 18

$$= \frac{\text{Coeff. of } x^{18} \text{ in } (x + x^2 + \ldots\ldots + x^6)^4}{6^4}$$

$$= \frac{\text{Coeff. of } x^{14} \text{ in } (1 + x + \ldots\ldots + x^5)^4}{6^4}$$

$$= \frac{\text{Coeff. of } x^{14} \text{ in } (1 - x^6)^4 (1 - x)^{-4}}{6^4}$$

$$= \frac{\text{Coeff. of } x^{14} \text{ in } (1 - 4x^6 + 6x^{12} \ldots\ldots) \times (1 + 4x + 10x^2 + \ldots\ldots + 16x^8 + \ldots\ldots + 680x^{14} + \ldots\ldots)}{6^4}$$

$$= \frac{680 - 660 + 60}{6^4} = \frac{80}{6^4} = \frac{5}{81}$$

Example 4: *A person throws two dice, one the common cube, and the other regular tetrahedron, the number in the lowest face being taken in the case of a tetrahedron. What is the chance that the sum of the numbers thrown is not less than 5.*

Solution: We know that the common cube has 6 faces and the regular tetrahedron has 4 faces. Hence, 6×4 is the total number of ways in which the cube and the tetrahedron can fall.

The favourable ways of getting a sum not less than 5

= the sum of the coeff. of $x^5, x^6, \ldots\ldots, x^{10}$ in $(x + x^2 + \ldots\ldots + x^6)(x + x^2 + x^3 + x^4)$

= the sum of the coeff. of $x^5, x^6, \ldots\ldots, x^{10}$ in $(x^2 + 2x^3 + 3x^4 + 4x^5 + 4x^6 + 4x^7 + 3x^8 + 2x^9 + x^{10})$

$= 4 + 4 + 4 + 3 + 2 + 1$

$\therefore$ The required chance $= \frac{18}{6 \times 4} = \frac{3}{4}$

Example 5: *Four tickets marked 00, 01, 10, 11 respectively are placed in a bag. A ticket is drawn at random five times, being replaced each time. Find the probability that the sum of the numbers on tickets thus drawn is 23.*

Solution: Given that four tickets marked 00, 01, 10, 11 respectively are placed in a bag.

The four tickets can be drawn five times in 4^5 ways.

The favourable number of ways for obtaining a sum

$$= \text{Coeff. of } x^{23} \text{ in } (x^0 + x^1 + x^{10} + x^{11})^5$$

$$= \text{Coeff. of } x^{23} \text{ in } (1 + x)^5 (1 + x^{10})^5 \qquad \text{(on factorization)}$$

$$= \text{Coeff. of } x^{23} \text{ in } (1 + 5x + 10x^2 + 10x^3 + 5x^4 + x^5)(1 + 5x^{10} + 10x^{20} + \ldots\ldots)$$

$$= 100$$

Hence, the required probability $= \dfrac{100}{4^5} = \dfrac{25}{256}$.

Example 6: *If m things are distributed among a men and b women, show that the chance that the number of things received by men is odd, is* $\dfrac{1}{2}\dfrac{(b+a)^m - (b-a)^m}{(a+b)^m}$.

Solution: Given that m things are distributed among a men and b women. The probability that men may get a thing is $\dfrac{a}{a+b}$.

The probability that a women may get a thing is $\dfrac{b}{a+b}$.

If out of m things 'a' men get only one thing and other things go to the women, then the probability for men is

$$= {}^mC_1 \left(\frac{a}{a+b}\right)\left(\frac{b}{a+b}\right)^{m-1}$$

Similarly, the probability that 'a' men may get three things, five things, ... are respectively

$${}^mC_3 \left(\frac{a}{a+b}\right)^3 \left(\frac{b}{a+b}\right)^{m-3}, {}^mC_5 \left(\frac{a}{a+b}\right)^5 \left(\frac{b}{a+b}\right)^{m-5}, \ldots\ldots$$

Let A be the event the number of things got by men is odd and A_i denotes the event 'men get ith thing'

$$i = 1, 3, 5, 7, \ldots\ldots$$

Thus, the probability that the number of things received by men is odd is,

$P(A) = P(A_1) + P(A_3) + P(A_5) + \ldots\ldots$

$$= {}^mC_1 \left(\frac{a}{a+b}\right)\left(\frac{b}{a+b}\right)^{m-1} + {}^mC_3 \left(\frac{a}{a+b}\right)^3 \left(\frac{b}{a+b}\right)^{m-3} + {}^mC_5 \left(\frac{a}{a+b}\right)^5 \left(\frac{b}{a+b}\right)^{m-5} + \ldots\ldots$$

$$= \frac{1}{(a+b)^m}\left[{}^mC_1\, ab^{m-1} + {}^mC_3\, a^3b^{m-3} + {}^mC_5\, a^5\, b^{m-5} + \ldots\ldots\right]$$

$$= \frac{1}{(a+b)^m}\left[\frac{1}{2}(b+a)^m - \frac{1}{2}(b-a)^m\right]$$

$$= \frac{1}{2}\cdot\frac{(b+a)^m - (b-a)^m}{(a+b)^m}.$$

Example 7: *From a bag containing 5 one rupee coins and 3 coins of 20 paisa each, a person is allowed to draw 2 coins indiscriminately. Find the value of his expectation.*

Solution: Given that a bag containing 5 one rupee coins and 3 coins of 20 paisa each.

Probability of drawing 2 rupees $= \frac{{}^5C_2}{{}^8C_2} = \frac{5}{14}$

Probability of drawing 1 rupee and one 20 paisa coin

$$= \frac{{}^5C_1 \times {}^3C_1}{{}^8C_2} = \frac{15}{28}$$

Probability of drawing two 20 paisa coins $= \frac{{}^3C_2}{{}^8C_2} = \frac{3}{28}.$

The values of draws in three cases are Rs 2, 1.20 and 0.40 rupees respectively.

$\therefore$ Expected value $= \frac{5}{14} \times 2 + \frac{15}{28} \times 1.20 + \frac{3}{28} \times 0.40$

$$= \frac{10}{14} + \frac{180}{280} + \frac{12}{280}$$

$$= \frac{200 + 180 + 12}{280}$$

$$= \frac{392}{280} = \frac{7}{5} = \text{Rs. } 1.40.$$

Example 8: *A person draws cards one by one from a pack until he draws all the aces. How many cards he may be expected to draw?*

Solution: Let a person has to make n draws for all aces. It means that in $n-1$ draws, three aces and in n^{th}, one ace. The probability of such an occurrence

$$= \frac{{}^4C_3 \times {}^{48}C_{n-4}}{{}^{52}C_{n-1}} \times \frac{1}{52-(n-1)}$$

$$= \frac{4 \times 48!(n-1)!(52-n+1)!}{(n-4)!(52-n)!52!} \times \frac{1}{52-n+1}$$

$$= \frac{4(n-1)(n-2)(n-3)}{49 \times 50 \times 51 \times 52}.$$

The least square number of draws he has to make is 4 and the maximum number 52. Hence, n range 4 to 52.

The expected number of draws

$$= \sum_{n=1}^{52} n \,.\, 4 \frac{(n-1)(n-2)(n-3)}{49 \times 50 \times 51 \times 52}$$

$$= \frac{4}{49 \times 50 \times 51 \times 52} \left[\sum_{n=1}^{52} n^4 - 6\sum_{n=1}^{52} n^3 - 11\sum_{n=1}^{52} n^2 - 6\sum_{n=1}^{52} n \right].$$

Example 9: *A makes a bet with B of 5S to 2S that in a single throw with two dice he will throw seven before B throws four. Each has a pair of dice and they throw simultaneously until one of them wins, equal throws being disregarded. Find B's expectation.*

Solution: The chance of 7 throwing with two dice

$$= \frac{\text{Coeff. of } x^7 \text{ in } (x + x^2 + x^3 + \ldots\ldots + x^6)^2}{6^2}$$

$$= \frac{\text{Coeff. of } x^5 \text{ in } (1 - x^6)^2 (1-x)^{-2}}{36} = \frac{6}{36} = \frac{1}{6}$$

Similarly, the chance of throwing 4 is $\frac{1}{12}$.

Hence, A's chance in each trail is double of B's.

Now let

$x = B$'s chance on the supposition

then, $2x = A$'s chance clearly and

A's chance + B's chance = 1

i.e. $2x + x = 1$

i.e. $3x = 1$

i.e. $x = \frac{1}{3}$

$\therefore$ B's expectation $= \frac{1}{3}$ of $5S - \frac{2}{3}$ of $2S = 4d$.

Example 10: *An urn contains n tickets numbered from 1 to n and m tickets are drawn one by one without replacements from the urn. What is the mathematical expectation of the sum of the numbers drawn?*

Solution: Suppose the variable associated with the jth ticket be x

$$j = 1, 2, 3, \ldots\ldots n.$$

Now,
$$E(x_j) = \Sigma p_j x_j = \frac{1}{n} \Sigma x_j$$

$$= \frac{1}{n} \frac{n(n+1)}{2} = \frac{n+1}{2}$$

Let S be the sum, then the expected value of the sum

$$E(S) = \Sigma\, E(x_j) = m\, \Sigma\, E(x_j)$$

$$= m\frac{(n+1)}{2}.$$

Example 11: *What is the expectation of the number of failures preceding the first success in an infinite series of independent trails with constant probability of successes.*

Solution: The probabilities of success in first, second, third, ... trails respectively are

$$p, qp, q^2p, q^3p, \ldots\ldots$$

Let X be the no. of failures

Then, the expected number of failures preceding the first success

$$E(X) = 0 \,.\, p + 1 \,.\, qp + 2q^2p + \ldots\ldots$$

$$= qp\, [1 + 2q + 3q^2 + \ldots\ldots \infty], q < 1$$

$$= \frac{qp}{(1-q)^2} \qquad [\because \; p = 1 - q]$$

$$= \frac{q}{p} = \frac{1-p}{p}.$$

EXERCISE 1.4

1. If on average 1 vessel in every 10 is wrecked find the probability that out of 5 vessels expected to arrive, 4 at least will arrive safely.
2. A teacher claims that he could often tell while his students were still in their first year whether they will obtain *I*, *II*, *III* divisions or fail in their final examinations. To demonstrate his claim, he forecasts the fates of 8 students. Find the probability of his being correct in 5 cases.
3. Five coins whose faces are marked, 2, 3 are thrown; what is the chance of obtaining a total of 12?
4. Determine the probability of throwing more than 8 with 3, perfectly symmetrical dice.
5. If three symmetrical dice are thrown, calculate the probability that the sum of numbers is 12.

6. A and B in turns toss an ordinary die for a prize of Rs. 44. The first to toss a 'six' wins. If A has first throw, what is his expectation?

7. A bag contains 2 white and 3 black balls. Four persons A, B, C and D in the order named each draw one ball and do not replace it. The person to draw a white ball receives Rs. 200. Determine their expectations.

8. A box containing 2^n tickets among which ^nC_i bear the number i, $(i = 0, 1, 2, ... n)$. A group of m tickets is drawn, what is the expectation of the sum of numbers?

9. A person draws 2 balls from a bag containing 3 white and 5 black balls. If he is to receive Rs. 10 for every white ball which he draws and Rs. 1 for every black ball. What is his expectation?

10. Two players of equal skill A and B play a set of games; they leave off playing when A wants 3 points and B wants 2. If the stake is Rs. 16, what share ought each to take?

ANSWERS

1. $\dfrac{45927}{50000}$ **2.** $\dfrac{189}{8192}$ **3.** $\dfrac{5}{6}$ **4.** $\dfrac{7}{27}, \dfrac{20}{27}$

5. $\dfrac{25}{216}$ **6.** 24 **7.** 80, 60, 40, 20 **8.** $\dfrac{mn}{2}$

9. 8.75 **10.** 5, 11

CHAPTER 2

Theoretical Distributions

In this chapter, we shall discuss some probability distribution such as Binomial, Poisson, Normal distribution with their applications. In theoretical distribution, normal distribution is most important distribution. Binomial and Poisson distributions are discrete probability distribution and Normal distribution is continuous probability distribution.

Before we discuss formal theoretical probability distribution, first we shall define certain terminologies and notations, which are used in defining theoretical distributions.

2.1. TERMINOLOGY AND NOTATIONS

1. **Random variable:** A variable, which takes a definite set of values with a definite probability associated with each value is called a random variable.
2. **Continuous random variable:** A random variable X is said to be a continuous random variable, if it takes all possible values between its limits. For example, height of a tree, weight of a school children etc.
3. **Discrete random variable:** A random variable X is said to be a discrete random variable, if it takes only finite values between its limits, for example, the number of student appearing in a festival consisting of 400 students is a discrete random variable which can assume values other than 0, 1, 2, ..., 400.
4. **Continuous probability distribution:** Let X be a random variable which assume values in the interval $[-\infty, \infty]$, then the probability in (a, b) is defined by

$$P(a \leq x \leq b) = \int_a^b f(x)\, dx$$

where, $f(x)$ is called probability density function which satisfies the following conditions:

$(i)\ f(x) \geq 0 \quad \forall\, x \in (-\infty, \infty)$ $\qquad$ $(ii) \int_{-\infty}^{\infty} f(x) = 1$

5. **Discrete probability distribution:** Let X be a random variable which assume values $x_1, x_2, ..., x_n$ with probability $p_1, p_2, ..., p_n$ respectively, then the probability is

$$P(X = x_i) = p(x_i) \quad \text{or} \quad f(x_i) \text{ for } i = 1, 2, \ldots\ldots, n$$

where, $p(x_i)$ is called the probability mass function which satisfies the following conditions:

(*i*) $f(x) \geq 0$ (*ii*) $\sum_{i=1}^{n} f(x) = 1$

6. **Mean and variance of random variables:** Let X be a random variable which assume values $x_1, x_2, \ldots\ldots, x_n$ with probability $p_1, p_2, \ldots\ldots, p_n$ respectively.

We denote mean (average or expected value) by μ ($\bar{x}$ or m)

$$\text{or} \qquad E(X) = \frac{\Sigma p_i x_i}{\Sigma p_i} = \Sigma p_i x_i \qquad [\because \Sigma p_i = 1]$$

And we denote variance by (σ^2 or μ_2 or $\mu'_2 - \mu'^2_1$ or $E(X^2) - [E(X)]^2$)

$$= \sum_i p_i (x_i - \bar{x})^2 \quad \text{or} \quad \sum_i p_i (x_i - m)^2$$

Standard deviation (σ) = + $\sqrt{\text{variance}}$.

2.2. BINOMIAL DISTRIBUTION

Binomial distribution is the extension of the theorem of probability. Binomial distribution was discovered by James-Bernoulli in 1700. Let X be the random variable denote the number of successes in these n trials. Let p and q be the probabilities of success and failure one any one trial. Then in the n independent trails the probability that there will be r successes and $n - r$ failures is given by

$$P(X = r) = {}^nC_r\, p^r q^{n-r}, \quad r = 0, 1, 2, \ldots\ldots, n$$

The probability distribution of the random variable X is therefore given by

X	0	1	2	.......r	.........n
$P(X)$	${}^nC_0\, p^0 q^n$	${}^nC_1\, p^1 q^{n-1}$	${}^nC_2\, p^2 q^{n-2}$	${}^nC_r\, p^r q^{n-r}$	${}^nC_n\, p^n q^{n-n}$

Hence, the probability distribution is called the binomial distribution because for $r = 0, 1, 2, \ldots\ldots, n$, are the probabilities of the successive terms of the binomial expansion of $(q + p)^n$.

Note. 1. The constant n, p, q are called parameters of the distribution.

2. Also denoted $b(r\,;\, n, p) = {}^nC_r\, p^r q^{n-r}, \quad r = 0, 1, 2, \ldots\ldots, n$

3. For N set of n trial the successes 0, 1, 2, ...r, ..., n are given by $N(q + p)^n$, which is called binomial distribution.

2.3. CONSTANTS OF BINOMIAL DISTRIBUTION

For the binomial distribution

$$P(X = r) = {}^nC_r\, p^r\, q^{n-r}, \qquad r = 0, 1, 2, \ldots\ldots, n \qquad \ldots(1)$$

$$, \qquad p + q = 1.$$

Taking out origin at 0 successes, we have

$$\mu_1' = E(X) = \sum_{r=0}^{n} r\, {}^nC_r\, p^r\, q^{n-r}$$

$$= 0.\, q^n + 1.\, {}^nC_1\, p\, q^{n-1} + 2.{}^nC_2\, p^2\, q^{n-2} + 3.\, {}^nC_3\, p^3\, q^{n-3} + \ldots\ldots.. + np^n$$

$$= 0 + npq^{n-1} + \frac{1}{2} n(n-1)\, 2p^2\, q^{n-2} + \frac{1}{6} n(n-1)(n-2)\, 3p^3\, q^{n-3} + \ldots\ldots. + np^n$$

$$= np[q^{n-1} + {}^{n-1}C_1\, pq^{n-2} + {}^{n-1}C_2\, p^2\, q^{n-3} + \ldots\ldots + p^{n-1}]$$

$$= np\,(q+p)^{n-1}, \text{ since } q + p = 1$$

$$= np$$

$\therefore$ $\boxed{\text{Mean } (\bar{X}) = np}$

$$\mu'_2 = E(X^2) = \sum_{r=0}^{n} r^2\, {}^nC_r\, p^r\, q^{n-r}$$

$$= \sum_{r=0}^{n} [r + r(r-1)]\, {}^nC_r\, p^r\, q^{n-r}$$

$$= \sum_{r=0}^{n} r\, {}^nC_r\, p^r\, q^{n-r} + \sum_{r=0}^{n} r(r-1)\, {}^nC_r\, p^r\, q^{n-r}$$

$$= np + \sum_{r=0}^{n} r(r-1)\, \frac{n(n-1)}{r(r-1)}\, {}^{n-2}C_{r-2}\, p^r\, q^{n-r}$$

$$= np + n(n-1)\, p^2 \sum_{r=0}^{n} {}^{n-2}C_{r-2}\, p^{r-2}\, q^{n-r}$$

$$= np + n(n-1)p^2\,(q+p)^{n-2}, \quad \text{since} \quad q + p = 1$$

$$= np + n(n-1)\, p^2$$

$$= np\,[1 + (n-1)\, p]$$

$$= np\,[1 + np - p]$$

$$= np\,[q + np]$$

$$= npq + n^2p^2$$

$\therefore$ Variance $(\mu_2) = \mu'_2 - \mu'^2_1$ or $E(X^2) - [E(X)]^2$

$$= npq + n^2p^2 - n^2p^2$$
$$= npq$$

	Variance (μ_2 or σ^2) = npq
and	Standard deviation (σ) = $\sqrt{npq}$

$$\mu'_3 = E(X^3) = \sum_{r=0}^{n} r^3\, {}^nC_r\, p^r q^{n-r}$$

$$= \sum_{r=0}^{n} [r(r-1)(r-2) + 3r(r-1) + r]\, {}^nC_r\, p^r q^{n-r}$$

On simplification in above similar manner, we have

$$= n(n-1)(n-2)p^3(q+p)^{n-3} + 3n(n-1)p^2(q+p)^{n-2} + np(q+p)^{n-1}$$
$$= n(n-1)(n-2)p^3 + 3n(n-1)p^2 + np$$
$$\mu_3 = \mu'_3 - 3\mu'_2\,\mu'_1 + 2\mu'^3_1 = npq(q-p)$$

$$\mu'_4 = E(X^4) = \sum_{r=0}^{n} r^4\, {}^nC_r\, p^r q^{n-r}$$

$$= \sum_{r=0}^{n} \{r(r-1)(r-2)(r-3) + 6r(r-1)(r-2) + 7r(r-1) + r\}\, {}^nC_r\, p^r q^{n-r}$$

On simplification in above similar manner, we have

$$= n(n-1)(n-2)(n-3)p^4 + 6n(n-1)(n-2)p^3 + 7n(n-1)p^2 + np$$
$$\mu_4 = \mu'_4 - 4\mu'_3\mu'_1 + 6\mu'_2\,\mu'^2_1 - 3\mu'^4_1 = npq\,[1 + 3(n-2)pq]$$

Now,
$$\beta_1 = \frac{\mu_3^2}{\mu_2^3} = \frac{[npq(q-p)]^2}{(npq)^3} = \frac{(q-p)^2}{npq}$$

$$\beta_2 = \frac{\mu_4}{\mu_2^2} = \frac{3n^2p^2q^2 + npq(1-6pq)}{n^2p^2q^2} = 3 + \frac{1-6pq}{npq} \quad \text{(Kurtosis)}$$

$$\gamma_1 = \sqrt{\beta_1} = \frac{q-p}{\sqrt{npq}} \quad \text{(Skewness)}$$

$$\gamma_2 = \beta_2 - 3 = \frac{1-6pq}{npq}.$$

Note. 1. The mean of binomial distribution is greater than the variance, since $0 < q < 1$.

2. If skewness is zero *i.e.*, $\beta_1 = 0$ so $p = q = \frac{1}{2}$.

3. If $p < \frac{1}{2}$, skewness is positive, if $p > \frac{1}{2}$, negative.

SOLVED EXAMPLES

Example 1: *Show that for the binomial distribution $(q + p)^n$, $\mu_{r+1} = pq\left(nr\,\mu_{r-1} + \frac{d\mu_r}{dp}\right)$ where, μ_r is the rth moment about the mean. Hence, obtain μ_2, μ_3 and μ_4.*

Solution: Using the definition of *r*th moment about mean, we have

$$\mu_r = E\,[X - E(X)]^r$$

$$= \sum_{x=0}^{n} (x - np)^r\, {}^nC_x p^x q^{n-x}$$

Differentiating with respect to p both sides, we get

$$\frac{d\mu_r}{dp} = \sum_{x=0}^{n} -nr\,(x - np)^{r-1}\, {}^nC_x p^x q^{n-x}$$

$$+ \sum_{x=0}^{n} (x - np)^r\, {}^nC_x\,[x\,p^{x-1} q^{n-x} - (n - x)\, p^x q^{n-x-1}]$$

$$= -nr \sum_{x=0}^{n} (x - np)^{r-1}\, {}^nC_x p^x q^{n-x}$$

$$+ \sum_{x=0}^{n} (x - np)^r\, {}^nC_x \left(\frac{1}{pq}\right) [xq - np + xp]\, p^x q^{n-x}$$

$$= -nr\,\mu_{r-1} + \frac{1}{pq} \sum_{x=0}^{n} (x - np)^r\, {}^nC_x\,[x\,(p + q) - np]\, p^x q^{n-x}$$

$$= -nr\,\mu_{r-1} + \frac{1}{pq} \sum_{x=0}^{n} (x - np)^{r+1}\, {}^nC_x\, p^x q^{n-x}$$

$$= -nr\,\mu_{r-1} + \frac{1}{pq}\,\mu_{r+1}$$

$$\Rightarrow \quad \boxed{\mu_{r+1} = pq\left(nr\mu_{r-1} + \frac{d\mu_r}{dp}\right)} \quad \text{...(1)}$$

Putting $r = 1, 2, 3$ in (1), we get

$$\mu_2 = pq\left(\frac{d\mu_1}{dp} + n\,.\,1\,.\,\mu_0\right)$$

$$= pq\,(0 + n), \quad \text{since } \mu_1 = 0,\ \mu_0 = 1$$

$$= npq$$

$$\mu_3 = pq\left(\frac{d\mu_2}{dp} + n\,.\,2\,\mu_1\right)$$

$$= pq\,(nq - np + 2n\,.\,0) = npq\,(q - p)$$

$$\mu_4 = pq\left(\frac{d\mu_3}{dp} + n\,.\,3\,\mu_2\right)$$

$$= pq\left[\frac{d}{dp}\{np(1-p)\,(1-2p)\} + 3n\,.\,npq\right]$$

$$= pq\left[\frac{d}{dp}\left\{np - 3np^2 + 2np^3\right\} + 3n^2pq\right]$$

$$= pq\,[n - 6np + 6np^2 + 3n^2pq]$$

$$= npq - 6np^2q + 6np^3q + 3n^2p^2q^2$$

$$= 3n^2p^2q^2 + npq + 6npq\,(p^2 - p)$$

$$= 3n^2p^2q^2 + npq + 6np^2q\,(p - 1)$$

$$= 3n^2p^2q^2 + npq - 6np^2q^2$$

$$= 3n^2p^2q^2 + npq\,(1 - 6pq).$$

Example 2: *A die is tossed thrice. Getting an even number is considered as success. What is variance of the binomial distribution ?*

Solution: Let p be the probability of getting an even number

i.e., $$p = \frac{3}{6} = \frac{1}{2} \text{ then } q = 1 - \frac{1}{2} = \frac{1}{2},\ n = 3$$

The variance of binomial distribution = npq, where $q = 1 - p$

$$= 3 \times \frac{1}{2} \times \frac{1}{2} = \frac{3}{4}.$$

Example 3: *The mean and variance of a binomial distribution are 4 and 3 respectively. Find the probability of getting exactly six successes in this distribution.*

Solution: The mean of binomial distribution $= np = 4$ (given) ...(1)

and the variance of binomial distribution $= npq = 3$ (given) ...(2)

Using (1) and (2), we have

$$npq = 3 \Rightarrow 4\,.\,q = 3 \Rightarrow q = \frac{3}{4}$$

and $$p = 1 - q = 1 - \frac{3}{4} = \frac{1}{4}$$

by (1), $$np = 4 \Rightarrow n \times \frac{1}{4} = 4 \Rightarrow n = 16$$

The probability of 6 success $= {}^{16}C_6 \left(\frac{1}{4}\right)^6 \left(\frac{3}{4}\right)^{10} = \frac{8008 \times 3^{10}}{4^{16}}$.

Example 4: *In 256 sets of 12 tosses of a coin, in how many cases one can expect 8 heads and 4 tails.*

Solution: Let p be the probability of head *i.e.*, $p = \frac{1}{2}$ and q be the probability of tail *i.e.*, $q = \frac{1}{2}$.

$$n = 12, N = 256.$$

The binomial distribution is $N(q + p)^n = 256 \left(\frac{1}{2} + \frac{1}{2}\right)^{12}$

The probability of 8 heads and 4 tails in 12 trial is

$$^{12}C_8 \left(\frac{1}{2}\right)^8 \left(\frac{1}{2}\right)^4 = \frac{495}{4096}$$

The expected number of such cases in 256 sets

$$= 256 \times \frac{495}{4096} = 30.9 \approx 31 \text{ (nearly)}.$$

Example 5: *During war, 1 ship out of 9 was sunk on an average in making a voyage. What was the probability that exactly 3 out of a convoy of 6 ships would arrive safely ?*

Solution: Let p be the probability of a ship arriving safely *i.e.*,

$$p = 1 - \frac{1}{9} = \frac{8}{9} \quad \text{then} \quad q = 1 - p = \frac{1}{9}$$

$$n = 6, N = 1$$

The binomial distribution is

$$N(q + p)^n = \left(\frac{1}{9} + \frac{8}{9}\right)^6$$

The probability that exactly 3 ship arrive safely

$$= {}^6C_3 \left(\frac{8}{9}\right)^3 \left(\frac{1}{9}\right)^3 = 20 \times \frac{512}{9^6} = \frac{10240}{9^6}.$$

Example 6: *The probability that a pen manufactured by a company will be defective is $\frac{1}{10}$. If 12 such pens are manufactured, find the probability that*

(*i*) *exactly two will be defective.*

(*ii*) *at least one will be defective.*

(*iii*) *none will be defective.*

Solution: Given that the probability of a defective pen = $\frac{1}{10} = 0.1$

Then the probability of a non-defective pen is = $1 - 0.1 = 0.9$

(*i*) The probability that exactly two pen will be defective

$$= {}^{12}C_2 \,(0.1)^2\,(0.9)^{10} = 0.2301$$

(*ii*) The probability that at least two will be defective

$$= 1 - \{\text{Probability that either none or one is non-defective}\}$$
$$= 1 - \{{}^{12}C_0\,(0.9)^{12} + {}^{12}C_1\,(0.1)\,(0.9)^{11}\}$$
$$= 1 - 0.6588 = 0.3412$$

(*iii*) The probability that none will be defective

$$= {}^{12}C_{12}\,(0.9)^{12} = 0.2833.$$

Example 7: *Six dice are thrown 729 times. How many times do you expect at least three dice to show a five or six ?*

Solution: Let p be the chance of getting 5 or 6 with one die

then $\quad p = \frac{1}{3}, \qquad q = 1 - \frac{1}{3} = \frac{2}{3}$

$\quad n = 6, \qquad N = 729$

Since, dice are in sets of 6 and there are 729 sets.

The binomial distribution is

$$N(q+p)^n = 729\left(\frac{2}{3}+\frac{1}{3}\right)^6$$

The required number is

$$= 729\left\{{}^6C_3\left(\frac{1}{3}\right)^3\left(\frac{2}{3}\right)^3 + {}^6C_4\left(\frac{1}{3}\right)^4\left(\frac{2}{3}\right)^2 + {}^6C_5\left(\frac{1}{3}\right)^5\left(\frac{2}{3}\right) + {}^6C_6\left(\frac{1}{3}\right)^6\right\}$$

$$= \frac{729}{3^6}\,\{160 + 60 + 12 + 1\} = 233.$$

Example 8: *A perfect cubical die is thrown a large number of times in set of 8. The occurrence of 5 or 6 is called a success. In what proportion of sets do you expect 3 successes.*

Solution: Let $\;p$ = the chance of occurrence 5 or 6 with one die = $\frac{2}{6} = \frac{1}{3}$

$$q = 1 - p = 1 - \frac{1}{3} = \frac{2}{3}, \quad n = 8$$

The binomial distribution is given by

$$N(q+p)^n = N\left(\frac{2}{3}+\frac{1}{3}\right)^8$$

The number of sets in which 3 successes are expected

$$N\left\{{}^8C_3\left(\frac{1}{3}\right)^3\left(\frac{2}{3}\right)^5\right\} = N\,\frac{8!}{3!5!}\cdot\frac{2^5}{3^8}$$

$$= N\,.\,\frac{8\times7\times6\times5!\times2^5}{3\times2\times5!\times3^8} = N\,\frac{56\times32}{243\times27} = N\,\frac{1792}{6591}$$

The percentage is $= N\,\dfrac{1792}{6591}\times\dfrac{100}{N} = 27.3\%.$

Example 9: *In 800 families with 4 children each how many families would be expected to have (i) 2 boys and 2 girls (ii) at least one boy (iii) no girl (iv) at least two girls ? Assuming that equal probabilities for boys and girls.*

Solutioin: Since, the probabilities for boys and girls are equal

Then let p = probability of having a boy $= \dfrac{1}{2}$

and q = probability of having a girl $= \dfrac{1}{2}$

Here $n = 4,\quad N = 800$

The binomial distribution is $N\,(q+p)^n = 800\left(\dfrac{1}{2}+\dfrac{1}{2}\right)^4.$

(*i*) The expected number of families having 2 boys and 2 girls

$$= 800\,{}^4C_2\left(\frac{1}{2}\right)^2\left(\frac{1}{2}\right)^2 = \frac{800\times6}{16} = 300.$$

(*ii*) The expected number of families having at least one boy

$$= 800\left\{{}^4C_1\left(\frac{1}{2}\right)^3\left(\frac{1}{2}\right)+{}^4C_2\left(\frac{1}{2}\right)^2\left(\frac{1}{2}\right)^2+{}^4C_3\left(\frac{1}{2}\right)^3\left(\frac{1}{2}\right)+{}^4C_4\left(\frac{1}{2}\right)^4\right\}$$

$$= \frac{800}{2^4}\,[4+6+4+1] = 750.$$

(*iii*) The expected number of families having no girl

$$= 800\,{}^4C_4\left(\frac{1}{2}\right)^4 = \frac{800}{16} = 50.$$

(*iv*) The expected number of families having at most two girls

$$= 800 \left\{ {}^4C_2 \left(\frac{1}{2}\right)^2 \left(\frac{1}{2}\right)^2 + {}^4C_3 \left(\frac{1}{2}\right)\left(\frac{1}{2}\right)^3 + {}^4C_4 \left(\frac{1}{2}\right)^4 \right\}$$

$$= \frac{800}{2^4} [6 + 4 + 1] = 550.$$

Example 10: *The probability that a men aged 60 years will live up to 70 years is 0.65. What is the probability that out of ten men now 60 years, at least 7 would live up to 70 years ?*

Solution: Given that the probability of a men aged 60 years will live up to 70 years is = 0.65

Then $p = 0.65, \quad q = 1 - p = 1 - 0.65 = 0.35$

$n = 10$

Let X be the number of men who live up to 70 years.

The probability that out of 10 men, r men will up to 70 years is

$$P(X = r) = {}^nC_r p^r q^{n-r} = {}^{10}C_r (0.65)^r (0.35)^{10-r}$$

The probability of at least 7 men would live up to 70 years.

$$P(X \geq 7) = P(X = 7) + P(X = 8) + P(X = 9) + P(X = 10)$$

$$= {}^{10}C_7 (0.65)^7 (0.35)^3 + {}^{10}C_8 (0.65)^8 (0.35)^2 + {}^{10}C_9 (0.65)^9 (0.35)^1 + {}^{10}C_{10} (0.65)^{10}$$

$$= 0.2523 + 0.1756 + 0.0725 + 0.0135 = 0.5140.$$

Example 11: *Assuming that on the average one telephone number out of 15 called between 2 p.m. and 3 p.m. on week days is busy. What is the probability that if 6 randomly selected telephone numbers are called (i) not more than three, (ii) at least three of them will be busy ?*

Solution: Let p be the probability of a telephone number out of 15 called between 2 p.m. and 3 p.m. on week days is busy,

i.e., $p = \frac{1}{15}$ then $q = 1 - \frac{1}{15} = \frac{14}{15}$, $n = 6, N = 1$

The binomial distribution is $N(q + p)^n = \left(\frac{14}{15} + \frac{1}{15}\right)^6$

(*i*) The probability that not more than three will be busy

$$P(X \leq 3) = P(X = 0) + P(X = 1) + P(X = 2) + P(X = 3)$$

$$= {}^6C_0 \left(\frac{14}{15}\right)^6 + {}^6C_1 \left(\frac{14}{15}\right)^5 \left(\frac{1}{15}\right) + {}^6C_2 \left(\frac{14}{15}\right)^4 \left(\frac{1}{15}\right)^2 + {}^6C_3 \left(\frac{14}{15}\right)^3 \left(\frac{1}{15}\right)^3$$

$$= \frac{(14)^3}{(15)^6} [2744 + 1176 + 210 + 20]$$

$$= \frac{2744 \times 4150}{(15)^6} = 0.9997$$

(*ii*) The probability that at least three of them will be busy

$$P(X \geq 3) = P(X = 3) + P(X = 4) + P(X = 5) + P(X = 6)$$

$$= {}^6C_3 \left(\frac{14}{15}\right)^3 \left(\frac{1}{15}\right)^3 + {}^6C_4 \left(\frac{14}{15}\right)^2 \left(\frac{1}{15}\right)^4 + {}^6C_5 \left(\frac{14}{15}\right)\left(\frac{1}{15}\right)^5 + {}^6C_5 \left(\frac{1}{15}\right)^6$$

$$= 0.005.$$

Example 12: *The following data are the number of seeds germinating out of 10 on damp filter for 80 sets of seeds. Fit a binomial distribution to these data:*

x	*0*	*1*	*2*	*3*	*4*	*5*	*6*	*7*	*8*	*9*	*10*	*Total*
f	*6*	*20*	*28*	*12*	*8*	*6*	*0*	*0*	*0*	*0*	*0*	*80*

Solution: Given that $n = 10$, $N = 80$ and $\Sigma f = 80$.

The arithmetic mean $= \dfrac{\Sigma fx}{\Sigma f}$

$$= \frac{1 \times 20 + 2 \times 28 + 3 \times 12 + 4 \times 8 + 5 \times 6 + 6 \times 0 + 7 \times 0 + 8 \times 0 + 9 \times 0 + 10 \times 0}{80}$$

$$= \frac{174}{80}$$

The mean of a binomial distribution $= np$

$$\therefore \quad np = \frac{174}{80} \Rightarrow p = \frac{174}{80 \times 10} = 0.2175$$

Then $q = 1 - p = 0.7825.$

Hence, the binomial distribution to be fitted to the data is

$$N(q + p)^n = 80\,(0.7825 + 0.2175)^{10}$$

Thus, the theoretical frequencies are

x	0	1	2	3	4	5	6	7	8	9	10
f	6.9	19.1	24.0	17.8	8.6	2.9	0.7	0.1	0	0	0

EXERCISE 2.1

1. The mean and variance of binomial distribution are 4 and $\frac{4}{3}$ respectively. Find $P(X \geq 1)$.

2. The items produced by a firm are supposed to contain 5% defective items. What is the probability that a sample of 8 items will contain less than 2 defective items ?

3. Find the binomial distribution for which the mean is 4 and variance is 3.

4. Bring out the fallcy in the statement:
The mean of a binomial distribution is 3 and variance is 4.

5. If 10% bolts produced by a machine are defective, determine the probability that out of 10 bolts chosen at random (*i*) 1 (*ii*) none (*iii*) at least one (*iv*) at most two bolts, will be defective.

6. An irregular six-faced die is thrown, and the expectation that in 100 throws it will give five even numbers is twice the expectation that it will give four even numbers. How many times in 10,000 sets of 10 throws would you expect it to give no even number ?

7. The incidence of occupational disease in an industry is such that the workmen have a 25% chance of suffering from it. What is the probability that out of six workmen 4 or more will contact the disease ?

8. A box contains '*a*' red and '*b*' black balls, '*n*' balls are drawn. Find the expected number of red balls drawn.

9. Assuming that half the population are consumers of rice so that the chance of an individual being a consumer is $\frac{1}{2}$ and assuming that 100 investigators, each take ten individuals to see whether they are consumers, how many investigators do you except to report that three people or less are consumers ?

10. Find the most probable number of heads in 99 tossing of a biased coin, given that the probability of a head in a single tossing is $\frac{3}{5}$.

11. In litters of 4 mice the number of litters which contained 0, 1, 2, 3, 4 females were noted. The figures are given in the table below:

Number of female mice	0	1	2	3	4	Total
Number of litters	8	32	34	24	5	103

In the chance of obtaining a female in a single trial is assumed constant, estimate this constant of unknown probability.

12. The probability of a man hitting a target in $\frac{1}{4}$. He fires 7 times. What is the probability of his hitting the target at least twice. How many times must he fires so that the probability of his hitting at least one is greater than $\frac{2}{3}$?

13. Eight coins are tossed at a time, 256 times. Number of heads observed at each throw are recorded and the results are as given below. Find the expected frequency and fit a binomial distribution. What are the theoretical values of the mean and standard deviation ? Calculate also the mean and standard deviation of the observed frequencies.

14. The probability that a bomb droped from a plane strike the target is $\frac{1}{5}$. If six such bombs are droped find the probability that at least two will strike the target, and exactly two will strike the target.

15. In a precision bombing attack there is a 50% chance that anyone bomb will strike the target. Two direct hits are required to destroy the target completely. How many bombs must be dropped to give a 99% chance or better of completely destroying the target ?

16. If a coin is tossed N times, where N is very large even number. Show that the probability of getting exactly $\frac{1}{2}N - p$ heads and $\frac{1}{2}N + p$ tails is approximately $\left(\frac{2}{\pi N}\right)^{1/2} e^{-2}p^{2/N}$.

ANSWERS

1. 0.99863
2. ${}^8C_0\,(0.05)^0\,(0.95)^8 + {}^8C_1\,(0.05)^1\,(0.95)^7$
3. $P(X = r) = {}^{16}C_r \left(\frac{1}{4}\right)^r \left(\frac{3}{4}\right)^{16-r}$, $r = 0, 1, 2, 3, \ldots\ldots, 16.$
4. $np > npq$, since $q < 1$
5. (*i*) 0.03874 (*ii*) 0.3487 (*iii*) 0.6513 (*iv*) 0.58114
6. 1 (approximately)
7. 0.0376
8. $\dfrac{na}{a+b}$
9. 17 (approximately)
10. 59 and 60
11. 103 $(0.534 + 0.466)^4$
12. $n > 3.8$, so the least value of n is 4.
13. .5, 4.0, 1.41, 4.0, 1.44
14. 0.345, 0.246
15. 11.

2.4. POISSON'S DISTRIBUTION

Poisson distribution was discovered by a french mathematician Simeon Denis Poisson in 1837. Poisson distribution is also a discrete probability distribution of a discrete random variable, which has no upper bound. Poisson distribution is a limiting form of the binomial distribution $(q + p)^n$ under the following conditions:

(*i*) $n \to \infty$, *i.e.*, the number of trials is indefinitely large.

(*ii*) $p \to 0$, *i.e.*, the constant probability of success for each trial is indefinitely small.

(*iii*) np is a finite quantity, say m.

Thus, $p = \dfrac{m}{n}$, $q = 1 - \dfrac{m}{n}$, where m is a positive real number.

Poisson distribution deals with situations explained below:

(*i*) Number of suicides or deaths by heart attack in 1 minute.

(*ii*) Number of accidents that take place on a busy road in time t.

(*iii*) Number of printing mistakes at each unit of the book.

(*iv*) Number of cars passing a certain street in time t.

(*v*) Emission of radioactive particles.

(*vi*) Number of faulty blades in a packet of 1000.

(*vii*) Number of person born blind per year in a certain village.

(*viii*) Number of telephone calls received at a particular switch board in 1 minute.

The probability distribution of a random variable X is said to be a Poisson distribution, if the random variable assumes only non-negative values and its probability distribution is given by

$$P(X=r)=\frac{m^r e^{-m}}{r!}, \quad r=0,1,2,\ldots\ldots$$

where, m is called the parameter of the distribution and $m>0$

The probability of r successes in a binomial distribution

$$P(X=r)={}^nC_r\,p^r q^{n-r}$$

$$=\frac{n!}{r!(n-r)!}p^r q^{n-r}=\frac{n(n-1)(n-2)\ldots\ldots(n-r+1)}{r!}p^r(1-p)^{n-r}$$

$$=\frac{1\left(1-\frac{1}{n}\right)\left(1-\frac{2}{n}\right)\ldots\ldots\left(1-\frac{r-1}{n}\right)p^r(1-p)^{n-r}n^r}{r!}$$

$$=\frac{\left(1-\frac{1}{n}\right)\left(1-\frac{2}{n}\right)\ldots\ldots\left(1-\frac{r-1}{n}\right)(np)^r(1-p)^{n-r}}{r!}$$

Taking limit $n\to\infty$, $p\to 0$ such that $np=m$

$$=\lim_{n\to\infty}\frac{\left(1-\frac{1}{n}\right)\left(1-\frac{2}{n}\right)\ldots\ldots\left(1-\frac{r-1}{n}\right)}{r!}m^r\left(1-\frac{m}{n}\right)^{n-r}$$

$$=\frac{m^r}{r!}\lim_{n\to\infty}\left(1-\frac{m}{n}\right)^{n-r}=\frac{m^r}{r!}\lim_{n\to\infty}\left(1-\frac{m}{n}\right)^{n}\left(1-\frac{m}{n}\right)^{-r}$$

$$=\frac{m^r}{r!}e^{-m}.1 \qquad \left[\because \lim_{n\to\infty}\left(1-\frac{m}{n}\right)^n=e^{-m} \text{ and } \lim_{n\to\infty}\left(1-\frac{m}{n}\right)^{-r}=1\right]$$

$$=\frac{m^r e^{-m}}{r!}, \quad r=0,1,2,\ldots\ldots$$

The probability of 0, 1, 2, ... r, ... successes are

$$e^{-m}, \frac{e^{-m}m}{1!}, \frac{e^{-m}m^2}{2!}, \frac{e^{-m}m^3}{3!}, \ldots\ldots \frac{e^{-m}m^r}{r!}, \ldots\ldots \text{ respectively.}$$

Note. It should be noted that $\sum_{r=0}^{\infty}P(X=r)=\sum_{r=0}^{\infty}\frac{m^r e^{-m}}{r!}=e^{-m}\sum_{r=0}^{\infty}\frac{m^r}{r!}=e^{-m}e^{m}=1.$

2.5. CONSTANTS OF POISSON DISTRIBUTION

$$\mu_1' = E(X) = \sum_{r=0}^{\infty} r \,.\, P(X=r)$$

$$= \sum_{r=0}^{\infty} r \,.\, \frac{e^{-m}\, m^r}{r\,!}$$

$$= e^{-m} \sum_{r=0}^{\infty} \frac{m^r}{(r-1)\,!} = e^{-m}\left(m + \frac{m^2}{1!} + \frac{m^3}{2!} + \ldots\right)$$

$$= m\, e^{-m}\left(1 + \frac{m}{1!} + \frac{m^2}{2!} + \ldots\right) = me^{-m} \,.\, e^{m} = m$$

$\therefore$ $\boxed{\text{Mean } (\bar{X}) = m}$

$$\mu_2' = E(X^2) = \sum_{r=0}^{\infty} r^2 \,.\, P(X=r)$$

$$= \sum_{r=0}^{\infty} \{r(r-1) + r\} \frac{e^{-m}\, m^r}{r\,!}$$

$$= \sum_{r=0}^{\infty} \frac{r(r-1)\, e^{-m}\, m^r}{r\,!} + \sum_{r=0}^{\infty} \frac{r \,.\, e^{-m}\, m^r}{r\,!}$$

$$= e^{-m} \sum_{r=0}^{\infty} \frac{m^r}{(r-2)\,!} + m$$

$$= e^{-m}\left(m^2 + \frac{m^3}{1!} + \frac{m^4}{2!} + \ldots\right) + m$$

$$= m^2 e^{-m}\left(1 + \frac{m}{1!} + \frac{m^2}{2!} + \ldots\right) + m$$

$$= m^2\, e^{-m} . e^{m} + m = m^2 + m$$

$\therefore$ Variance $(\mu_2) = \mu'_2 - \mu_1'^2$ or $E(X^2) - [E(X)]^2$

$$= m^2 + m - m^2 = m$$

$\boxed{\text{Variance } (\mu_2 \text{ or } \sigma^2) = m}$

Hence, mean of Poisson distribution = Variance of Poisson distribution

$\boxed{\text{Standard deviation } \quad (\sigma) = \sqrt{m}}$

$$\mu_3' = E(X^3) = \sum_{r=0}^{\infty} r^3\, P(X=r)$$

$$= \sum_{r=0}^{\infty} \{r(r-1)(r-2) + 3r(r-1) + r\} \frac{e^{-m} m^r}{r!}$$

$$= \sum_{r=0}^{\infty} \frac{r(r-1)(r-2)e^{-m} m^r}{r!} + 3\sum_{r=0}^{\infty} \frac{r(r-1)e^{-m} m^r}{r!} + \sum_{r=0}^{\infty} \frac{r e^{-m} m^r}{r!}$$

$$= e^{-m} m^3 \sum_{r=0}^{\infty} \frac{m^{r-3}}{(r-3)!} + 3e^{-m}m^2 \sum_{r=0}^{\infty} \frac{m^{r-2}}{(r-2)!} + e^{-m} . m \sum_{r=0}^{\infty} \frac{m^{r-1}}{(r-1)!}$$

$$= m^3 + 3m^2 + m.$$

$$\mu_3 = \mu'_3 - 3\mu'_2 \mu'_1 + 2\mu'^3_1$$

$$= m$$

$$\mu'_4 = E(X^4) = \sum_{r=0}^{\infty} r^4 P(X = r)$$

$$= \sum_{r=0}^{\infty} \{r(r-1)(r-2)(r-3) + 6r(r-1)(r-2) + 7r(r-1) + r\} \frac{e^{-m}m^r}{r!}$$

$$= e^{-m} m^4 \sum_{r=0}^{\infty} \frac{m^{r-4}}{(r-4)!} + 6e^{-m}m^3 \sum_{r=0}^{\infty} \frac{m^{r-3}}{(r-3)!} + 7e^{-m}m^2 \sum_{r=0}^{\infty} \frac{m^{r-2}}{(r-2)!}$$

$$+ e^{-m} \cdot m \sum_{r=0}^{\infty} \frac{m^{r-1}}{(r-1)!}$$

$$= m^4 + 6m^3 + 7m^2 + m.$$

$$\mu_4 = \mu'_4 - 4\mu'_3 \mu'_1 + 6\mu'_2\mu'^2_1 - 3\mu'^4_1$$

$$= (m^4 + 6m^3 + 7m^2 + m) - 4(m^3 + 3m^2 + m)(m) + 6(m^2 + m)m - 3m^4$$

$$= 3m^2 + m$$

Now $\beta_1 = \dfrac{\mu_3^2}{\mu_2^3} = \dfrac{m^2}{m^3} = \dfrac{1}{m}$

$$\beta_2 = \frac{\mu_4}{\mu_2^2} = \frac{3m^2 + m}{m^2} = 3 + \frac{1}{m}$$

$$\gamma_1 = \sqrt{\beta_1} = \frac{1}{\sqrt{m}}$$

$$\gamma_2 = \beta_2 - 3 = \frac{1}{m}$$

When $m \to \infty$, $\beta_1 = 0$ and $\beta_2 = 3$

Hence, Poisson distribution is always positively skewed.

Note. $\mu_1 = \mu'_1 - \mu'_0\, m = m - m = 0$, since
$\mu_n = \mu'_n - {}^nC_1\, \mu'_{n-1} \cdot m + {}^nC_2\, \mu'_{n-2} m^2 - \ldots\ldots$

2.6. RECURRENCE FORMULA FOR POISSON DISTRIBUTION

Poisson distribution for r successes, we have

$$P(X = r) = \frac{m^r e^{-m}}{r!}, \quad r = 0, 1, 2, 3 \ldots\ldots \qquad \ldots(1)$$

where m is the mean of successes.

Also $$P(X = r + 1) = \frac{m^{r+1} e^{-m}}{(r+1)!}, \quad r = 0, 1, 2, \ldots\ldots \qquad \ldots(2)$$

divide (2) by (1), we get

$$\frac{P(r+1)}{P(r)} = \frac{m}{r+1}$$

$$\Rightarrow \qquad \boxed{P(r + 1) = \frac{m}{r+1} P(r)}$$

This is the recurrence formula for Poisson distribution.

2.7. MODE OF THE POISSON DISTRIBUTION

The value of r which gives the greatest probability is the mode of the Poisson distribution. Thus

$$\frac{m^{r-1} e^{-m}}{(r-1)!} \le \frac{m^r e^{-m}}{(r)!} \ge \frac{m^{r+1} e^{-m}}{(r+1)!}$$

Where, $m > r > m - 1$

i.e., $m - 1 < r < m$

Thus, if m is an integer then there are two modes $m - 1$ and m. If m is not an integer then the mode is the integral value between $m - 1$ and m.

SOLVED EXAMPLES

Example 1: *Show that for the Poisson distribution with mean m*

$$\mu_{r+1} = r\, m\, \mu_{r-1} + m \frac{d\mu_r}{dm}, \text{ where } \mu_r = \sum_{x=0}^{\infty} (x - m)^r \frac{e^{-m} m^x}{x!}.$$

Solution: Given that $\mu_r = \sum_{x=0}^{\infty} (x-m)^r e^{-m} \dfrac{m^x}{x!}$

Differentiating with respect to m both sides, we get

$$\frac{d\mu_r}{dm} = \sum_{x=0}^{\infty} (-r)(x-m)^{r-1} \frac{e^{-m} m^x}{x!} + \sum_{x=0}^{\infty} (x-m)^r \cdot \frac{1}{x!} \{e^{-m} . x m^{x-1} - m^x e^{-m}\}$$

$$= -r \sum_{x=0}^{\infty} (x-m)^{r-1} \frac{e^{-m} m^x}{x!} + \sum_{x=0}^{\infty} \frac{(x-m)^r}{x!} e^{-m} m^{x-1} \{x-m\}$$

$$= -r \mu_{r-1} + \sum_{x=0}^{\infty} (x-m)^{r+1} \frac{e^{-m} m^{x-1}}{x!}$$

$$= -r \mu_{r-1} + \frac{1}{m} \sum_{x=0}^{\infty} (x-m)^{r+1} \frac{e^{-m} m^x}{x!}$$

$$= -r \mu_{r-1} + \frac{1}{m} \mu_{r+1}$$

$$\Rightarrow \qquad \boxed{\mu_{r+1} = rm\, \mu_{r-1} + m \frac{d\mu_r}{dm}}$$

Example 2: *Six coins are tossed 6400 times. Using the Poisson distribution, what is approximate probability of getting six heads x times.*

Solution: Let p be the probability of getting all the six head in a throw of six coins. *i.e.*, $p = \dfrac{1}{2^6} = \dfrac{1}{64}$.

Here, $n = 6400$

The mean of Poisson distribution $= np = 6400 \times \dfrac{1}{64} = 100$

The probability of getting all six heads x times according to Poisson distribution is $\dfrac{e^{-100} (100)^x}{x!}$.

Example 3: *Show that for a Poisson's distribution $\gamma_1 \gamma_2 \sigma m = 1$, when σ and m are the standard deviation and mean respectively.*

Solution: In Poisson distribution $\gamma_1 = \dfrac{1}{\sqrt{m}}$ and $\gamma_2 = \dfrac{1}{m}$

The standard deviation $(\sigma) = \sqrt{m}$

and the mean $= m$

Now, we have $\gamma_1 \gamma_2 \sigma m = \dfrac{1}{\sqrt{m}} \times \dfrac{1}{m} \times \sqrt{m} \times m = 1$

Example 4: *If X is a Poisson variate such that P(X = 2) = 9P (X = 4) + 90 P (X = 6). Find the mean and variance of X.*

Solution: The Poisson distribution is

$$P(X = r) = \frac{m^r e^{-m}}{r!}. \quad r = 0, 1, 2, \ldots\ldots, m > 0$$

Given that $\quad P(X = 2) = 9\,P(X = 4) + 90P\,(X = 6)$

$$\Rightarrow \quad \frac{m^2 e^{-m}}{2!} = 9\,\frac{m^4 e^{-m}}{4!} + 90\,\frac{m^6 e^{-m}}{6!}$$

$$\Rightarrow \quad \frac{m^2}{2} = \frac{3m^4}{8} + \frac{m^6}{8}$$

$$\Rightarrow \quad 4 = 3m^2 + m^4 \;\Rightarrow\; m^4 + 3m^2 - 4 = 0$$

$$\Rightarrow \quad (m^2 - 1)\,(m^2 + 4) = 0 \;\Rightarrow\; (m^2 - 1) = 0 \;\Rightarrow\; m = 1 \qquad \text{(Taking only real values)}$$

Hence, the mean and variance of X are each equal to 1.

Example 5: *In a Poisson distribution P(x) for x = 0 is 0.1. Find the mean given that $\log_e 10 = 2.3026$.*

Solution: The Poisson distribution is

$$P(x) = \frac{m^x e^{-m}}{x!}, \quad x = 0, 1, 2, \ldots\ldots., m > 0 \qquad \ldots(1)$$

Given that for $x = 0$ $\;P(x) = 0.1$

Then by (1), we have

$$0.1 = \frac{m^0 e^{-m}}{0!} \;\Rightarrow\; e^{-m} = 0.1 \;\Rightarrow\; e^m = 10$$

$$\Rightarrow \quad m = \log_e 10 = 2.3026.$$

Example 6: *If 2% electric bulbs manufactured by a company be defective, find the probability of (i) 0 (ii) 1 (iii) 2 (iv) 3 defectives in a lot of 100 bulbs.*

Solution: Let p be the probability of electric bulbs manufactured by a company be defective.

i.e., $$p = \frac{2}{100}.$$

Given that, $\quad n = 100$

we have, $$np = 100 \times \frac{2}{100} = 2 = m \text{ (mean)}$$

Let X be the number of electric bulbs manufactured by a company be defective.

Then, $$P(X = r) = \frac{m^r\, e^{-m}}{r!} = \frac{2^r\, e^{-2}}{r!}, \qquad r = 0, 1, 2, 3, \ldots..$$

(i) $P(X = 0) = \dfrac{2^0 e^{-2}}{0!} = \dfrac{1}{e^2} = \dfrac{1}{(2.718)^2} = 0.135$

(ii) $P(X = 1) = \dfrac{2^1 e^{-2}}{1!} = \dfrac{2}{e^2} = \dfrac{2}{(2.718)^2} = 2 \times (0.135) = 0.2706$

(iii) $P(X = 2) = \dfrac{2^3 e^{-2}}{2!} = \dfrac{2}{e^2} = \dfrac{2}{(2.718)^2} = 2 \times (0.135) = 0.2706$

(iv) $P(X = 3) = \dfrac{2^3 e^{-2}}{3!} = \dfrac{4}{3} \times \dfrac{1}{e^2} = \dfrac{4}{3} \times \dfrac{1}{(2.718)^2} = (1.33) \times (0.135) = 0.18.$

Example 7: *A Poisson distribution has a double mode at x = 4 and x = 5. Find the probability that x will have either of these values P(x = 4) = P(x = 5).*

Solution: The Poisson distribution is

$$P(X = r) = \frac{m^r e^{-m}}{r!}, \quad r = 0, 1, 2, ...$$

Given that $P(x = 4) = P(x = 5)$

$$\Rightarrow \quad \frac{m^4 e^{-m}}{4!} = \frac{m^5 e^{-m}}{5!} \quad \Rightarrow \quad m = 5$$

We know that, $e^{-5} = 0.006738$

Then $P(x = 4) = \dfrac{5^4 \times 0.006738}{4!}$

The required probability is

$$P(x = 4) + P(x = 5) = 2P(x = 4)$$

$$= 2 \times \frac{5^4 \times 0.006738}{4!} = 0.35.$$

Hence, it is easy to show that $P(4) > P(3)$ and also greater than $P(6)$.

Example 8: *Fit a Poisson's distribution to the following data and calculate theoretical frequencies:*

Deaths	*0*	*1*	*2*	*3*	*4*
Frequencies	*122*	*60*	*15*	*2*	*1*

Solution: The Poisson distribution is

$$P(X = r) = \frac{m^r e^{-m}}{r!}, \qquad r = 0, 1, 2, 3, ... \qquad ...(1)$$

Here, $$\text{mean} = \frac{\Sigma f(x)}{\Sigma f} = \frac{0 \times 122 + 1 \times 60 + 2 \times 15 + 3 \times 2 + 4 \times 1}{122 + 60 + 15 + 2 + 1} = \frac{100}{200} = 0.5 = m$$

Now, $e^{-0.5} = 0.61$ (approximately)

By equation (1), the number of r deaths is given by

$$= 200 \times \frac{(0.5)^r e^{-0.5}}{r!}, \quad r = 0, 1, 2, 3, 4.$$

for $r = 0, 1, 2, 3, 4$ the theoretical frequencies are 121, 61, 15, 2 and 0.

Example 9: *If the variance of the Poisson distribution is 2, find the probabilities for r = 1, 2, 3,4 from the reccurrence relation of the Poisson distribution. Also find P(r ≥ 4).*

Solution: We know that for Poisson's distribution mean and variance both are equal.

i.e., mean = variance = 2.

Recurrence relation for Poisson distribution

$$P(r+1) = \frac{m}{r+1} P(r)$$

$$\Rightarrow \quad P(r+1) = \frac{2}{r+1} P(r) \qquad ...(1)$$

The Poisson's distribution is $P(r) = \dfrac{e^{-m} m^r}{r!}, \quad r = 0, 1, 2, 3, ...$

$$P(0) = e^{-m} = e^{-2} = 0.1353$$

By (1), if $r = 0$, $$P(1) = \frac{2}{0+1} P(0) = 2 \times (0.1353) = 0.2706$$

if $r = 1$, $$P(2) = \frac{2}{1+1} P(1) = 0.2706$$

if $r = 2$, $$P(3) = \frac{2}{2+1} P(2) = \frac{2}{3} \times (0.2706) = 0.1804$$

if $r = 3$, $$P(4) = \frac{2}{3+1} P(3) = \frac{1}{3} \times (0.1804) = 0.0902$$

Now, we have
$$\begin{aligned} P(r \geq 4) &= P(4) + P(5) + P(6) + ... \\ &= 1 - [P(0) + P(1) + P(2) + P(3)] \\ &= 1 - [0.1353 + 0.2706 + 0.2706 + 0.1804] \\ &= 1 - 0.8569 = 0.1431. \end{aligned}$$

Example 10: *If the probability that an individual suffers a bad reaction from injection of a given serum is 0.001. Determine the probability that out of 2000 individual (i) exactly 3, (ii) more than 2 individuals suffer from bad reaction, (iii) none.*

Solution: Let p be the probability that an individual suffers a bad reaction from injection of a given serum

i.e., $$p = 0.001$$

Since, p is very small so we used Poisson's distribution.

Given that $$n = 2000$$

we have $$np = 0.001 \times 2000 = 2$$

Let X be the number of individuals who suffer from bad reaction.

Then $$P(X = r) = \frac{m^r e^{-m}}{r!} = \frac{2^r e^{-2}}{r!}$$

(*i*) The probability for exactly 3

$$P(X = 3) = \frac{2^3 e^{-2}}{3!} = \frac{4}{3e^2} = \frac{4}{3 \times (2.718)^2} = 0.18$$

(*ii*) The probability for more than 2 individuals suffer from bad reaction

$$P(X > 2) = 1 - [P(X = 0) + P(X = 1) + P(X = 2)]$$

$$= 1 - \left[\frac{2^0 e^{-2}}{0!} + \frac{2^1 e^{-2}}{1!} + \frac{2^2 e^{-2}}{2!}\right]$$

$$= 1 - \left[\frac{1}{e^2} + \frac{2}{e^2} + \frac{2}{e^2}\right] = 1 - \frac{5}{e^2} = 1 - \frac{5}{(2.718)^2} = 0.323$$

(*iii*) The probability for none

$$P(X = 0) = \frac{2^0 e^{-2}}{0!} = e^{-2} = \frac{1}{e^2} = \frac{1}{(2.718)^2} = 0.135.$$

Example 11: *In a certain factory turning Lexi blades, there is a small chance, 0.002 for any blade to be defective. The blades are in packets of 10. Use Poisson distribution to calculate the approximate number of packet containing no defective, one defective and two defective blades respectively in a consignment of 10,000 packets.*

Solution: Let p be the probability that a blade is defective

i.e., $$p = 0.002$$

Since, p is very small so Poisson's distribution is used. Here, $n = 10$ then

$$m = np = 10 \times 0.002 = 0.02.$$

Let X denote the number of defective blades in a packet of 10. Then

$$P(X = r) = \frac{m^r e^{-m}}{r!} = \frac{(0.02)^r e^{-0.2}}{r!}, \quad r = 0, 1, 2, \ldots\ldots$$

$$e^{-0.02} = 1 - (0.2) + \frac{1}{2}(0.02)^2 + \ldots\ldots = 0.9802 \text{ (nearly)}$$

Now the number of packets containing no defective blade is

$$= 10{,}000 \times 0.9802 = 9802$$

The number of packets containing one defective blade

$$= 10{,}000 \times m\, e^{-m}$$
$$= 10{,}000 \times (0.2)\, e^{-0.2}$$
$$= 10{,}000 \times (0.2)\,(.9802)$$
$$= 196$$

The number of packets containing two defective blades

$$= 10{,}000 \times m^2\, e^{-m}$$
$$= 10{,}000 \times (0.2)^2\, e^{-0.2}$$
$$= 10{,}000 \times (0.2)^2\,(0.9802)$$
$$= 2 \text{ (nearly).}$$

Example 12: *If X and Y be independent Poisson variates, show that the conditional distribution of X, given X + Y = n, is binomial.*

Solution: Let X and Y be independent Poisson variates with parameters m_1 and m_2.

We have, $$P\left[X = \frac{r}{X+Y=n}\right] = \frac{P(X=r,\, X+Y=n)}{P(X+Y=n)}$$

$$= \frac{P(X=r)\times P(Y=n-r)}{P(X+Y=n)}$$

$$= \frac{\dfrac{m_1^{\,r}\, e^{-m_1}}{r\,!} \times \dfrac{m_2^{\,n-r}\, e^{-m_2}}{(n-r)\,!}}{\dfrac{(m_1+m_2)^n\, e^{-(m_1+m_2)}}{n\,!}} = \frac{n\,!}{r\,!\,(n-r)\,!} \times \frac{m_1^{\,r}\, m_2^{\,n-r}}{(m_1+m_2)^n}$$

Letting $$p = \frac{m_1}{m_1+m_2} \text{ and } q = \frac{m_2}{m_1+m_2} \text{ then}$$

$$= {}^nC_r\, p^r\, q^{n-r} \quad (0 \le r \le n)$$

Hence, the conditional distribution of X, given $X + Y$, is binomial.

EXERCISE 2.2

1. If X be a Poisson variate such that $3P(X = 3) = 4P(X = 4)$, find $P(X = 7)$.
2. Criticise the following statement:

   The mean of a Poisson distribution is 7, while the standard deviation is 6.
3. The probability that a man aged 50 years will die within a year is 0.01125. What is the probability that of 12 such men at least 11 will reach their fifty first birthday ?

4. A car-hire-firm has two cars, which if hires, out day by day. The number of demands for a car on each day is distributed as Poisson distribution with mean 1.5. Calculate the proportion of days on which neither car is used and the proportion of days on which some demand is refused $\{e^{-1.5} = 0.2231\}$.

5. If m is the parameter of a Poisson variate, show that the probabilities that the value of the variate taken at random is even or odd are $e^{-m} \cosh m$ and $e^{-m} \sinh m$.

6. Letters were received in an office on each of 100 days. Assuming the following data to form a random sample from a Poisson's distribution. Fit the distribution and calculate the expected frequencies taking $e^{-4} = 0.183$.

Number of letters	0	1	2	3	4	5	6	7	8	9	10
Frequencies	1	4	15	22	21	20	8	6	2	0	1

7. The frequency of accidents per shift in a factory is given in the following table. Calculate the mean number of accidents per shift. Find the corresponding Poisson distribution and compare with actual observations.

Accidents per shift	0	1	2	3	4	Total
Frequencies	192	100	24	3	1	320

8. A manufacturer of coffer pins knows that 5 per cent of his product is defective. If he sells coffer pins in boxes of 100 and guarantees that not more than 4 pins will be defective, what is the approximate probability that a box will fail to meet the guaranteed quality ? $(e^{-5} = 0.0067)$.

9. A telephone switch board handles 600 calls on the average during a rush hour. The board can make a maximum of 20 connections per minute. Use Poisson distribution to estimate the probability that the board will be over during any given minute $[e^{-10} = 0.00004539]$.

10. Suppose that a book of 600 pages contains 40 printing mistakes. Assume that these errors are randomly distributed throughout the book and x, the number of errors per page has a Poisson distribution. What is the probability that 10 pages selected at random will be free of errors ?

11. Find the probability that at most defective fuses will be found in a box of 200 fuses, if experience show that 2 per cent of such fuses are defective.

12. An insurance company found that only 0.01% of the population is involved in a certain type of accident each year. If its 1000 policy holders were randomly selected from the population. What is the probability that not more than two of its clients are involved in such an accident next year ? $\{e^{-0.1} = 0.9048\}$.

13. Red blood cell deficiency may be determined by examining a specimen of the blood under a microscope. Suppose a certain small fixed volume contains on the average 20 red cells for normal persons. Using Poisson distribution, obtain the probability that a specimen from a normal person will contain less than 15 red cells.

14. Show how the Poisson distribution

$$\frac{m^r e^{-m}}{r!}, \quad (r = 0, 1, 2, 3, \ldots\ldots)$$

can be regarded as the limiting case of the binomial distribution. Hence or otherwise obtain the mean and the variance of the Poisson distribution, assuming the variance of the binomial distribution.

ANSWERS

1. $\frac{3^7 e^{-3}}{7!}$.
2. Wrong, because $\sigma = \sqrt{m}$.
3. 0.9916.

4. 0.1912625.

6. 1.83, 7.32, 14.64, 9.22, 19.52, 15.62, 10.41, 5.95, 2.975, 1.322 and 5.99.

7. 194, 97, 24, 4, 1.
8. 0.5620.
9. $1 - 0.00004539 \sum_{r=0}^{20} \frac{10^r}{r!}$.

10. 0.51.
11. 0.785.
12. 0.9998.

13. $\sum_{r=0}^{14} \frac{e^{-20}(20)^r}{r!}$.

2.8. NORMAL DISTRIBUTION

Normal distribution was discovered by a English mathematician De-Moivre in 1733. Normal distribution is a continuous probability distribution. Normal distribution has got wide application in the theory of statistics. The normal distribution also known as error function. Normal distribution as the limiting case of binomial distribution $(q + p)^n$ as $n \to \infty$, neither p nor q being very small, it is given by

$$N(m, \sigma^2) = f(x) = y(x) = \frac{1}{\sigma\sqrt{2\pi}} e^{-\frac{1}{2}(x-m)^2/\sigma^2}; \quad -\infty < x < \infty, -\infty < m < \infty, \quad \sigma > 0 \qquad ...(1)$$

where m = Arithmetic mean (μ)

σ = Standard deviation are two parameters of the continuous distribution. Equation (1) is also called the normal curve. It is also denoted by $X \sim N(m, \sigma^2)$.

2.9. CONSTANTS OF NORMAL DISTRIBUTION

About the origin, we have

$$\mu'_n = \int_{-\infty}^{\infty} \frac{1}{\sqrt{2\pi}\,\sigma} e^{-x^2/2\sigma^2} x^n \, dx \qquad ...(1)$$

$$= \frac{1}{\sqrt{2\pi}} \int_{-\infty}^{\infty} \frac{x^n e^{-x^2/2\sigma^2}}{\sigma} dx$$

Put $\frac{x}{\sigma} = t \Rightarrow \frac{dx}{\sigma} = dt = \frac{1}{\sqrt{2\pi}} \int_{-\infty}^{\infty} \sigma^n t^n e^{-t^2/2} dt$

$$= \frac{\sigma^n}{\sqrt{2\pi}} \int_{-\infty}^{\infty} t^n e^{-\frac{1}{2}t^2} dt$$

$$= 0, \text{ when } n \text{ is odd} \quad \{\text{property of definite integrals}\}$$

Hence, all odd moment about the origin is vanish.

i.e., $$\mu'_1 = \mu'_3 = \mu'_5 = ... = 0$$

If n is even, by equation (1), we have

$$\mu'_n = \frac{2}{\sqrt{2\pi}} \frac{1}{\sigma} \int_0^{\infty} x^n e^{-x^2/2\sigma^2} dx$$

Putting, $\frac{x^2}{2\sigma^2} = u \Rightarrow \frac{x}{\sigma^2} dx = du$

$$= \frac{2}{\sqrt{2\pi}\,\sigma} \int_0^{\infty} (2u\sigma^2)^{(n-1)/2} e^{-u} \sigma^2 du$$

$$= \frac{\sigma^n 2^{n/2}}{\sqrt{\pi}} \int_0^{\infty} u^{(n-1)/2} e^{-u} du$$

$$= \frac{\sigma^n 2^{n/2}}{\sqrt{\pi}} \Gamma\left(\frac{n+1}{2}\right) \text{d} \qquad \left[\because \Gamma n = \int_0^{\infty} x^{n-1} e^{-x} dx\right]$$

When $n = 2$

$$\mu'_2 = \frac{\sigma^2 2^{2/2}}{\sqrt{\pi}} \Gamma 3/2 = \frac{\sigma^2 . 2}{\sqrt{\pi}} . \frac{1}{2} \sqrt{\pi} = \sigma^2$$

When $n = 4$

$$\mu'_4 = \frac{\sigma^2 2^2}{\sqrt{\pi}} . \Gamma \frac{5}{2} = \frac{\sigma^4 . 4}{\sqrt{\pi}} . \frac{3}{2} . \frac{1}{2} \sqrt{\pi} = 3\sigma^4$$

When $n = 0$

$$\mu'_0 = \frac{\sigma^0 2^0}{\sqrt{\pi}} \Gamma \frac{1}{2} = \frac{1.1}{\sqrt{\pi}} \sqrt{\pi} = 1$$

Here mean is at the origin itself, these are also undashed

μ'_s *i.e.,* $\mu_2 = \mu'_2$, $\mu_3 = \mu'_3$ etc.

Now, $$\beta_1 = \frac{\mu_3^{\,2}}{\mu_2^{\,3}} = 0\,; \quad \beta_2 = \frac{\mu_4}{\mu_2^{\,2}} = \frac{3\sigma^4}{\sigma^4} = 3$$

$$\gamma_1 = \sqrt{\beta_1} = 0\,; \quad \gamma_2 = 0$$

Hence, the normal curve has zero kurtosis.

2.10. MOMENT ABOUT THE MEAN M

$$\mu_{2n+1} = \int_{-\infty}^{\infty} (x-m)^{2n+1} \frac{1}{\sqrt{2\pi}\,\sigma} e^{-\frac{1}{2}\left(\frac{x-m}{\sigma}\right)^2} dx$$

$$= \frac{\sigma^{2n+1}}{\sqrt{2\pi}} \int_{-\infty}^{\infty} \left(\frac{x-m}{\sigma}\right)^{2n+1} \frac{e^{-\frac{1}{2}\left(\frac{x-m}{\sigma}\right)^2}}{\sigma} dx$$

Putting $\frac{x-m}{\sigma} = t \Rightarrow \frac{dx}{\sigma} = dt$

$$= \frac{\sigma^{2n+1}}{\sqrt{2\pi}} = \int_{-\infty}^{\infty} t^{2n+1} e^{-\frac{1}{2}t^2} dt$$

= 0, since it is integral of an odd function of t.

Now we have $$\mu_{2n} = \int_{-\infty}^{\infty} (x-m)^{2n} \frac{1}{\sqrt{2\pi}\,\sigma} e^{-\frac{1}{2}\left(\frac{x-m}{\sigma}\right)^2} dx$$

$$= \frac{\sigma^{2n}}{\sqrt{2\pi}} \int_{-\infty}^{\infty} \left(\frac{x-m}{\sigma}\right)^{2n} \frac{e^{-\frac{1}{2}\left(\frac{x-m}{\sigma}\right)^2}}{\sigma} dx$$

Putting $\frac{x-m}{\sigma} = t \Rightarrow \frac{dx}{\sigma} = dt$

$$= \frac{\sigma^{2n}}{\sqrt{2\pi}} \int_{-\infty}^{\infty} t^{2n} e^{-\frac{1}{2}t^2} dt$$

$$= \frac{\sigma^{2n}}{\sqrt{2\pi}} \int_{-\infty}^{\infty} (2T)^n e^{-T} \frac{dT}{\sqrt{2T}}$$

Putting $t^2 = 2T, t = \sqrt{2T}$

$\Rightarrow \quad 2t\,dt = 2dT$

or $\quad dt = \frac{dT}{\sqrt{2T}}$

$$= \frac{2^{n-1}\sigma^{2n}}{\sqrt{\pi}} \int_{-\infty}^{\infty} T^n e^{-T} dT$$

$$= \frac{2^n \sigma^{2n}}{\sqrt{\pi}} \int_{0}^{\infty} T^n e^{-T} dT$$

$$\Rightarrow \quad \mu_{2n} = \frac{2^n \sigma^{2n}}{\sqrt{\pi}} \Gamma\left(n + \frac{1}{2}\right) \quad ...(1)$$

Putting $n = n - 1$ in (1), we get

$$\mu_{2n-2} = \frac{2^{n-1}\,\sigma^{2n-2}}{\sqrt{\pi}}\,\Gamma\left(n - \frac{1}{2}\right) \qquad \text{...(2)}$$

Now equation (1) divide by (2), we get

$$\frac{\mu_{2n}}{\mu_{2n-2}} = \frac{2\sigma^2\,\Gamma\left(n + \frac{1}{2}\right)}{\Gamma\left(n - \frac{1}{2}\right)} \qquad [\because \quad \Gamma n = (n-1)\,\Gamma(n-1)]$$

$$= 2\sigma^2\left(n - \frac{1}{2}\right)$$

$$\Rightarrow \qquad \boxed{\mu_{2n} = \sigma^2\,(2n-1)\,\mu_{2n-2}}$$

which is the recurrence relation for the moments of normal distribution.

2.11. AREA UNDER THE NORMAL CURVE

The probability density function of the normal distribution is given by

$$f(x) = \frac{1}{\sqrt{2\pi}\,\sigma}\,e^{\frac{-1}{2}\left(\frac{x-m}{\sigma}\right)^2}, \qquad -\infty < x < \infty.$$

If x is a normal random variable with mean m and standard deviation σ, then the random variable $z = \left(\frac{x-m}{\sigma}\right)$ has the normal distribution with mean 0 and the standard deviation 1. The random variable z is called the standard normal random variable.

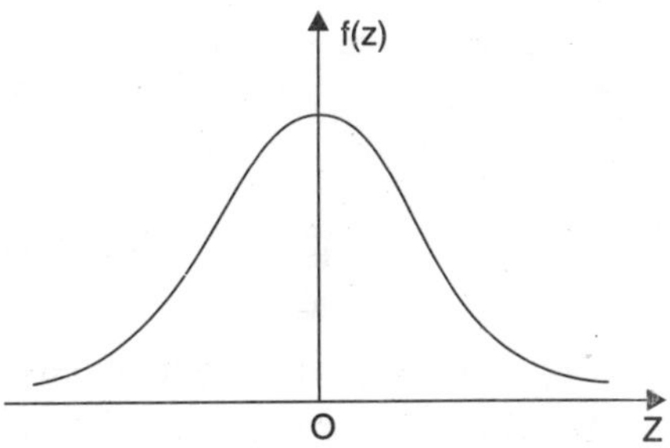

The probability density function of standard normal random variable z is given by

$$f(z) = \frac{1}{\sqrt{2\pi}}\,e^{-\frac{1}{2}z^2}, \qquad -\infty < z < \infty$$

Note 1. The graph of $f(z)$ is famous 'bell shaped' curve.

2. If $f(z)$ is the probability density function for the normal distribution, then

$$P(z_1 \le z \le z_2) = \int_{z_1}^{z_2} f(z)\,dz = \int_{z_1}^{z_2} \frac{1}{\sqrt{2\pi}}\,e^{-\frac{1}{2}z^2}\,dz$$

2.12. PROPERTIES OF THE NORMAL DISTRIBUTION AND NORMAL CURVE

The normal probability curve is

$$f(x) = \frac{1}{\sqrt{2\pi}\,\sigma} e^{-\frac{1}{2}\left(\frac{x-m}{\sigma}\right)^2}, \quad -\infty < x < \infty$$

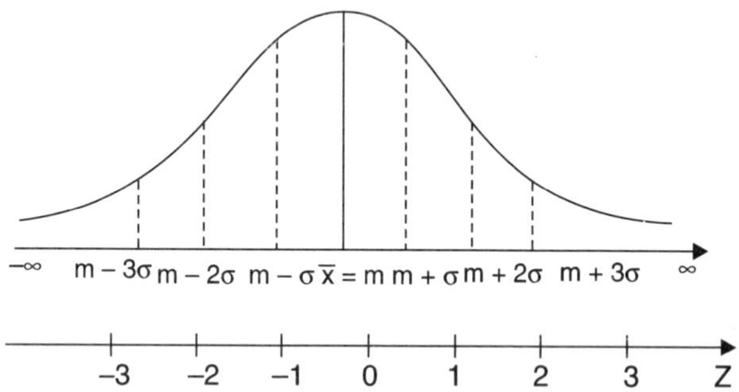

(1) Mean, Meadian and Mode of the Normal distribution coincide.

(2) Area under the normal curve is unity.

(3) The probability of the continuous random variable X $(X_1 \le X \le X_2)$ is denoted by

$$P(X_1 \le X \le X_2) = \int_{X_1}^{X_2} \frac{1}{\sqrt{2\pi}\,\sigma} e^{-\frac{1}{2}\left(\frac{x-m}{\sigma}\right)^2} dx$$

Where m is the mean and σ is standard deviation.

(4) The curve is bell-shaped and symmetry about the line $x = m$ *i.e.*, y-axis.

(5) As x increases then $f(x)$ decreases rapidly, the maximum probability at point $x = \mu$ is given by

$$f(x) \text{ or } [P(x)]_{max} = \frac{1}{\sqrt{2\pi}\,\sigma}$$

(6) Since, $f(x)$ is probability density function, so it never be negative, *i.e.*, no portion of the curve lies below the x-axis.

(7) x-axis is an asymptote of the curve.

(8) The point of inflexion of the normal curve are $x = \pm\,\sigma$ when the origin is taken at the mean.

(9) Mean deviation about mean $= \sqrt{\frac{2}{\pi}}\,\sigma \simeq \frac{4}{5}\,\sigma$ (approximate).

(10) Area under the normal curve is distributed as following 68.27% area lies between $m - \sigma$ to $m + \sigma$ or $(\bar{x} - \sigma$ to $\bar{x} + \sigma)$ *i.e.*, between $-1 \le z \le 1$

94.45% area lies between $m - 2\sigma$ to $m + 2\sigma$, *i.e.*, between $-2 \le z \le 2$

99.73% area lies between $m - 3\sigma$ to $m + 3\sigma$, *i.e.*, between $-3 \le z \le 3$

50% area lies in between $-0.745 \le z \le 0.745$

99% area lies in between $-2.58 \le z \le 2.58$.

SOLVED EXAMPLES

Example 1: *Prove that for normal distribution the mean deviation from the mean equal to $\frac{4}{5}$ of standard deviation nearly.*

Solution: Let m and σ be the mean and standard deviation of the normal distribution respectively.

By definition of mean deviation

$$= \int_{-\infty}^{\infty} |x-m| f(x)\, dx$$

$$= \frac{1}{\sqrt{2\pi}\,\sigma}\int_{-\infty}^{\infty} |x-m|\, e^{-\frac{1}{2}\left(\frac{x-m}{\sigma}\right)^2} dx$$

$$= \frac{1}{\sqrt{2\pi}\,\sigma}\int_{-\infty}^{\infty} \sigma |z| e^{-\frac{1}{2}z^2}\, \sigma\, dz \qquad \text{Putting } z = \frac{x-m}{\sigma} \Rightarrow dz = \frac{dx}{\sigma}$$

$$= \sigma\sqrt{\frac{2}{\pi}}\int_{0}^{\infty} z\, e^{-\frac{1}{2}z^2}\, dz$$

$$= \sigma\sqrt{\frac{2}{\pi}}\left[-e^{-\frac{z^2}{2}}\right]_0^{\infty}$$

$$= \sigma\sqrt{\frac{2}{\pi}}\,.\,1 = 0.7979\,\sigma \approx 0.8\,\sigma$$

$$\because \frac{8}{10}\,\sigma = \frac{4}{5}\sigma.$$

Example 2: *If X is a normal variate with mean12 and standard deviation 4. Find*

(i) $P(X \geq 20)$ *(ii) $P(X \leq 20)$* *(iii) $P(0 \leq X \leq 12)$.*

Solution: Given that $m = 12$ and $\sigma = 4$

Let $$Z = \frac{X-m}{\sigma} = \frac{X-12}{4}$$

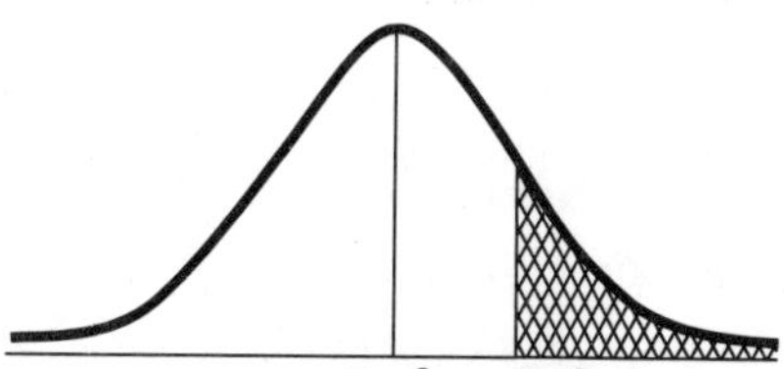

(*i*) When $X = 20$, $$Z = \frac{20-12}{4} = 2$$

$$P(X \geq 20) = P(Z \geq 2)$$

$$= 0.5 - P(0 \leq Z \leq 2)$$

$$= 0.5 - 0.4772 \qquad [\because P(0 \leq Z \leq 2) + P(Z \geq 2) = 0.5]$$

$$= 0.0228. \qquad \text{(Using table)}$$

(ii) $$P(X \le 20) = P(Z \le 2)$$
$$= 0.5 + P(0 \le Z \le 2)$$
$$= 0.5 + 0.4772$$
$$= 0.9772$$

(iii) When $X = 0, Z = \frac{-12}{4} = -3$ and for $X = 12, Z = 0$

$\therefore$ $$P(0 \le X \le 12) = P(-3 \le Z \le 0) = P(0 \le Z \le 3) = 0.4987.$$

Example 3: *Assume the mean height of soldiers to be 68.22 inches with a variance of 10.8 (inches)^2. How many soldiers in a regiment of 1000 would you expect to be over 6 feet tall? Given that the area under the standard normal curve between Z = 0 and Z = 0.35 is 0.1368; and between Z = 0 and Z = 1.15 is 0.3746.*

Solution: Given that mean $(m) = 68.22$, variance $(\sigma^2) = 10.8$

and standard deviation $(\sigma) = \sqrt{10.8}$

We have $$Z = \frac{X - m}{\sigma}$$

$\Rightarrow$ $$Z = \frac{X - 68.22}{\sqrt{10.8}}$$

$\Rightarrow$ $$Z = \frac{X - 68.22}{3.286}$$

When $X = 72$, $$Z = \frac{72 - 68.22}{3.286} = 1.15$$

Now $$P(X \ge 72) = P(Z \ge 1.15) = 0.5 - 0.3749 = 0.1251$$

The probability of getting a soldier above six feet is 0.1251. Hence, the number of soldiers who are over 6 feet tall.

$$= 1000 \times 0.1251 = 125.1$$
$$= 125 \text{ (nearly)}.$$

Example 4: *Assuming the resistance of the resistors to be normally distributed with mean 100 ohms and standard deviation 2 ohms, what percentage of resistors will have resistance between 98 ohms to 102 ohms.*

Solution: Given that mean $(m) = 100$ ohms

standard deviation $(\sigma) = 2$ ohms

We have $$Z = \frac{X - m}{\sigma}$$

$$\Rightarrow \qquad Z = \frac{X - 100}{2}$$

When $X = 102$, $\quad Z = \frac{102 - 100}{2} = 1$

When $X = 98$, $\quad Z = \frac{98 - 100}{2} = -1$

Now, $\quad P(98 \le X \le 102) = P(-1 \le Z \le 1)$

$$= 2P(0 \le Z \le 1)$$

$$= 2 \times (0.3413) = 0.6826$$

Hence, 68.26% of resistors will have resistance between 98 ohms to 102 ohms.

Example 5: *In a Normal distribution 31% items are under 45 and 8% are over 64. Find the mean and standard deviation of the distribution.*

Solution: Let m and σ be the mean and standard deviation respectively.

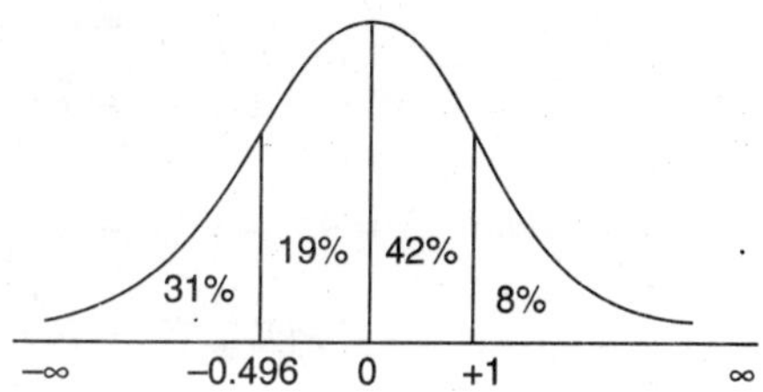

If $x = 45$, $\quad Z = \frac{45 - m}{\sigma}$

If $x = 64$, $\quad Z = \frac{64 - m}{\sigma}$

The area between 0 and $\frac{45 - m}{\sigma} = 0.50 - 0.31 = 0.19$

From the table, for the area 0.19,

$$Z = 0.496$$

$$\frac{45 - m}{\sigma} = -0.496 \qquad ...(1)$$

Area between Z = 0 and $Z = \frac{64 - m}{\sigma} = 0.5 - 0.08 = 0.42$

From the table, for the area 0.42, $Z = 1.405$

$$\frac{64 - m}{\sigma} = 1.405 \qquad ...(2)$$

Solving equation (1) and (2), we get

$$m = 50 \quad \text{and} \quad \sigma = 10.$$

Example 6: *The distribution of weekly wages for 500 workers in a factory is approximately normal with the mean and students derivation of Rs. 75 and Rs. 15. Find the number of workers who receive weekly wages:*

(i) more than Rs. 90 *(ii) less than Rs. 45.*

Solution: The normal distribution is

$$f(x) = \frac{1}{\sqrt{2\pi}\,\sigma} e^{-\frac{1}{2}\left(\frac{x-m}{\sigma}\right)^2}, \quad -\infty < x < \infty \qquad \text{...(1)}$$

Given that $N = 500$, $m = 75$ and $\sigma = 15$.

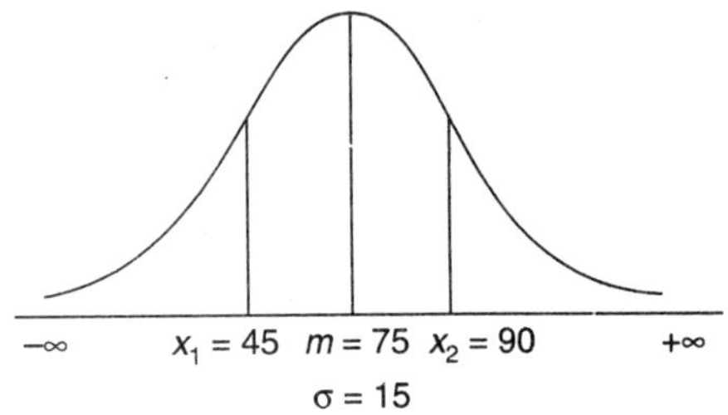

we have $$Z = \frac{X-m}{\sigma} = \frac{X-75}{15}$$

(*i*) when $X = 90$, $$Z = \frac{90-75}{15} = 1$$

$$P(X > 90) = P(Z > 1) = 0.5 - P(0 < Z < 1)$$
$$= 0.5 - 0.3413 = 0.1587$$

The number of workers who receive weekly wages more than Rs. 90

$$= 0.1587 \times 500$$
$$= 79.35 = 79 \text{ (nearly) workers.}$$

(*ii*) When $X = 45$, $$Z = \frac{45-75}{15} = -2$$

$$P(X < 45) = P(Z < -2) = 0.5 - P(-2 < Z < 0)$$
$$= 0.5 - P(0 < Z < 2)$$
$$= 0.5 - 0.4772 = 0.0228$$

Hence, the number of workers who receive weekly wages more than Rs. 45

$$= 0.0228 \times 500 = 11.4 \quad \textit{i.e.}, 11 \text{ (nearly) workers.}$$

Example 7: *The life of army shoes is 'normally' distributed with mean 8 months and standard deviation 2 months. If 5000 pairs are issued how many pair would be expected to need replacement after12 months?* $\left\{\textit{Given that} \quad P(Z \geq 2) = 0.0288 \textit{ and } Z = \left(\frac{x-m}{\sigma}\right)\right\}$

Solution: Given that

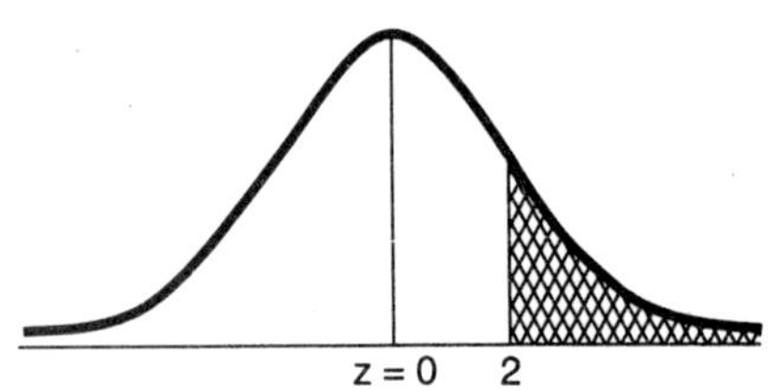

Mean $(m) = 8$,

Standard deviation $(\sigma) = 2$

Number of pairs of shoes $= 5000$

Total month $(x) = 12$

We have $$Z = \frac{x-m}{\sigma} = \frac{12-8}{2} = 2$$

Area $(Z \geq 2) = 0.0228$

The number of pairs whose life is more than 12 months $(Z > 2)$

$= 5000 \times 0.0228 = 114$

Replacement after 12 months

$= 5000 - 114 = 4886$ pairs of shoes.

Example 8: *If the height of 300 students are normally distributed with mean 64.5 inches and standard deviation 3.3 inches, how many students have height*

(i) less than 5 feet

(ii) between 5 feet and 5 feet 9 inches.

Also, find the height below which 99% of the students lie.

Solution: Given that $m = 64.5$, $\sigma = 3.3$ and $N = 300$.

We have
$$Z = \frac{X - m}{\sigma} = \frac{X - 64.5}{3.3}$$

(*i*) For $X = 60$,
$$Z = \frac{60 - 64.5}{3.3} = -\frac{4.5}{3.3} = -1.36$$

$$P(X < 60) = P(Z < -1.36) = 0.5 - (-1.36 < Z < 0)$$
$$= 0.5 - (0 < Z < 1.36)$$
$$= 0.5 - 0.4131 \qquad \text{(Using table)}$$
$$= 0.0869.$$

The number of student having height less than 5 feet

$= 0.0869 \times 300$

$= 26.07$ *i.e.*, 26 students.

(*ii*)
$$P(60 < X < 69) = P(-1.36 < Z < 1.36)$$
$$= 2P(0 < Z < 1.36) = 2(0.4131) = 0.8262$$

The number of student having heights lie between 60 inches and 69 inches $= 0.8262 \times 300 = 247.86$ *i.e.*, 248 students.

Now let $P(X < X_1) = 0.99$

$\Rightarrow \quad P(Z < Z_1) = 0.99$

where
$$Z_1 = \frac{X_1 - m}{\sigma}$$

$\Rightarrow \quad P(0 < Z < Z_1) = 0.99 - 0.5 = 0.49$

Using the table giving areas under normal curve $Z_1 = 2.33$

$$\therefore \quad Z_1 = \frac{X_1 - m}{\sigma} = \frac{X_1 - 64.5}{3.3}$$

$\Rightarrow \quad X_1 = 64.5 + 3.3 \times 2.33$

$\Rightarrow \quad X_1 = 72.18$

Hence, 99% of the students have height below 72.18 inches.

Example 9: *A coin is tossed 12 times. Find the probability, both exactly and by fitting a normal distribution of getting*

(*i*) *4 heads,* (*ii*) *atmost 4 heads.*

Solution: The binomial distribution is

$$P(X = r) = {}^nC_r\, p^r\, q^{n-r},\ r = 0, 1, 2, \ldots\ldots \qquad \ldots(1)$$

Here, $p = \frac{1}{2}, q = 1 - \frac{1}{2} = \frac{1}{12}$ and $n = 12$. Let X denote the probability of head occurrence.

The probability of 4 heads

$$P(X = 4) = {}^{12}C_4 \left(\frac{1}{2}\right)^4 \left(\frac{1}{2}\right)^8 = \frac{(12)!}{4! \times 8! \times 2^{12}} = \frac{11 \times 5 \times 9}{2^{12}} = 0.121$$

Now, $P(X \leq 4) = P(X = 0) + P(X = 1) + P(X = 2) + P(X = 3) + P(X = 4)$

$$= {}^{12}C_0 \left(\frac{1}{2}\right)^0 \left(\frac{1}{2}\right)^{12} + {}^{12}C_1 \left(\frac{1}{2}\right)^1 \left(\frac{1}{2}\right)^{11} + {}^{12}C_2 \left(\frac{1}{2}\right)^2 \left(\frac{1}{2}\right)^{10} + {}^{12}C_3 \left(\frac{1}{2}\right)^3 \left(\frac{1}{2}\right)^9 + {}^{12}C_4 \left(\frac{1}{2}\right)^4 \left(\frac{1}{2}\right)^8$$

$$= \frac{1}{2^{12}} [{}^{12}C_0 + {}^{12}C_1 + {}^{12}C_2 + {}^{12}C_3 + {}^{12}C_4]$$

$$= \frac{1}{2^{12}} [1 + 12 + 66 + 220 + 495]$$

$$= \frac{794}{2^{12}} = 0.194$$

The normal distribution is

$$N(m, \sigma^2) = \frac{1}{\sqrt{2\pi}\,\sigma} e^{-\frac{1}{2}\left(\frac{x-m}{\sigma}\right)^2}, \quad -\infty < x < \infty,\ -\infty < m < \infty,\ \sigma > 0 \qquad \ldots(2)$$

$$\text{Mean } (m) = np = 12 \times \frac{1}{2} = 6$$

$$\text{Variance } (\sigma^2) = npq = 12 \times \frac{1}{2} \times \frac{1}{2} = 3$$

For $n \leq 4$

$$Z = \frac{X - m}{\sigma} = \frac{4.5 - 6}{\sqrt{3}} = \frac{-1.5}{\sqrt{3}} = -0.5 \times \sqrt{3} = -0.5 \times 1.732 = -0.866$$

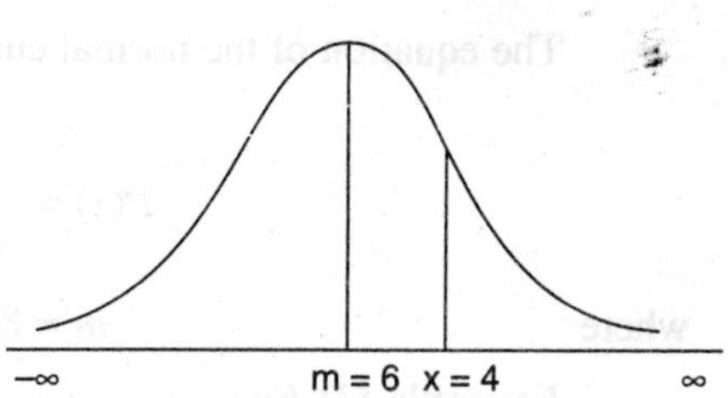

The probability

$$P(-\infty < Z < -0.866) = 0.5 - P(0 < Z < 0.866)$$

Probability number of heads $\leq 4 = 0.5 - 0.3068$

$$= 0.1932$$

Probability of 4 heads by normal distribution

$$= P(3.5 < x < 4.5)$$

$$= P\left(\frac{3.5-6}{\sqrt{3}} < Z < \frac{4.5-6}{\sqrt{3}}\right)$$

$$= P(-1.231 < Z < -0.7385)$$

$$= P(0 < Z < 1.231) - P(0 < Z < 0.7385)$$

$$= 0.3907 - 0.2700 = 0.120.$$

Example 10: *Fit a normal curve to the following data:*

Length of the line (in Cms.)	*8.60*	*8.59*	*8.58*	*8.57*	*8.56*	*8.55*	*8.54*	*8.53*	*8.52*
Frequencies	*2*	*3*	*4*	*9*	*10*	*8*	*4*	*1*	*1*

Solution: Here $a = 8.56$

x	f	$d = (x - a)$	fd	fd^2
8.60	2	0.04	0.08	0.0032
8.59	3	0.03	0.09	0.0027
8.58	4	0.02	0.08	0.0016
8.57	9	0.01	0.09	0.0009
8.56	10	0	0	0
8.55	8	– 0.01	– 0.08	0.0008
8.54	4	– 0.02	– 0.08	0.0016
8.53	1	– 0.03	– 0.03	0.0009
8.52	1	– 0.04	– 0.04	0.0016
	$\Sigma f = 42$		$\Sigma fd = 0.11$	$\Sigma fd^2 = 0.0133$

$$\text{Mean } (m) = a + \frac{\Sigma fd}{\Sigma f} = 8.56 + \frac{0.11}{42} = 8.56262$$

$$\text{Standard deviation } (\sigma) = \sqrt{\frac{\Sigma fd^2}{\Sigma f} - \left(\frac{\Sigma fd}{\Sigma f}\right)^2} = \sqrt{\frac{0.0133}{42} - \left(\frac{0.11}{42}\right)^2} = 0.0176.$$

The equation of the normal curve fitted to the given data is

$$P(x) = \frac{1}{\sqrt{2\pi}\,\sigma} e^{-\frac{1}{2}\left(\frac{x-m}{\sigma}\right)^2}, \quad -\infty \le x \le \infty$$

where $m = 8.56262, \quad \sigma = 0.0176.$

Example 11: *For* $-\infty < x < \infty$, *and probability density*

$$f(x) = \frac{1}{\sqrt{2\pi}\,\sigma} e^{-\frac{1}{2}\left(\frac{x-m}{\sigma}\right)^2}$$

Show that the total probability is 1.

Solution: The total probability

$$= \int_{-\infty}^{\infty} f(x)\,dx = \frac{1}{\sqrt{2\pi}\,\sigma}\int_{-\infty}^{\infty} e^{-\frac{1}{2}\left(\frac{x-m}{\sigma}\right)^2} dx$$

Let $\frac{x-m}{\sqrt{2}\,\sigma} = t$

$\Rightarrow \quad dx = \sqrt{2}\,\sigma\, dt$

$$= \frac{\sigma\sqrt{2}}{\sqrt{2\pi}\,\sigma}\int_{-\infty}^{\infty} e^{-t^2}\, dt$$

$$= \frac{1}{\sqrt{\pi}}\int_{-\infty}^{\infty} e^{-t^2}\, dt$$

$$= \frac{2}{\sqrt{\pi}}\int_{0}^{\infty} e^{-t^2}\, dt \qquad \{\text{As } f(-t) = f(t)\}$$

Now let $\quad t^2 = y \quad \text{or} \quad t = \sqrt{y}$

$\Rightarrow \quad 2t\,dt = dy$

$$= \frac{2}{\sqrt{\pi}}\int_{0}^{\infty} e^{-y}\,\frac{dy}{2\sqrt{y}}$$

$$= \frac{1}{\sqrt{\pi}}\int_{0}^{\infty} y^{-1/2}\, e^{-y}\, dy$$

$$= \frac{1}{\sqrt{\pi}}\int_{0}^{\infty} y^{\frac{1}{2}-1}\, e^{-y}\, dy \qquad \left[\text{By Gamma integral } \int_0^{\infty} e^{-x}\, x^{n-1}\, dx = \Gamma n\right]$$

$$= \frac{1}{\sqrt{\pi}}\Gamma(1/2) = \frac{1}{\sqrt{\pi}} \times \sqrt{\pi} = 1$$

EXERCISE 2.3

1. Let X be a normal variate with mean 30 and standard deviation 5, find the probabilities that
(*i*) $26 \le X \le 40$, (*ii*) $X \ge 45$ and (*iii*) $|X - 30| > 5$
2. Students of a class were given anaptitute test. Their marks were found to be normally distributed with mean 60 and standard deviation 5. What percentage of students scored more than 60 marks?
3. 1000 light bulbs with a mean life of 120 days are installed in a new factory, their length of life is normally distributed with standard deviation 20 days. How many bulbs will expire in less than 90 days?
4. On a final examination in mathematics, the mean was 72, and the standard deviation was 15. Determine the standard scores of students receiving grades.
(*i*) 60 (*ii*) 72 (*iii*) 93.
5. In a male population of 1000, the mean height is 68.16 inches and standard deviation is 3.2 inches. How many men may be more than 6 feet (72 inches)?
6. For a certain normal distribution the first moment about 10 is 40 and the fourth moment about 50 is 48. What is the arithmetic mean and variance of the normal distribution.
7. In a normal distribution whose mean is 2 and standard deviation 3, find a value of the variate such that the probability of the interval from the mean to the value is 0.4115. Find another value such that the probability for the interval from $x = 3.5$ to that value is 0.2307.
8. A random variable x is normally distributed with $m = 12$ and standard deviation 2. Find $P(9.6 < x < 13.8)$ given that for $\frac{x}{\sigma} = 0.9, A = 0.3159$ and for $\frac{x}{\sigma} = 1.2, A = 0.3849$.
9. There are six hundred students in P.G. class and the probability for any student to need a particular book on statistics from the University Library on one day is 0.05. How many copies of the book should be kept in the University Library, so that the probability may be greater than 0.90 that none of the student needing a copy from the Library has to come back disappointed.
10. If shells are classified as A, B, C according as the length breadth index as under 75, between 75 and 80, or over 80, find approximately (assuming that distribution is normal) the mean and standard deviation of a series in which A are 58%, B are 38%, and C are 4% being given that

$$f(t) = \frac{1}{\sqrt{2\pi}} \int_0^t e^{-\frac{1}{2}x^2}\, dx \quad \text{then} \quad f(0.20) = 0.08 \quad \text{and} \quad f(1.75) = 0.46.$$

11. Prove that for the normal distribution, the quartile deviation, the mean deviation and the standard deviation are approximately in the ratio 10 : 12 : 15.

ANSWERS

1. (*i*) 0.7653 (*ii*) 0.0014 (*iii*) 0.3174. **2.** 50%. **3.** 67.
4. (*i*) – 0.8 (*ii*) 0 (*iii*) 1.4. **5.** 115. **6.** 6.05, 6.26.
7. 0.7008. **8.** 37. **9.** 74.3, 3.23.

CHAPTER 3

Correlation and Regression

Modern age is an age of planning. Economic planning means direction and control of economic resources both current and potential to meet the social and economic objectives of the state. In absence of statistics planning cannot be imagined. All economic plans of a country are based on statistical data of economic activity of that country. Statistics is the branch of scientific method which deals with data obtained by counting or measuring the properties of population of natural phenomena.

Statistics is a branch of science which deals with the collection of data. Classification and tabulation of numerical facts as the basis for explanation, description and comparison of phenomena. Modern statistics takes into consideration however, biological, astronomical and physical as well as social phenomena. Statistics methods are used in every department of human activity. Psychology, education, public health, agriculture, business, economics and administration.

Statistics methods are also useful to stock-exchange brokers, bankers, speculators and public utility. Concerns like water works, electric supply companies, railways etc. Statistics analysis of railway working is very important in their expansion programmes. The records on the basis of demand in several parts of the year, railways authorities make adequate provision for services by increasing the length of trains providing special trains such as Jammu, Banglore, Goa, Mumbai etc. in summer. Therefore, in every business (big or small) used of statistical data in any form.

3.1. FREQUENCY DISTRIBUTION

Frequency distribution is a way to represent the large amount of data according to quantitative characteristics frequency distribution helps the management of the bank for announcing new saving policies for the benefit to its customers.

3.2. BIVARIATE FREQUENCY DISTRIBUTION

The study of relationship between two or more variables is one of the most important problems in life. In a bivariate distribution, we find the relationship between the two variables. For example, if we find the heights and weights of 10th class of student, we shall get a bivariate distribution. Here, one variable is related to height and other variable is related to weight.

Let the pairs $(x_1, y_1), (x_2, y_2), \ldots\ldots, (x_n, y_n)$ of two variable x and y with frequencies $f_1, f_2, \ldots\ldots, f_n$, $\sum_{i}^{n} f_i = N$, then it define a frequency distribution which is called a bivariate frequency distribution.

3.3. CORRELATION

The relationship between two variables such that a change in one is accompanied by change in the other in such a way that an increase in one is accompanied by an increase or decrease in the other, is called a correlation. For example, the number of policemen and the number of crimes, the number of trains and the number of passengers.

If there is no relationship indicated between two variables then they are said to uncorrelated or statistically independent. For example, marks in statistics of a child and weights of her mother.

It is not necessary that one unit of change in one variable is followed by one unit of change in other variable. It is also not essential that the ratio should be the same in all cases.

3.4. POSITIVE CORRELATION

If an increase or decrease in the values of one variable corresponds to an increase or decrease in other variable respectively, then the correlation is said to be positive correlation. For example, age of husbands and wives are known to have a positive correlation.

If an increase or decrease in the values of one variable is always followed by a corresponding and proportional increase or decrease in other variable, then the correlation is said to be perfect positive correlation.

3.5. NEGATIVE CORRELATION

If an increase or decrease in the values of one variable corresponds to a decrease or increase in other variable respectively, then the correlation is said to be negative correlation. For example, the number of fourth class workers in a college and the number of duty room in a college.

If an increase or decrease in the values of one variable is always followed by a corresponding and proportional decrease or increase in other variable, then the correlation is said to be perfect positive correlation.

3.6. LINEAR OR NON-LINEAR CORRELATION

When the variations in the values of two variables are in a constant ratio, then correlation is said to be linear correlation. This type of relationship represented by a linear equation of the form $Y = aX + b$.

Otherwise, if the ratio of variation in the values of two variables are not constant, then correlation is said to be non-linear correlation. This type of relationship represented by $y = a + bx + cx^2$ etc.

Note: If the change in one variable has a very little effect in the other variable then there is no correlation.

3.7. COEFFICIENT OF CORRELATION

The coefficient of correlation measures the degree of relationship between two or more variable. The coefficient of correlation varies between –1 to +1. If the coefficient of correlation reaches unity then it is perfect. The +1 indicates perfect positive correlation and –1 indicates perfect negative correlation. Whenever the value of coefficient of correlation is zero then there is no correlation.

3.8. MEASUREMENT OF CORRELATION

Correlation in two or more variables can be studied by the following methods :

(*i*) Karl Pearson's coefficients of correlation.

(*ii*) Rank correlation or Spearman's coefficient of rank correlation.

(*iii*) Scatter diagram or dot diagram.

(*iv*) Correlation table.

(*v*) Graphic method.

(*vi*) Concurrent deviations method.

In this chapter, we shall discuss Karl Pearson's coefficient of correlation, Rank correlation and Scatter diagram method only.

3.9. KARL PEARSON'S COEFFICIENTS OF CORRELATION

Karl Pearson (1867–1936), a great statistician and biologist, developed a formula for the calculation of coefficient of correlation. The correlation coefficient between two variables X and Y denoted by γ, defined as

$$\gamma(X, Y) = \frac{\text{Cov}(X, Y)}{\sigma_X \sigma_Y} \qquad ...(1)$$

If (x_i, y_i) ; $i = 1, 2, 3, ... n$ be the set of n pairs of values of variable X and Y then

$$\text{Cov}(X, Y) = E[(X - \overline{X})(Y - \overline{Y})]$$

$$= \frac{1}{n} \sum_{i=1}^{n} (x_i - \overline{X})(y_i - \overline{Y})$$

where, $$\overline{X} = \frac{1}{n}\sum_{i=1}^{n} X_i,\quad \overline{Y} = \frac{1}{n}\sum_{i=1}^{n} Y_i$$

$$\sigma_X^2 = E[X - E(X)]^2 = E[X - \overline{X}]^2$$

$$= \frac{1}{n}\sum_{i=1}^{n} (x_i - \overline{X})^2$$

$$\sigma_Y^2 = E[Y - E(Y)]^2 = E[Y - \overline{Y}]^2$$

$$= \frac{1}{n}\sum_{i=1}^{n} (y_i - \overline{Y})^2$$

Putting these values of Cov (X, Y), σ_X and σ_Y in (1), we get

$$\gamma(X, Y) = \frac{\frac{1}{n}\sum_{i=1}^{n} (x_i - \overline{X})(y_i - \overline{Y})}{\left[\frac{1}{n}\sum_{i=1}^{n} (x_i - \overline{X})^2\right]^{1/2}\left[\frac{1}{n}\sum_{i=1}^{n} (y_i - \overline{Y})^2\right]^{1/2}} \quad \text{...(2)}$$

$$\Rightarrow \quad \gamma(X, Y) = \frac{\sum_{i=1}^{n} (x_i - \overline{X})(y_i - \overline{Y})}{\left[\sum_{i=1}^{n} (x_i - \overline{X})^2\right]^{1/2}\left[\sum_{i=1}^{n} (y_i - \overline{Y})^2\right]^{1/2}} \quad \text{...(3)}$$

where $\overline{X}$ and $\overline{Y}$ are the means of X and Y respectively.

Note. 1. Cov $(X, Y) = \sigma_{XY}$ (also write)

2. Karl Pearson's coefficient of correlation is also called product moment correlation coefficient, because

$$\text{Cov}(X, Y) = E[\{X - E(X)\}\ \{Y - E(Y)\}]$$

3. Simplifying equation (2), we have

$$\frac{1}{n}\sum_{i=1}^{n} (x_i - \overline{X})(y_i - \overline{Y}) = \frac{1}{n}\sum_{i=1}^{n} \{x_i y_i - x_i\overline{Y} - \overline{X}y_i + \overline{X}\overline{Y}\}$$

$$= \frac{1}{n}\sum_{i=1}^{n} x_i y_i - \overline{Y}\frac{1}{n}\sum_{i=1}^{n} x_i - \overline{X}\frac{1}{n}\sum_{i=1}^{n} y_i + \overline{X}\overline{Y}$$

$$= \frac{1}{n}\sum_{i=1}^{n} x_i y_i - \overline{X}\overline{Y}$$

and $$\frac{1}{n}\sum_{i=1}^{n}(x_i-\overline{X})^2=\frac{1}{n}\sum x_i^2-\overline{X}^2$$

$$\frac{1}{n}\sum_{i=1}^{n}(y_i-\overline{Y})^2=\frac{1}{n}\sum y_i^2-\overline{Y}^2$$

Now by (2), we get

$$\gamma(X, Y)=\frac{\frac{1}{n}\sum_{i=1}^{n}x_iy_i-\overline{X}\overline{Y}}{\sqrt{\left\{\frac{1}{n}\sum x_i^2-\overline{X}^2\right\}\left\{\frac{1}{n}\sum y_i^2-\overline{Y}^2\right\}}}.$$

Theorem 1: Prove that the coefficient of correlation $|\gamma|\leq 1$ *i.e.*, $-1<\gamma<1$.

Proof: We know that by schwartz inequality, if a_i, b_i, $i = 1, 2, 3, \ldots\ldots, n$ are real quantities then we have

$$\left(\sum_{i=1}^{n}a_ib_i\right)^2\leq\left(\sum_{i=1}^{n}a_i^2\right)\left(\sum_{i=1}^{n}b_i^2\right) \qquad \ldots(1)$$

the sign of equality is satisfied iff $\frac{a_1}{b_1}=\frac{a_2}{b_2}=\frac{a_3}{b_3}=\ldots\ldots=\frac{a_n}{b_n}$

Put $a_i = X-\overline{X}$ and $b_i = Y-\overline{Y}$

where, X and Y are two variables and $\overline{X}$ and $\overline{Y}$ are their means. By (1), we have

$$\left[\sum_{i=1}^{n}(X-\overline{X})(Y-\overline{Y})\right]^2\leq\sum_{i=1}^{n}(X-\overline{X})^2\sum_{i=1}^{n}(Y-\overline{Y})^2$$

or $$\frac{\left[\sum_{i=1}^{n}(X-\overline{X})(Y-\overline{Y})\right]^2}{\sum_{i=1}^{n}(X-\overline{X})^2\sum_{i=1}^{n}(Y-\overline{Y})^2}\leq 1$$

or $$\gamma^2(X, Y)\leq 1$$

$$\Rightarrow \qquad |\gamma|\leq 1$$

$$\Rightarrow \qquad -1\leq\gamma\leq 1.$$

Note: 1. If $\gamma = -1$, then perfect negative correlation.

2. If $\gamma = 1$, then perfect positive correlation.

Theorem 2: If X and Y are independent variable, then they are uncorrelated but converse is not true.

Proof: We know that the covariance between two variable X and Y is

$$\begin{aligned}\text{Cov}(X, Y) &= E[\{X - E(X)\}\{Y - E(Y)\}] \\ &= E[(X - \bar{X})(Y - \bar{Y})] \\ &= E(XY) - E(X) \,.\, E(Y)\end{aligned}$$

Since, it is given X and Y are independent variables then

$$E(XY) = E(X)\, E(Y)$$

$$\therefore \quad \text{Cov}(XY) = 0$$

The coefficient of correlation

$$\gamma(X, Y) = \frac{\text{Cov.}(XY)}{\sqrt{\text{variance } X} \,.\, \sqrt{\text{variance } Y}} = 0$$

Hence, two variables X and Y are uncorrelated.

Now let two random variables X and Y with probability density function

$$f(x) = \begin{cases} \dfrac{1}{4}, & -2 < x < 2 \\ 0, & \text{otherwise} \end{cases} \quad \text{and } Y = X^2$$

Now,
$$E(X) = \int_{-2}^{2} x\, f(x)\, dx = \frac{1}{4}\int_{-2}^{2} x\, dx = \frac{1}{4}\left[\frac{x^2}{2}\right]_{-2}^{2} = \frac{1}{8}[4 - 4] = 0$$

and
$$E(XY) = E(X^3) = \int_{-2}^{2} x^3\, f(x)\, dx = \frac{1}{4}\int_{-2}^{2} x^2\, dx$$

$$= \frac{1}{4}\left[\frac{x^4}{4}\right]_{-2}^{2} = \frac{1}{16}[16 - 16] = 0$$

$$\therefore \quad \text{Cov}(XY) = E(XY) - E(X) \,.\, E(Y) = 0$$

i.e., the correlation coefficient is zero but the variables are dependent, by the relation $Y = X^2$.

Theorem 3: The correlation coefficient is independent of change of origin and scale.

Proof: The correlation coefficient, between two variables X and Y is

$$\gamma(X, Y) = \frac{\text{Cov}(X, Y)}{\sigma_X\, \sigma_Y} \qquad \text{...(1)}$$

Let $X = a + hU$ and $Y = b + kV$, where a, b, h, k are constants; $n > 0, k > 0$.

To show that $\gamma(X, Y) = \gamma(U, V)$

Since, $X = a + hU \Rightarrow E(X) = a + hE(U)$

and $Y = b + kV \Rightarrow E(Y) = b + k\,E(V)$

$\therefore$ $X - E(X) = h[U - E(U)]$

and $Y - E(Y) = k[V - E(V)]$.

Now, $\text{Cov}(X, Y) = E[\{X - E(X)\}\{Y - E(Y)\}]$

$$= E[h\{U - E(U)\}\,k\,\{V - E(V)\}]$$

$$= hk\,E[\{U - E(U)\}\{V - E(V)\}] = hk\,\text{Cov}(UV) \quad ...(2)$$

$$\sigma_X^2 = E[\{X - E(X)\}^2] = E[h^2\{U - E(U)\}^2]$$

$$= h^2\,E[U - E(U)] = h^2\,\sigma_U^2 \quad ...(3)$$

and

$$\sigma_Y^2 = E[\{Y - E(Y)\}^2] = E[k^2\{V - E(V)\}^2]$$

$$= k^2\,E[V - E(V)] = k^2\sigma_V^2 \quad ...(4)$$

Using equation (1), (2), (3) and (4), we have

$$\gamma(X, Y) = \frac{\text{Cov}(X, Y)}{\sigma_X\,\sigma_Y} = \frac{hk\,\text{Cov}(U, V)}{h\sigma_U k\sigma_V} = \frac{\text{Cov}(U, V)}{\sigma_U\sigma_V} = \gamma(U, V).$$

3.10. PROBABLE ERROR

The coefficient of correlation also has a probable error. It is that amount when added to and subtracted from the coefficient of correlation would give such limits within which we can reasonably expect the values of coefficient of correlation to vary. The formula for probable error is

$$\text{P.E.} = (0.6745) \times \frac{1 - \gamma^2}{\sqrt{n}}.$$

The following rules are observed for probable error:

(*i*) If $\gamma < 0.30$, then it is insignificant.

(*ii*) If $\gamma <$ P.E, then it is not at all significant.

(*iii*) If $0.4 < \gamma < 0.6$, then correlation cofficient is normal.

(*iv*) If $\gamma > 0.9$, then it is highly significant.

(*v*) If $\gamma > 6$ P.E, then it is significant *i.e.*, correlation is certain.

3.11. CORRELATION COEFFICIENT FOR A BIVARIATE FREQUENCY DISTRIBUTION

For a bivariate frequency distribution

$$\gamma = \frac{\Sigma f\,UV - \dfrac{\Sigma f\,U\,\Sigma f\,V}{\Sigma f}}{\left[\left\{\Sigma f\,U^2 - \dfrac{(\Sigma f\,U)^2}{\Sigma f}\right\}\left\{\Sigma f\,V^2 - \dfrac{(\Sigma f\,V)^2}{\Sigma f}\right\}\right]^{1/2}}$$

where, f is the frequency of a particular rectangle in the correlation table, and

$$U = \frac{X - \overline{X}}{h},\ V = \frac{Y - \overline{Y}}{k}.$$

SOLVED EXAMPLES

Example 1: *Calculate the coefficient of correlation between X and Y using the following data.*

X	*– 10*	*– 5*	*0*	*5*	*10*
Y	*5*	*9*	*7*	*11*	*13*

Solution:

X	Y	X^2	Y^2	XY
– 10	5	100	25	– 50
– 5	9	25	81	– 45
0	7	0	49	0
5	11	25	121	55
10	13	100	169	130
$\Sigma X = 0$	$\Sigma Y = 45$	$\Sigma X^2 = 250$	$\Sigma Y^2 = 445$	$\Sigma XY = 90$

We have,
$$\overline{X} = \frac{1}{n}\Sigma X = 0,$$

$$\overline{Y} = \frac{1}{n}\Sigma Y = \frac{1}{5} \times 45 = 9$$

The correlation coefficient

$$\gamma(X, Y) = \frac{\frac{1}{n}\Sigma XY - \overline{X}\,\overline{Y}}{\sqrt{\left\{\frac{1}{n}\Sigma X^2 - \overline{X}^2\right\}\left\{\frac{1}{n}\Sigma Y^2 - \overline{Y}^2\right\}}}$$

$$= \frac{\frac{1}{5}\times 90 - 0\times 9}{\sqrt{\left\{\frac{1}{5}\times 250-(0)^2\right\}\left\{\frac{1}{5}\times 445-(9)^2\right\}}}$$

$$= \frac{18}{\sqrt{50\times 8}} = \frac{18}{20} = 0.9.$$

Example 2: *Calculate the correlation coefficient for the following heights (in inches) of fathers (X) and their sons (Y).*

X	65	66	67	67	68	69	70	72
Y	67	68	65	68	72	72	69	71

Solution:

X	Y	X^2	Y^2	XY
65	67	4225	4489	4355
66	68	4356	4624	4488
67	65	4489	4225	4355
67	68	4489	4624	4556
68	72	4624	5184	4896
69	72	4761	5184	4968
70	69	4900	4761	4830
72	71	5184	5041	5112
$\Sigma X = 544$	$\Sigma Y = 552$	$\Sigma X^2 = 37028$	$\Sigma Y^2 = 38132$	$\Sigma XY = 37560$

We have $\quad \bar{X} = \frac{1}{n}\Sigma X = \frac{544}{8} = 68.$

$$\bar{Y} = \frac{1}{n}\Sigma Y = \frac{552}{8} = 69.$$

The correlation coefficient

$$\gamma(X, Y) = \frac{\frac{1}{n}\Sigma XY - \bar{X}\bar{Y}}{\sqrt{\left\{\frac{1}{n}\Sigma X^2 - \bar{X}^2\right\}\left\{\frac{1}{n}\Sigma Y^2 - \bar{Y}^2\right\}}}$$

$$= \frac{\frac{1}{8}\times 37560 - 68\times 69}{\sqrt{\left\{\frac{1}{8}\times 37028-(68)^2\right\}\left\{\frac{1}{8}\times 38132-(69)^2\right\}}} = \frac{4695-4692}{\sqrt{4.5\times 5.5}} = 0.603.$$

Example 3: *Calculate the coefficient of correlation between X and Y using the following data.*

X	*1*	*2*	*3*	*4*	*5*	*6*	*7*	*8*	*9*
Y	*9*	*8*	*10*	*12*	*11*	*13*	*14*	*16*	*15*

Solution: Let us assume mean for *X* be 5 and for *Y* be 12

X	Y	$U = X - \overline{X}$	$V = Y - \overline{Y}$	U^2	V^2	UV
1	9	−4	−3	16	9	12
2	8	−3	−4	9	16	12
3	10	−2	−2	4	4	4
4	12	−1	0	1	0	0
5	11	0	−1	0	1	0
6	13	1	1	1	1	1
7	14	2	2	4	4	4
8	16	3	4	9	16	12
9	15	4	3	16	9	12
$n = 9$		0	0	$\Sigma U^2 = 60$	$\Sigma V^2 = 60$	$\Sigma UV = 57$

$$\overline{U} = \frac{1}{n}\Sigma U = 0, \qquad \overline{V} = \frac{1}{n}\Sigma V = 0$$

$$\text{Cov}(U, V) = \frac{1}{n}\Sigma UV - \overline{U}\,\overline{V} = \frac{1}{9} \times 57 - 0 = \frac{57}{9} = 6.33$$

$$\sigma_{V^2} = \frac{1}{n}\Sigma U^2 - \overline{U}^2 = \frac{1}{9} \times 60 - 0 = \frac{60}{9} = 6.66$$

$$\sigma_{V^2} = \frac{1}{n}\Sigma V^2 - \overline{V}^2 = \frac{1}{9} \times 60 - 0 = \frac{60}{9} = 6.66$$

The correlation coefficient

$$\gamma(U, V) = \frac{\text{Cov}(U, V)}{\sigma_U\,\sigma_V} = \frac{6.33}{\sqrt{6.66 \times 6.66}} = \frac{6.33}{6.66} = 0.950.$$

Example 4: *Calculate the coefficient of correlation from the following data.*

X	*1*	*3*	*5*	*7*	*8*	*10*
Y	*8*	*12*	*15*	*17*	*18*	*20*

Solution: Let us assume mean for X be 7 and for Y be 15

X	Y	$U = X - \overline{X}$	$V = Y - \overline{Y}$	U^2	V^2	UV
1	8	−6	−7	36	49	42
3	12	−4	−3	16	9	12
5	15	−2	0	4	0	0
7	17	0	2	0	4	0
8	18	1	3	1	9	3
10	20	3	5	9	25	15
$n = 6$		−8	0	$\Sigma U^2 = 66$	$\Sigma V^2 = 96$	$\Sigma UV = 72$

$$\overline{U} = \frac{1}{n}\Sigma U = \frac{-8}{6} = -1.33, \quad \overline{V} = \frac{1}{n}\Sigma V = \frac{1}{6}\times 0 = 0$$

$$\text{Cov}(U, V) = \frac{1}{n}\Sigma UV - \overline{U}\,\overline{V} = \frac{1}{6}\times 72 - (-1.33)\times 0 = 12$$

$$\sigma_{U^2} = \frac{1}{n}\Sigma U^2 - \overline{U}^2 = \frac{1}{6}\times 66 - (-1.33)^2$$

$$= 11 - 1.77 = 9.23$$

$$\sigma_{V^2} = \frac{1}{n}\Sigma V^2 - \overline{V}^2 = \frac{1}{6}\times 96 - (0)^2 = 16$$

The correlation coefficient

$$\gamma(U, V) = \frac{\text{Cov}(U, V)}{\sigma_U\,\sigma_V} = \frac{12}{\sqrt{9.23\times 16}} = \frac{12}{12.15} = 0.9876.$$

Example 5: *Calculate the value of the Pearson's coefficient of correlation for the following table.*

X \ *Y*	*16–18*	*18–20*	*20–22*	*22–24*
10–20	*2*	*1*	*1*	–
20–30	*3*	*2*	*3*	*2*
30–40	*3*	*4*	*5*	*6*
40–50	*2*	*2*	*3*	*4*
50–60	–	*1*	*2*	*2*
60–70	–	*1*	*2*	*1*

Solution: Let us assume $\overline{X} = 30–40$, and $\overline{Y} = 18–20$,

And we have $$U = \frac{X - (30–40)}{10} \text{ and } V = \frac{Y - (18–20)}{2}$$

Y / X	16–18	18–20	20–22	22–24	f	U	fU	fU^2	ΣfV	$U\Sigma fV$
10–20	2	1	1		4	– 2	– 8	16	– 1	2
20–30	3	2	3	2	10	– 1	– 10	10	4	– 4
30–40	3	4	5	6	18	0	0	0	14	0
40–50	2	2	3	4	11	1	11	11	9	9
50–60		1	2	2	5	2	10	20	6	12
60–70		1	2	1	4	3	12	36	4	12
f	10	11	16	15	52	Total	15	93	36	31
V	– 1	0	1	2	Total					
fV	– 10	0	16	30	36					
fV^2	10	0	16	60	86					
ΣfU	– 5	3	8	9	15					
$V\Sigma fU$	5	0	8	18	31					

From the table,

$$\Sigma fV = 36$$
$$\Sigma fV^2 = 86$$
$$\Sigma fU = 15$$
$$\Sigma fU^2 = 93$$
$$\Sigma f = 52$$
$$\Sigma fUV = 31.$$

The Pearson coefficient of correlation is

$$\gamma = \frac{\Sigma f\,UV - \dfrac{\Sigma f\,U \times \Sigma f\,V}{\Sigma f}}{\sqrt{\left\{\Sigma f\,U^2 - \dfrac{(\Sigma f\,U)^2}{\Sigma f}\right\}\left\{\Sigma f\,V^2 - \dfrac{(\Sigma f\,V)^2}{\Sigma f}\right\}}}$$

$$= \frac{31 - \dfrac{15 \times 36}{52}}{\sqrt{\left(93 - \dfrac{15^2}{52}\right)\left(86 - \dfrac{36^2}{52}\right)}} = \frac{31 \times 52 - 15 \times 36}{\sqrt{(93 \times 52 - 225)\,(86 \times 52 - 1296)}}$$

$$= \frac{1072}{\sqrt{14644536}} = \frac{1072}{3826.8} = 0.28.$$

3.12. RANK CORRELATION OR SPEARMAN'S COEFFICIENT OF RANK CORRELATION

Charles Edward Spearman (1906) a great psychologist and statistician, developed a formula for the calculation of coefficient of correlation, which is called the "Rank correlation coefficient or Spearman's correlation coefficient.

Let us suppose that sometimes the condition is such type which cannot be measured quantitatively, but can be ranked among themselves. For example, it is possible for a class-teacher to arrange his students in ascending or descending order of intelligence, even though intelligence cannot be measured quantitatively. There are so many attributes which are incapable of quantitative measurement such as honesty, character, morality etc.

Suppose (x_i, y_i), $i = 1, 2, 3, \ldots\ldots, n$ be the ranks of n individuals of a group corresponding to two characteristics A and B. Assuming that no two individuals are equal in either classification, each individual takes the values 1, 2, 3, ... n.

Now,
$$\bar{X} = \frac{1}{n}(1 + 2 + 3 + \ldots\ldots + n) = \frac{n+1}{2} = \bar{Y}$$

$$\sigma_X^2 = \frac{1}{n}\Sigma x_i^2 - \bar{X}^2 = \frac{1}{n}(1^2 + 2^2 + 3^2 + \ldots\ldots + n^2) - \left(\frac{n+1}{2}\right)^2$$

$$= \frac{(n+1)(2n+1)}{6} - \frac{(n+1)^2}{4} = \frac{n^2 - 1}{12} = \sigma_Y^2$$

Let
$$d_i = x_i - y_i = (x_i - \bar{X}) - (y_i - \bar{Y})$$

$$\frac{1}{n}\Sigma d_i^2 = \frac{1}{n}\Sigma\{(x_i - \bar{X}) - (y_i - \bar{Y})\}^2$$

$$= \frac{1}{n}\{\Sigma(x_i - \bar{X})^2 + \Sigma(y_i - \bar{Y})^2 - 2\,\Sigma(x_i - \bar{X}) - (y_i - \bar{Y})\}$$

$$= \sigma_x^2 + \sigma_y^2 - 2\,\text{Cov}\,(X, Y)$$

or
$$\text{Cov}\,(X, Y) = \frac{1}{2}\left[\sigma_x^2 + \sigma_y^2 - \frac{1}{n}\Sigma d_i^2\right] = \frac{1}{2}\left[\frac{n^2-1}{12} + \frac{n^2-1}{12} - \frac{1}{n}\Sigma d_i^2\right]$$

$$= \frac{1}{12}(n^2 - 1) - \frac{1}{2n}\Sigma d_i^2$$

The correlation coefficient, between two variable X and Y is

$$\gamma\,(X, Y) = \frac{\text{Cov}\,(X, Y)}{\sigma_X \sigma_Y} = \frac{\frac{1}{12}(n^2 - 1) - \frac{1}{2n}\Sigma d_i^2}{\sqrt{\frac{n^2 - 1}{12}}\sqrt{\frac{n^2 - 1}{12}}}$$

$$= \frac{\dfrac{(n^2-1)}{12} - \dfrac{1}{2n}\Sigma d_i^2}{\dfrac{n^2-1}{12}}$$

or
$$\boxed{\gamma = 1 - \frac{\sigma \Sigma d_i^2}{n(n^2-1)}}$$

The method of measuring association between two sets of ranks is known as the rank correlation method.

3.13. RANK CORRELATION COEFFICIENT FOR REPEATED RANKS

Suppose given the following data:

X	68	64	75	50	64	80	75	40	55	64
Y	62	58	68	45	81	60	68	48	50	70

Here, we see that in X-series, 75 occurs 2 times. 64 occurs 3 times and in Y-series, 68 occurs 2 times. If any two or more individuals have same value with respect to characteristics X and Y, then Spearman's rank formula is fails. For these common ranking the correct formula for γ, to Σd^2 we add $\dfrac{m(m^2-1)}{12}$ for each value repeated, where m is the number of times a value occurs. As in given above example,

For the X-series $$= \frac{2(2^2-1)}{12} + \frac{3(3^2-1)}{12} = \frac{1}{2} + 2 = \frac{5}{2}$$

For the Y-series $$= \frac{2(2^2-1)}{12} = \frac{1}{2}$$

and $$\gamma = 1 - \frac{6\left(\Sigma d^2 + \dfrac{5}{2} + \dfrac{1}{2}\right)}{n(n^2-1)}.$$

3.14. SCATTER DIAGRAM OR DOT DIAGRAM

In this method the values of two variables (series) are plotted on a graph, taking one on X-axis and the other along Y-axis. There is one dot for one pair of series, there being as many dots on the graph as there pairs. This collection of dots is called a scatter diagram or dot diagram.

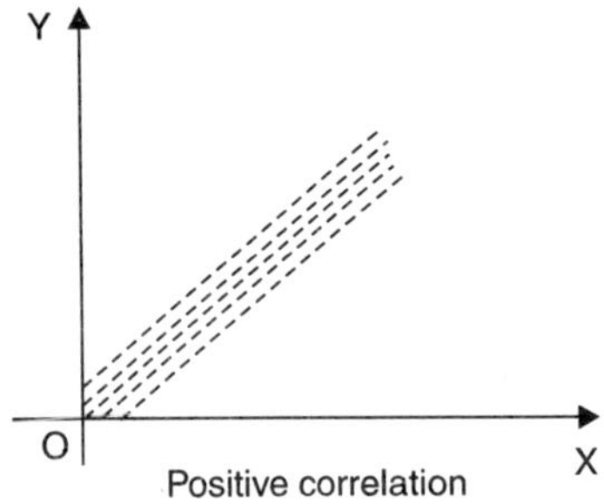

Positive correlation

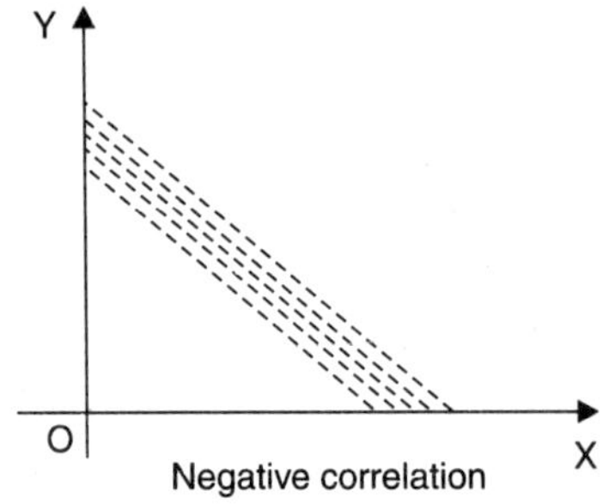

Negative correlation

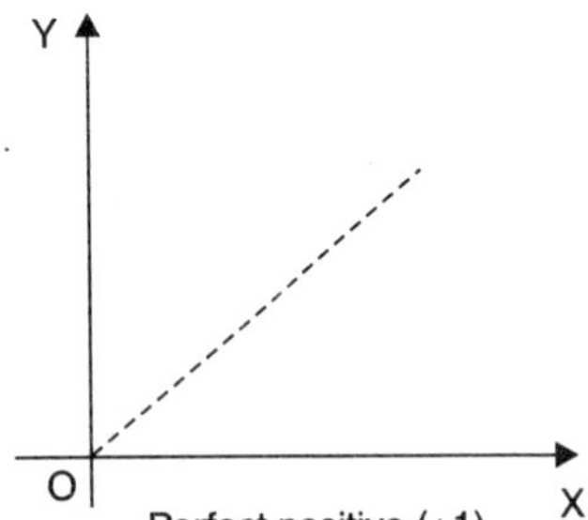

Perfect positive (+1)

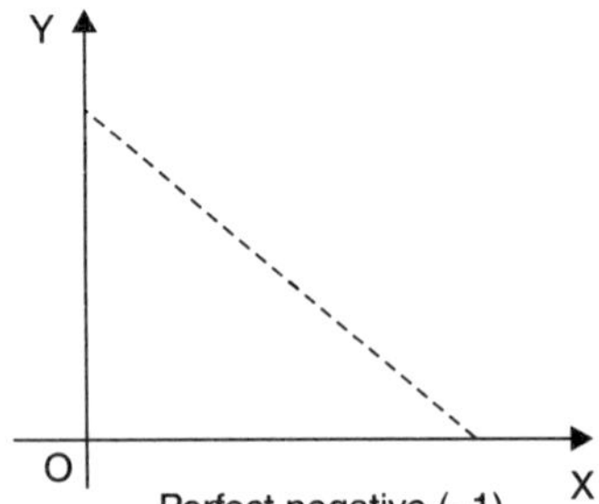

Perfect negative (–1)

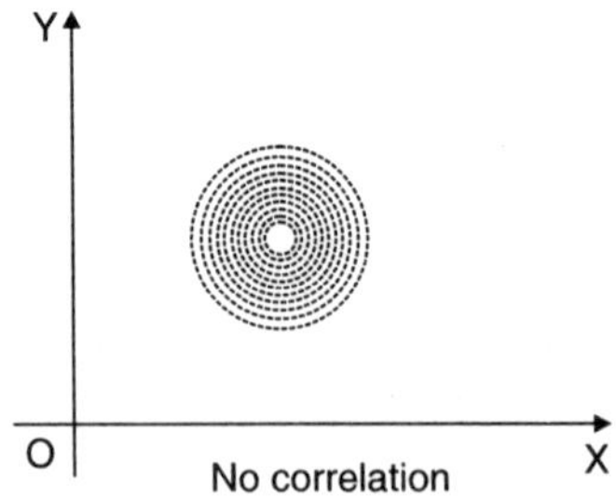

No correlation

Example 6: *The marks secured by students in mathematics and statistics are given below:*

Mathematics (X)	*10*	*15*	*12*	*17*	*13*	*16*	*25*	*14*	*22*
Statistics (Y)	*30*	*42*	*45*	*46*	*33*	*34*	*40*	*35*	*39*

Calculate the rank correlation coefficient.

Solution: Calculation of the coefficient of rank correlation

X	Rank of X R_1	Y	Rank of Y R_2	$R_1 - R_2 = d$	$(R_1 - R_2)^2 = d^2$
10	9	30	9	0	0
15	5	42	3	2	4
12	8	45	2	6	36
17	3	46	1	2	4
13	7	33	8	– 1	1
16	4	34	7	– 3	9
25	1	40	4	– 3	9
14	6	35	6	0	0
22	2	39	5	– 3	9
$n = 9$		$n = 9$		$\Sigma d = 10$	$\Sigma d^2 = 72$

The coefficient of rank correlation

$$\gamma = 1 - \frac{6\,\Sigma d^2}{n(n^2-1)} = 1 - \frac{6 \times 72}{9(9^2-1)} = 1 - \frac{432}{720} = 0.4.$$

Example 7: *Ten competitors in beauty contest are ranked by three judges in the following order:*

First Judge by R_1	*1*	*6*	*5*	*10*	*3*	*2*	*4*	*9*	*7*	*8*
Second Judge by R_2	*3*	*5*	*8*	*4*	*7*	*10*	*2*	*1*	*6*	*9*
Third Judge by R_3	*6*	*4*	*9*	*8*	*1*	*2*	*3*	*10*	*5*	*7*

Use the rank correlation method to discuss which pair of judges have the nearest approach to common tastes in beauty.

Solution: Calculation of the coefficient of rank correlation between the ranking of first, second and third Judges.

R_1	R_2	R_3	$R_1 - R_2 = d_1$	$R_1 - R_3 = d_2$	$R_2 - R_3 = d_3$	d_1^2	d_2^2	d_3^2
1	3	6	– 2	– 5	– 3	4	25	9
6	5	4	1	2	1	1	4	1
5	8	9	– 3	– 4	– 1	9	16	1
10	4	8	6	2	– 4	36	4	16
3	7	1	– 4	2	6	16	4	36
2	10	2	– 8	0	8	64	0	64
4	2	3	2	1	– 1	4	1	1
9	1	10	8	– 1	– 9	64	1	81
7	6	5	1	2	1	1	4	1
8	9	7	– 1	1	2	1	1	4
			$\Sigma d_1 = 0$	$\Sigma d_2 = 0$	$\Sigma d_3 = 0$	$\Sigma d_1^2 = 200$	$\Sigma d_2^2 = 60$	$\Sigma d_3^2 = 214$

Here, $n = 10$.

$$\gamma(1, 2) = 1 - \frac{6\Sigma d_1^2}{n(n^2-1)} = 1 - \frac{6 \times 200}{10 \times 99} = 1 - \frac{40}{33} = -\frac{7}{33} = -0.212$$

$$\gamma(1, 3) = 1 - \frac{6\Sigma d_2^2}{n(n^2-1)} = 1 - \frac{6 \times 60}{10 \times 90} = 1 - \frac{4}{11} = \frac{7}{11} = 0.636$$

$$\gamma(2, 3) = 1 - \frac{6\Sigma d_3^2}{n(n^2-1)} = 1 - \frac{6 \times 214}{10 \times 90} = 1 - \frac{214}{165} = -\frac{49}{165} = -0.297$$

Here, we see that the first and third Judges have the nearest approach to their tastes for beauty.

Example 8: *Calculate the coefficient of rank correlation from the following data.*

X	*48*	*33*	*40*	*9*	*16*	*16*	*65*	*24*	*16*	*57*
Y	*13*	*13*	*24*	*6*	*15*	*4*	*20*	*9*	*6*	*19*

Solution: Calculation of the coefficient of rank correlation

X	*Rank of X* R_1	*Y*	*Rank of Y* R_2	$R_1 - R_2 = d$	$R_1 - R_2 = d^2$
48	3	13	5.5	– 2.5	6.25
33	5	13	5.5	– 0.5	0.25
40	4	24	1	3	9.00
9	10	6	8.5	1.5	2.25
16	8	15	4	4	16.00
16	8	4	10	– 2	4.00
65	1	20	2	– 1	1.00
24	6	9	7	– 1	1.00
16	8	6	8.5	– 0.5	0.25
57	2	19	3	– 1	1.00
$n = 10$		$n = 10$		$\Sigma d = 0$	$\Sigma d^2 = 41.00$

Here, we see that in *X*-series, 16 occurs 3 times and in *Y*-series 13 occurs 2 times and 6 occurs 2 times.

The total correlation for *X*-series is

$$= \frac{3(3^2 - 1)}{12} = 2$$

The total correlation for *Y*-series is

$$= \frac{2(2^2 - 1)}{12} + \frac{2(2^2 - 1)}{12} = \frac{1}{2} + \frac{1}{2} = 1$$

The coefficient of rank correlation

$$\gamma = 1 - \frac{6\left[\Sigma d^2 + 2 + \frac{1}{2} + \frac{1}{2}\right]}{10(10^2 - 1)} = 1 - \frac{6[41 + 3]}{990} = 1 - \frac{264}{990}$$

$$= 1 - 0.2666 = 0.7334.$$

Example 9: *Calculate the coefficient of rank correlation from the following data.*

X	*68*	*64*	*75*	*50*	*64*	*80*	*75*	*40*	*55*	*64*
Y	*62*	*58*	*68*	*45*	*81*	*60*	*68*	*48*	*50*	*74*

Solution: Calculation of the coefficient of rank correlation.

X	Rank of X R_1	Y	Rank of Y R_2	$R_1 - R_2 = d$	$R_1 - R_2 = d^2$
68	4	62	5	– 1	1
64	6	58	7	– 1	1
75	2.5	68	3.5	– 1	1
50	9	45	10	– 1	1
64	6	81	1	5	25
80	1	60	6	– 5	25
75	2.5	68	3.5	– 1	1
40	10	48	9	1	1
55	8	50	8	0	0
64	6	74	2	4	16
$n = 10$		$n = 10$		$\Sigma d = 0$	$\Sigma d^2 = 72$

Here, we see that in X-series, 75 occurs 2 times, 64 occurs 3 times and in Y-series, 68 occurs 2 times.

The total correlation for X-series is

$$= \frac{2(2^2 - 1)}{12} + \frac{3(3^2 - 1)}{12} = \frac{1}{2} + 2 = \frac{5}{2}$$

The total correlation for Y-series is

$$= \frac{2(2^2 - 1)}{12} = \frac{1}{2}$$

The rank correlation coefficient

$$\gamma = 1 - \frac{6\left[\Sigma d^2 + \frac{5}{2} + \frac{1}{2}\right]}{n(n^2 - 1)} = 1 - \frac{6[72 + 3]}{10 \times 90} = 1 - \frac{450}{900} = 0.5.$$

Example 10: *If $X_i + Y_i = n + 1$, show that $\gamma = -1$.*

Solution: Given $X_i + Y_i = n + 1$

Let $X_i - Y_i = d_i$

$\Rightarrow$ $d_i = 2X_i - (n + 1)$

$\therefore$ $\Sigma d^2 = 4\Sigma X^2 - 4(n + 1)\,\Sigma X + n(n + 1)^2$

$$= 4\,\frac{n(n+1)(2n+1)}{6} - 4\,(n+1)\,.\,\frac{n(n+1)}{2} + n(n+1)^2$$

$$= \frac{1}{3}\,n(n^2 - 1)$$

$$\therefore \qquad \gamma = 1 - \frac{6 \cdot \frac{1}{3} n(n^2-1)}{n(n^2-1)} = 1 - 2 = -1.$$

Example 11: *If X and Y are uncorrelated random variables, find the coefficient of correlation between X + Y and X – Y.*

Solution: Let $U = X + Y$ and $V = X - Y$

Then correlation coefficient is

$$\gamma = \frac{\Sigma(U - \bar{U})(V - \bar{V})}{n\,\sigma_U\,\sigma_V} \qquad ...(1)$$

Let $\bar{U} = \bar{X} + \bar{Y}, \quad \bar{V} = \bar{X} - \bar{Y}$

and $X - \bar{X} = x, \quad Y - \bar{Y} = y$

We have $\Sigma(U - \bar{U})(V - \bar{V}) = \Sigma[\{(X+Y) - (\bar{X} + \bar{Y})\}\{(X - Y) - (\bar{X} - \bar{Y})\}]$

$$= \Sigma[(x+y)(x-y)] = \Sigma x^2 - \Sigma y^2 = n\sigma_x^2 - n\sigma_y^2$$

Also we have, $\sigma_U^2 = \frac{1}{n}\Sigma(U - \bar{U})^2 = \frac{1}{n}\Sigma[(X+Y) - (\bar{X} + \bar{Y})]^2$

$$= \frac{1}{n}\Sigma(x+y)^2 = \frac{1}{n}[\Sigma x^2 + \Sigma y^2 + 2\Sigma xy]$$

$$= \sigma_x^2 + \sigma_y^2$$

Here, we have $\Sigma xy = 0$, becuase X and Y are not correlated

Similarly, $\sigma_V^2 = \sigma_x^2 + \sigma_y^2$

from equation (1), we have

$$\gamma = \frac{n(\sigma_x^2 - \sigma_y^2)}{\sqrt{\{n(\sigma_x^2 + \sigma_y^2)\}\{n(\sigma_x^2 + \sigma_y^2)\}}}$$

$$\gamma = \frac{\sigma_x^2 - \sigma_y^2}{\sigma_x^2 + \sigma_y^2}.$$

Example 12: *Establish the formula* $\sigma_{X-Y}^2 = \sigma_X^2 + \sigma_Y^2 - 2\gamma\sigma_X\sigma_Y$, *where* γ *is the correlation coefficient between X and Y.*

Solution: The correlation coefficient between X and Y is

$$\gamma(X, Y) = \frac{\text{Cov}(X, Y)}{\sigma_x \sigma_y} \qquad ...(1)$$

We know that, $\sigma_X^2 = E(X^2)$ and $\text{Cov}(X, Y) = E(X, Y)$

$$\Rightarrow \qquad \sigma_{X-Y}^2 = E(X-Y)^2 = E(X^2) + E(Y^2) - 2E(X, Y)$$

$$= \sigma_X^2 + \sigma_Y^2 - 2\,\text{Cov}\,(X, Y)$$

$$\Rightarrow \qquad \text{Cov}\,(X, Y) = \frac{1}{2}[\sigma_X^2 + \sigma_Y^2 - \sigma_{X-Y}^2]$$

from equation (1), we have

$$\gamma = \frac{\frac{1}{2}[\sigma_X^2 + \sigma_Y^2 - \sigma_{X-Y}^2]}{\sigma_X \sigma_y}$$

$$\Rightarrow \qquad \gamma = \frac{\sigma_X^2 + \sigma_Y^2 - \sigma_{X-Y}^2}{2\sigma_X \sigma_Y}$$

$$\Rightarrow \qquad \sigma_{X-Y}^2 = \sigma_X^2 + \sigma_Y^2 - 2\gamma\sigma_X\sigma_Y.$$

EXERCISE 3.1

1. Calculate the coefficient of correlation between X and Y from the table of their values:

X	1	3	4	6	8	9	11	14
Y	1	2	4	4	5	7	8	9

2. Calculate the coefficient of correlation between the values of X and Y given below:

X	78	89	97	69	59	79	68	61
Y	125	137	156	112	107	136	123	108

3. Calculate the coefficient of correlation for the following ages of husband and wife:

Husband's age	23	27	28	29	30	31	33	35	36	39
Wife's age	18	22	23	24	25	26	28	29	30	32

4. Calculate of coefficient of correlation for X and Y from the following data:

X	45	55	56	58	60	65	68	70	75	80	85
Y	56	50	48	60	62	64	65	70	74	82	90

5. From the following data calculate the coefficient of correlation and find its probable error:

Mean Annual birth rate per 1000 of population	35.3	33.5	31.4	30.5	29.3	28.2	26.3	23.6	20.1	19.9	16.7
Mean Annual death rate per 1000 of population	30.8	19.4	18.9	18.7	17.7	16.0	14.7	14.3	14.4	12.2	12.1

6. Find the value of the Pearson's coefficient of correlation for the following table:

X \ *Y*	*0–5*	*5–10*	*10–15*	*15–20*	*20–25*
0–4	1	2	—	—	—
4–8	—	4	5	8	—
8–12	—	—	3	4	—
12–16	—	—	—	2	1

7. Find Pearson's coefficient of correlation from the following table :

X \ *Y*	*94.5*	*96.5*	*98.5*	*100.5*	*102.5*	*104.5*	*106.5*	*108.5*	*110.5*
29.5	—	—	4	3		4	1	—	1
59.5	1	3	6	18	6	9	2	3	1
89.5	7	3	16	16	4	4	1	—	1
119.5	5	9	10	9	2	—	1	2	—
149.5	3	5	8	1	—	1	—	—	—
179.5	4	2	3	1	—	—	—	—	—
209.5	4	4	—	1	—	—	—	—	—
239.5	1	1	—	—	—	—	—	—	—

8. Ten students got the following percentage of marks in mathematics and statistics:

S.No. of students	1	2	3	4	5	6	7	8	9	10
Marks in mathematic	78	36	98	25	75	82	90	62	65	39
Marks in statistics	84	51	91	60	68	62	86	58	53	47

Find the coefficient of correlation and rank correlation of the above.

9. The figures in the following table give the number of criminal convictions and the number unemployed (in millions) for the years 1998–2007. Find the coefficient of rank correlation.

Years	1998	1999	2000	2001	2002	2003	2004	2005	2006	2007
Number convicted of crime	7.88	8.12	7.86	7.25	7.44	7.22	8.28	8.83	10.54	9.46
Number of unemployed	1.26	1.24	1.43	1.19	1.33	1.34	2.50	2.67	2.78	2.26

10. Calculate the coefficient of correlation from the following data by the method of rank differences:

X	81	78	73	73	69	68	62	58
Y	10	12	18	18	18	22	20	24

11. The following are the numbers of hours which 10 students studied for an examination and the scores they obtained:

No. of hour studied (X)	8	5	11	13	10	5	18	15	2	8
Scores (Y)	56	44	79	72	70	54	94	85	33	65

Calculate the rank correlation coefficient.

12. A sample of 12 fathers and their eldest sons gave the following data about their height in inches:

Father	65	63	67	64	68	62	70	66	68	67	69	71
Son	68	66	68	65	69	66	68	65	71	67	68	70

Calculate the coefficient of rank correlation.

ANSWERS

1. 0.977. **2.** 0.957. **3.** 0.95. **4.** 0.92. **5.** 0.98, 0.0805.

6. 0.6. **7.** – 0.49. **8.** 0.78, 0.818. **9.** 0.09. **10.** – 0.928.

11. 0.975. **12.** 0.722.

3.15. REGRESSION

The study of regression has a great importance in physical sciences, where the data are generally in functional relationship. The word 'Regression' was used by sir Francis Galton (1822–1911) in a paper entitled 'Regression towards Mediocrity in Hereditary stature' in his studies connected with the relation between the heights of fathers and heights of sons. Knowledge of regression analysis comes in handy in finding the probable value of one variable for a given value of the other, when the two variables are known to be correlated. Thus, if we know that two series is relating to supply and price are correlated we can find out what would be the effect on price, if the supply of commodity is increased or decreased to a particular level.

3.16. LINE OF REGRESSION

Regression is the estimation for prediction of unknown values of one variable from known values of another variable. The line which describes the average relationship between two variables is known as line of regression. The equations describing the regression lines are called regression equations. A regression equation is the algebric expression of regression line.

A regression line is one which shows that the best mean values of one variable corresponding to mean values of the other variable. With two series x and y, if the two regression lines range themselves along two straight lines, the correlation between x and y is linear. If the two straight lines of regression coincide, correlation is perfect. If the two lines cut each other at right angles, correlation is zero.

Let $Y = aX + b$ be the equation of the line of the best fit of x. Changing the origin to $(\overline{X}, \overline{Y})$, where $\overline{X}$ is the mean of x-series and $\overline{Y}$ is the mean of y-series. Let x, y be the deviations from the respective means $\overline{X}$ and $\overline{Y}$.

$$\therefore \qquad x = X - \overline{X}, \qquad y = Y - \overline{Y}$$

Then, equation $Y = aX + b$ is changed in the form

$$y = ax + b$$

where, $x = X - \overline{X}$ and $y = Y - \overline{Y}$

Let (x_c, y_c) be any point. Then the difference between this point and the above line is

$$y_c - ax_c - b$$

Hence U, the sum of the squares of such distance is

$$U = \Sigma\,(y - ax - b)^2, \text{ for all values of } c.$$

Now for U is minimum, we choose a and b such that

$$\frac{\partial U}{\partial a} = -2\,\Sigma\, x(y - ax - b) = 0$$

and $$\frac{\partial U}{\partial b} = -2\Sigma(y - ax - b) = 0$$

$$\Rightarrow \qquad \Sigma xy - a\Sigma x^2 - b\Sigma x = 0$$

and $$\Sigma y - a\Sigma x - nb = 0$$

Since, $\Sigma x = 0 = \Sigma y$, we get

$$\Rightarrow \qquad b = 0 \quad \text{and} \quad a = \frac{\Sigma xy}{\Sigma x^2}$$

$$\Rightarrow \qquad a = \frac{\Sigma xy}{\Sigma x^2} = \frac{\text{Cov}(x, y)}{\sigma_x^2} = \gamma \frac{\sigma_y}{\sigma_x}$$

where, γ is the coefficient of correlation between x and y.

Hence, the equation to the line of regression

$$y = \gamma \frac{\sigma_y}{\sigma_x} x.$$

$$\Rightarrow \qquad \boxed{Y - \overline{Y} = \gamma \frac{\sigma_y}{\sigma_x}(X - \overline{X})} \text{ where, } x = X - \overline{X} \text{ and } y = Y - \overline{Y},$$

which is called the regression line of Y on X.

Similarly, the regression line of X on Y is

$$\boxed{X - \overline{X} = \gamma \frac{\sigma_x}{\sigma_y}(Y - \overline{Y})}$$

Therefore, if the straight line is so chosen that the sum of squares of deviation parallel to the axis of x is minimum, it is called the line of regression of x on y.

3.17. STANDARD ERROR OF ESTIMATE

The standard error measures the accuracy of estimate.

It is given by $\qquad E_{yx} = \sqrt{\dfrac{\Sigma(y - y_c)^2}{n}}$

where, y is the observed value, y_c is the estimated value and n is the number of pairs of items.

Note.

1. The coefficient $\gamma \frac{\sigma_y}{\sigma_x}$ and $\gamma \frac{\sigma_x}{\sigma_y}$ are called the regression coefficient of y on x, and of x on y respectively.

2. If the two lines of regression coincide, the correlation between the variables is perfect *i.e.*,

$$\gamma \frac{\sigma_y}{\sigma_x} = \frac{1}{\gamma} \frac{\sigma_y}{\sigma_x}$$

$$\Rightarrow \qquad \gamma^2 = 1$$

$$\Rightarrow \qquad \gamma = \pm 1.$$

3. If the variables x and y are independent *i.e.*, the coefficient of correlation between them is zero.
4. Some diagrams for differents values of γ.

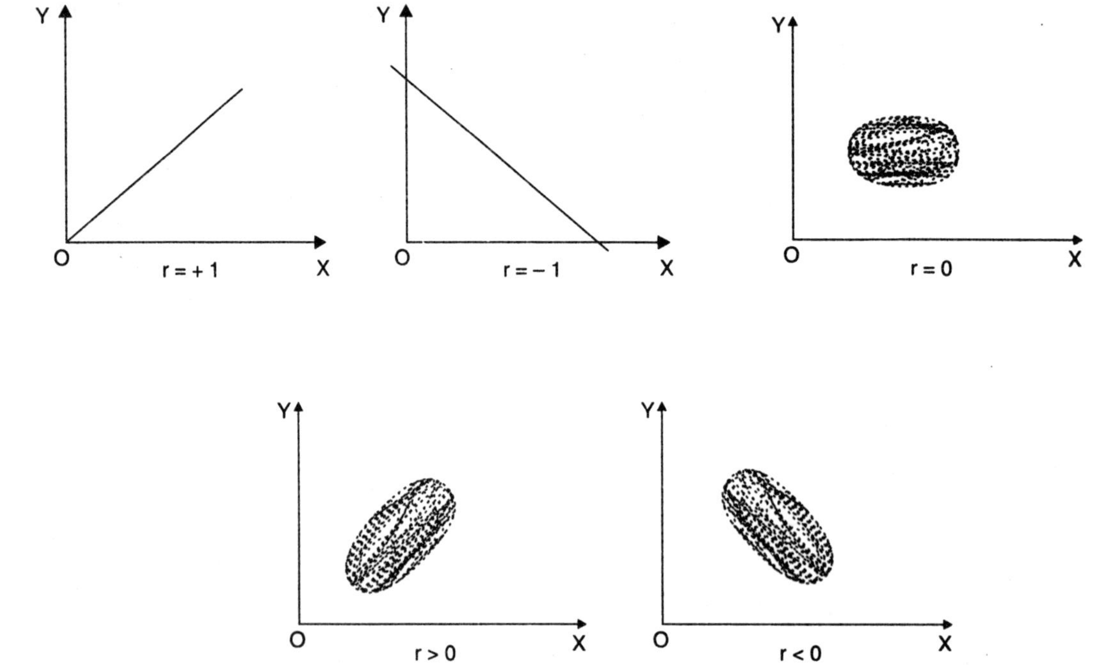

5. Also denotes $\gamma \dfrac{\sigma_y}{\sigma_x} = b_{yx}$ and $\gamma \dfrac{\sigma_x}{\sigma_y} = b_{xy}$.

SOLVED EXAMPLES

Example 1: *If θ be the angle between the lines of regression, show that*

$$\tan\theta = \frac{(1-\gamma^2)}{\gamma} \cdot \frac{\sigma_x \sigma_y}{\sigma_x^2 + \sigma_y^2}.$$

Explain the significance of the formula, when $\gamma = 0$, $\gamma = \pm 1$.

Solution: The regression line of y on x

$$y - \bar{y} = \gamma \frac{\sigma_y}{\sigma_x}(x - \bar{x}) \qquad \text{...(1)}$$

Here slope $\quad m_1 = \gamma \dfrac{\sigma_y}{\sigma_x}$ ...(2)

The regression line of x on y

$$x - \bar{x} = \gamma \frac{\sigma_x}{\sigma_y}(y - \bar{y}) \qquad \text{...(3)}$$

The slope $$m_2 = \frac{1}{\gamma}\frac{\sigma_y}{\sigma_x}$$

If m_1 and m_2 are the slops of the lines and θ is the angle between them, we have

$$\tan\theta = \frac{m_2 - m_1}{1 + m_1 m_2} = \frac{\frac{1}{\gamma}\frac{\sigma_y}{\sigma_x} - \gamma\frac{\sigma_y}{\sigma_x}}{1 + \gamma\frac{\sigma_y}{\sigma_x}\cdot\frac{1}{\gamma}\frac{\sigma_y}{\sigma_x}}$$

$$= \frac{\frac{\sigma_y}{\sigma_x}\left(\frac{1-\gamma^2}{\gamma}\right)}{1 + \frac{\sigma_y^2}{\sigma_x^2}} = \left(\frac{1-\gamma^2}{\gamma}\right)\cdot\frac{\sigma_x\sigma_y}{\sigma_x^2 + \sigma_y^2}.$$

when $\gamma = 0$, $\tan\theta = \infty$, $\theta = \frac{\pi}{2}$ *i.e.,* the two lines of regression are perpendicular to each other or variables are uncorrelated.

When $\gamma = \pm 1$, $\tan\theta = 0$, $\theta = 0$ or π *i.e.,* the lines coincide and therefore is a perfect correlation between the two variables x and y. The estimated value of y is the same for all values of x or vice-versa.

Example 2: *For two variables x and y with the same mean, the two regression equations are $y = ax + b$ and $x = \alpha y + \beta$. Show that $\frac{b}{\beta} = \frac{1-a}{1-\alpha}$. Find also the common mean.*

Solution: Suppose the common mean is m.

The given two regression equations are $y = ax + b$ ...(1)

and $x = \alpha y + \beta$ ...(2)

$y - m = a(x - m)$ ...(3)

and $x - m = \alpha(y - m)$ ...(4)

Comparing equation (3) and (4) with (1) and (2), we get

$b = m(1 - a)$ ...(5)

and $\beta = m(1 - \alpha)$ ...(6)

Dividing (5) by (6), we get

$$\frac{b}{\beta} = \frac{1-a}{1-\alpha}$$

Also (5) and (6), we have

$$m = \frac{b}{1-a} = \frac{\beta}{1-\alpha}.$$

Example 3: *Prove that the Pearson's coefficient of correlation γ lies between – 1 and + 1.*

Solution: Let $$U = \Sigma(y - ax - b)^2 \quad ...(1)$$

For U is minimum when $\frac{\partial U}{\partial a} = 0$ and $\frac{\partial U}{\partial b} = 0$

differentiate partially (1) with respect to a and b respectively, we get

$$\frac{\partial U}{\partial a} = -2\,\Sigma\, x\,(y - ax - b) = 0$$

and
$$\frac{\partial U}{\partial b} = -2\,\Sigma(y - ax - b) = 0.$$

Since, $\Sigma x = 0 = \Sigma y$, then we get

$$b = 0 \quad \text{and} \quad a = \frac{\Sigma\, xy}{\Sigma\, x^2}$$

Now by (1), we have

$$U = \Sigma(y - ax)^2 = \Sigma y^2 - 2a\Sigma\, xy + a^2\Sigma\, x^2$$

$$= \Sigma y^2 - 2\,\frac{\Sigma\, xy}{\Sigma\, x^2}\,.\,\Sigma xy + \left(\frac{\Sigma\, xy}{\Sigma\, x^2}\right)^2 \Sigma x^2$$

$$= \Sigma y^2 - 2\,\frac{(\Sigma\, xy)^2}{\Sigma x^2} + \frac{(\Sigma\, xy)^2}{\Sigma x^2} = \Sigma y^2 - \frac{(\Sigma\, xy)^2}{\Sigma x^2}$$

$$= \Sigma y^2\left[1 - \frac{(\Sigma\, xy)^2}{\Sigma\, x^2 \Sigma y^2}\right]$$

$$= \Sigma y^2\,[1 - \gamma^2], \quad \text{where } \gamma = \frac{\Sigma xy}{\sqrt{\Sigma x^2\, \Sigma y^2}}$$

$$= (1 - \gamma)^2\, \Sigma\, y^2$$

Since, U be the sum of squares then $\Sigma\, y^2$ will not be negative.

$$\Rightarrow \quad 1 - \gamma^2 \geq 0 \quad \text{or} \quad \gamma^2 \leq 1$$

Hence, γ lies between – 1 and + 1.

Example 4: *Show that the coefficient of correlation is the geometric mean between the two regression coefficients.*

Solution: The regression coefficient of y on $x = \gamma\,\frac{\sigma_y}{\sigma_x}$

The regression coefficient of x on $y = \gamma\,\frac{\sigma_x}{\sigma_y}$

$$\text{Geometric mean} = \sqrt{\gamma\,\frac{\sigma_y}{\sigma_x}\cdot\gamma\,\frac{\sigma_x}{\sigma_y}}$$

$= \gamma =$ coefficient of correlation.

Example 5: *For a bivariate distribution*

$$n = 18, \Sigma X^2 = 60, \Sigma Y^2 = 96, \Sigma X = 12, \Sigma Y = 18, \Sigma XY = 48,$$

find the equation of the lines of regression and γ.

Solution: We have,

$$\bar{X} = \frac{1}{n}\Sigma X = \frac{12}{18} = \frac{2}{3} \quad ...(1)$$

$$\bar{Y} = \frac{1}{n}\Sigma Y = \frac{18}{18} = 1 \quad ...(2)$$

$$\sigma_X^2 = \frac{1}{n}\Sigma X^2 - \bar{X}^2 = \frac{60}{18} - \frac{4}{9} = \frac{26}{9} \quad ...(3)$$

$$\sigma_Y^2 = \frac{1}{n}\Sigma Y^2 - \bar{Y}^2 = \frac{96}{18} - 1 = \frac{13}{3} \quad ...(4)$$

$$\text{Cov}(X, Y) = \frac{1}{n}\Sigma XY - \overline{XY}$$

$$= \frac{48}{18} - \left(\frac{2}{3}\right)(1) = \frac{48 - 12}{18} = \frac{36}{18} = 2$$

Now
$$\gamma(X, Y) = \frac{\text{Cov}(X, Y)}{\sqrt{\sigma_X^2 \sigma_Y^2}} = \frac{2}{\sqrt{\frac{26}{9} \times \frac{13}{3}}} = \frac{6}{13}\sqrt{\frac{3}{2}}$$

The regression line of Y on X is

$$Y - \bar{Y} = \gamma \frac{\sigma_Y}{\sigma_X}(X - \bar{X})$$

$$Y - 1 = \frac{6}{13}\sqrt{\frac{3}{2}}\sqrt{\frac{3}{2}}\left(X - \frac{2}{3}\right)$$

$$Y - 1 = (0.692)(X - 0.666)$$

$$Y = 0.692X - 0.461 + 1$$

$$Y = 0.692X + 0.538$$

The regression line of X on Y is

$$(X - \bar{X}) = \gamma \frac{\sigma_X}{\sigma_Y}(Y - \bar{Y})$$

$$\Rightarrow \quad X - \frac{2}{3} = \frac{6}{13}\sqrt{\frac{3}{2}} \times \sqrt{\frac{2}{3}}(Y - 1)$$

$$\Rightarrow \quad X - \frac{2}{3} = \frac{6}{13}(Y - 1)$$

$$\Rightarrow \quad X = 0.46\,Y - 0.46 + 0.66$$

$$\Rightarrow \quad X = 0.46Y + 0.20.$$

Example 6: *Two random variables have the least square regression lines with equations*

$$3x + 2y - 26 = 0 \quad \text{and} \quad 6x + y - 31 = 0,$$

find the mean values and the coefficient of correlation between x and y.

Solution: Given equations are

$$3x + 2y - 26 = 0 \qquad ...(1)$$

and

$$6x + y - 31 = 0 \qquad ...(2)$$

Since, the two lines of regression pass through the point $(\bar{x}, \bar{y})$, we have

$$3\bar{x} + 2\bar{y} = 26 \qquad ...(3)$$

and

$$6\bar{x} + \bar{y} = 31 \qquad ...(4)$$

Solving (3) and (4), we get

$$\bar{x} = 4 \quad \text{and} \quad \bar{y} = 7$$

From (1), we have $\quad y = -\dfrac{3}{2}x + 13$

$$\therefore \quad b_{yx} = -\frac{3}{2} \qquad ...(5)$$

Now from (2), we have

$$x = -\frac{1}{6}y + \frac{31}{6}$$

$$\therefore \quad b_{xy} = -\frac{1}{6} \qquad ...(6)$$

Multiplying (5) and (6), we get

$$b_{yx} \cdot b_{xy} = \left(-\frac{3}{2}\right) \times \left(-\frac{1}{6}\right) \qquad \left[\because \; b_{yx} = \gamma \frac{\sigma_y}{\sigma_x} \text{ and } b_{xy} = \gamma \frac{\sigma_x}{\sigma_y}\right]$$

$$\gamma^2 = \frac{1}{4}$$

$$\Rightarrow \quad \gamma = \pm \frac{1}{2} = \pm 0.5$$

Since, both regression coefficient b_{yx} and b_{xy} are negative.

Then $\quad \gamma = -0.5.$

Example 7: *Calculate the coefficient of correlation and obtain the line of regression for the following data.*

X	*1*	*2*	*3*	*4*	*5*	*6*	*7*	*8*	*9*
Y	*9*	*8*	*10*	*12*	*11*	*13*	*14*	*16*	*15*

find also an estimate for Y which would correspond to X = 6.2.

Solution: Let $\overline{X} = 4$ and $\overline{Y} = 13$.

X	Y	$U = X - \overline{X}$	$V = Y - \overline{Y}$	U^2	V^2	UV
1	9	−3	−4	9	16	12
2	8	−2	−5	4	25	10
3	10	−1	−3	1	9	3
4	12	0	−1	0	1	0
5	11	1	−2	1	4	−2
6	13	2	0	4	0	0
7	14	3	1	9	1	3
8	16	4	3	16	9	12
9	15	5	2	25	4	10
		$\Sigma U = 9$	$\Sigma V = -9$	$\Sigma U^2 = 69$	$\Sigma V^2 = 69$	$\Sigma UV = 48$

Hence,
$$\overline{U} = \frac{1}{n}\Sigma U = \frac{1}{9} \times 9 = 1,$$

$$\overline{V} = \frac{1}{n}\Sigma V = \frac{1}{9} \times (-9) = -1$$

$$\text{Cov}\,(U, V) = \frac{1}{n}\Sigma UV - \overline{U}\,\overline{V} = \frac{1}{9} \times 48 - 1(-1) = 5.33 + 1 = 6.33$$

$$\sigma_U^2 = \frac{1}{n}\Sigma U^2 - \overline{U}^2 = \frac{1}{9} \times 69 - 1 = \frac{60}{9} = 6.66$$

$$\sigma_V^2 = \frac{1}{n}\Sigma V^2 - \overline{V}^2 = \frac{1}{9} \times 69 - 1 = \frac{60}{9} = 6.66$$

$$\gamma = \frac{\text{Cov}\,(UV)}{\sqrt{\sigma_U^2\,\sigma_V^2}} = \frac{6.33}{\sqrt{6.66 \times 6.66}} = \frac{6.33}{6.66} = 0.95.$$

So $\bar{X} = 4, \quad \bar{Y} = 13,$

$$\sigma_X = \sigma_Y = \sqrt{\frac{60}{9}}$$

The regression line of Y on X is

$$Y - \bar{Y} = \gamma \frac{\sigma_Y}{\sigma_X}(X - \bar{X})$$

$$\Rightarrow \quad Y - 13 = (0.95)\frac{\sqrt{60/9}}{\sqrt{60/9}}(X - 4)$$

$$\Rightarrow \quad Y - 13 = (0.95)(X - 4)$$

$$\Rightarrow \quad Y = 0.95X - 3.80 + 13$$

$$\Rightarrow \quad Y = 0.95X + 9.20.$$

Example 8: *Show that the A.M. of the coefficient of regression is greater than the coefficient of correlation.*

Solution: The regression line of Y on X is

$$Y - \bar{Y} = \gamma \frac{\sigma_Y}{\sigma_X}(X - \bar{X}) \quad ...(1)$$

The regression line of X on Y is

$$X - \bar{X} = \gamma \frac{\sigma_X}{\sigma_Y}(Y - \bar{Y}) \quad ...(2)$$

Now to show that, we have

$$\frac{\gamma \frac{\sigma_Y}{\sigma_X} + \gamma \frac{\sigma_X}{\sigma_Y}}{2} > \gamma$$

$$\Rightarrow \quad \frac{\sigma_Y^2 + \sigma_X^2}{2\sigma_X \sigma_Y} > 0 \quad \Rightarrow \quad \sigma_Y^2 + \sigma_X^2 - 2\sigma_X\sigma_Y > 0$$

$$\Rightarrow \quad (\sigma_Y - \sigma_X)^2 > 0$$

Which is true.

EXERCISE 3.2

1. Find the correlation coefficient and the equation of regression lines for the following values of X and Y:

X	1	2	3	4	5
Y	2	5	3	8	7

2. Find Cov (X, Y), when $\Sigma X = 50$, $\Sigma Y = -30$, $\Sigma XY = -115$, $n = 10$.

3. Given the following data:

	X	Y
Arithmetic mean	36	85
Standard deviation	11	8

Correlation coefficient between X and Y is 0.66.

(*a*) Find two regression lines.

(*b*) Estimate value of X, when $Y = 75$.

4. Find the coefficient of correlation for X and Y from the following data:

X	45	55	56	58	60	65	68	70	75	80	85
Y	56	50	48	60	62	64	65	70	74	82	90

Also find the equation of the lines of regression.

5. The lines of regression for a data are given as under:

$$2y - x - 50 = 0, \quad 3y - 2x - 10 = 0.$$

Show that the regression estimate of y for $x = 150$ is 100, whereas the regression estimate of x corresponding to $y = 100$ is 145. Explain the difference.

6. The variables x and y are connected by the equations $ax + by + c = 0$, show that the correlation between them is -1, if the sign of a and b are like, and $+1$ if they are different.

7. A study of prices of a certain commodity at Hapur and Kanpur yield the following data:

	Hapur Rs	*Kanpur Rs*
Average price per kilo	2.463	2.797
Standard deviation	0.326	0.207
γ between prices at Hapur and Kanpur	0.774	

Estimate from the above data the most likely price

(*i*) At Kanpur corresponding to the price of Rs. 3.052 per kilo at Hapur.

(*ii*) At Hapur corresponding to the price of Rs. 2.334 per kilo at Kanpur.

8. Heights of fathers and sons are given in the inches

Height of Father (X)	65	66	67	67	68	69	71	73
Height of Son (Y)	67	68	64	68	72	70	69	70

From the two lines of regression and calculate the expected average height of the son when the height of the father is 67.5 inches.

9. The equations of two regression lines obtained in a correlation analysis of 60 observations are $5x = 6y + 24$ and $1000\,y = 768x - 3608$. What is the correlation coefficient and what is its probable error ?

Show that the ratio of the coefficient of variability of x to that of y is $\frac{5}{24}$. What is the ration of variances of x and y?

10. In a partially destroyed laboratory record of an analysis correlation data, the following results only are legible:

variance of $x = 9$. Regression equations:

$$8x - 10y + 66 = 0, \qquad 40x - 18y = 214.$$

(*i*) What are the mean values of x and y,

(*ii*) The standard deviation of y,

(*iii*) The coefficient of correlation between x and y?

11. A student has obtained the following answers to certain problems. Discuss the criticise them

(*i*) Mean = 3, variance = 5 for a binomial distribution.

(*ii*) Mean = 4, variance = 3 for a poisson distribution.

(*iii*) Coefficient of regression of y on $x = 3.2$.

Coefficient of regression of x on $y = 0.8$ for a bivarate distribution.

12. Prove that:

(*a*) $E_{yx} = \sigma_y (1 - \gamma^2)^{1/2}$ (*b*) $E_{xy} = \sigma_x (1 - \gamma^2)^{1/2}$.

ANSWERS

1. 0.81, $X = 0.5Y + 0.5$, $Y = 1.3X + 1.1$

2. 3.5

3. (*a*) $X = 0.9075Y - 41.1375$, $Y = 0.48X + 67.72$, (*b*) 26.925

4. 0.92, $Y = 0.99X + 1.02$, $X = 0.85\,Y + 9.47$

7. (*i*) 3.086 (*ii*) 1.899

8. $X = 0.524\,Y + 32.29$, $Y = 0.421X + 39.77$, 68.19.

9. $\gamma = \pm\,0.96$, 0.0068, 5/4

10. (*i*) 13,17 (*ii*) 4 (*iii*) 0.6

11. (*i*) Wrong, because here $q = \frac{5}{3} > 1$

(*ii*) Mean = variance (in Poisson distribution) wrong

(*iii*) Contradiction; because $\gamma^2 = 2.56 > 1$, which is never greater than unity.

UNIT V

CALCULUS OF VARIATIONS AND TRANSFORMS

In this chapter, we deal with calculus of variations and problems of determining the maximum or minimum values of a definite integral involving in a certain function, strong variation, weak variation, simple variation problems and Euler's equation are also discussed.

Calculus of variations has a great importance to solving the problems of dynamics of rigid bodies, as a unifying influence in mechanics, vibration problems and as a guide in the mathematical interpretation of many physical phenomena. First in 1744 Euler, discovered the basic differential equation for a minimizing curve. The calculus of variations also deals with minimum problems depending upon surfaces, analytical mechanics, optimization of orbits and the study of extrema of functionals.

In differential calculus, we discuss the problems of maxima and minima of functions. The calculus of variations is concerned with maximizing or minimizing functionals.

CHAPTER 1

Calculus of Variations

1.1. FUNCTIONAL

A real valued function f whose domain is the set of real functions is called a functional. Let us consider the problem to determine a curve $y = y(x)$ through two points $P(x_1, y_1)$ and $Q(x_2, y_2)$ for which $y(x_1) = y_1$, $y(x_2) = y_2$ such that

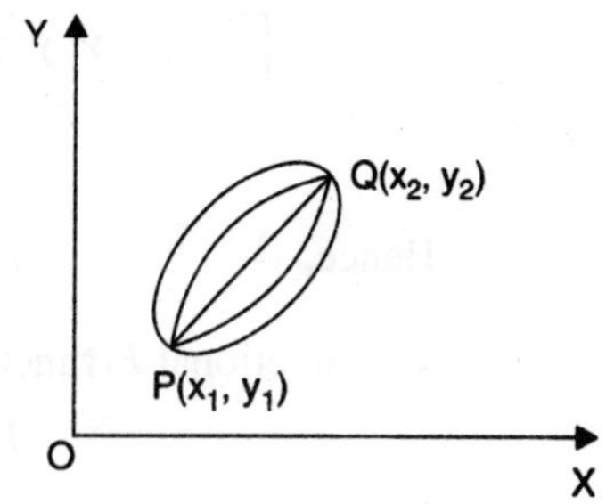

$$\int_{x_1}^{x_2} \sqrt{1 + \left(\frac{dy}{dx}\right)^2}\, dx \qquad ...(1.1)$$

is a minimum.

In general terms, to determine the curve $y = y(x)$, where $y(x_1) = y_1$ and $y(x_2) = y_2$ such that for a given function $f\left(x, y, \frac{dy}{dx}\right)$ or $f(x, y, y')$,

$$\int_{x_1}^{x_2} f(x, y, y')\, dx \text{ is maximum or minimum} \qquad ...(1.2)$$

Hence, an integral such as (1.1) and (1.2), which assumes a definite value for functions of the type $y = y(x)$ is known as functional.

Note. A function $y = y(x)$, which extremizes a functional is known as extremal or extremizing function.

1.2. EULER'S EQUATION

A necessary condition for

$$I = \int_{x_1}^{x_2} f(x, y, y')\, dx \qquad \text{with } y(x_1) = y_1 \text{ and } y(x_2) = y_2$$

to be an extremum (maximum or minimum) is that

$$\frac{\partial f}{\partial y} - \frac{d}{dx}\left(\frac{\partial f}{\partial y'}\right) = 0, \quad \text{where } y' = \frac{dy}{dx}$$

This is called Euler's equation.

Proof: Given that $$I = \int_{x_1}^{x_2} f(x, y, y')\, dx \qquad ...(1.3)$$

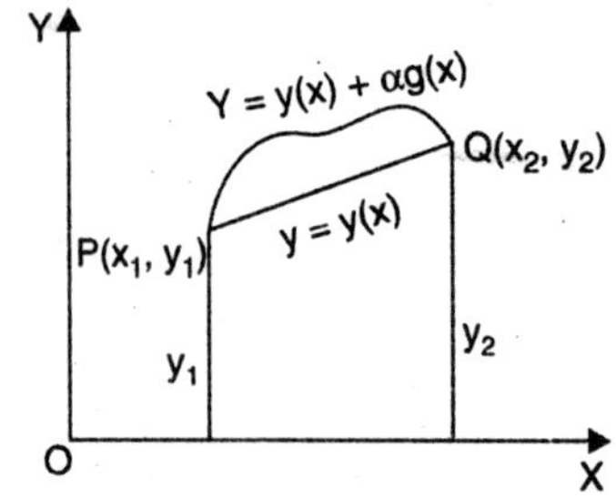

Let $y = y(x)$ be the curve joining two points $P(x_1, y_1)$ and $Q(x_2, y_2)$ which makes the given function I an extremum.

Let $$Y = y(x) + \alpha g(x) \qquad ...(1.4)$$

be a neighbouring curve joining these point P and Q, where α is a small parameter but independent of x and $g(x)$ is an arbitrary differentiable function.

At the end point P and Q, $g(x_1) = 0 = g(x_2)$ ...(1.5)

When $\alpha = 0$, neighbouring curve (1.4) becomes

$Y = y(x)$, which is extremal

If we replace y by Y in equation (1.3) then, we get

$$\int_{x_1}^{x_2} f(x, Y, Y')\, dx = \int_{x_1}^{x_2} f\{(x, y(x) + \alpha g(x), y'(x) + \alpha g'(x)\}\, dx$$

$$= I(\alpha) \quad \text{which is a function of } \alpha$$

Hence, $$I(\alpha) = \int_{x_1}^{x_2} f(x, Y, Y')\, dx \qquad ...(1.6)$$

The functional I (function of α) is a maximum or minimum for $\alpha = 0$, when

$$I'(\alpha) = 0 \qquad ...(1.7)$$

Differentiating equation (1.6) under the integral sign by Leibnitz's rule, we have

$$I'(\alpha) = \int_{x_1}^{x_2} \left(\frac{\partial f}{\partial x}\frac{\partial x}{\partial \alpha} + \frac{\partial f}{\partial Y}\frac{\partial Y}{\partial \alpha} + \frac{\partial f}{\partial Y'}\frac{\partial Y'}{\partial \alpha} \right) dx$$

Since α is independent of x, so $\frac{\partial x}{\partial \alpha} = 0$, we have

$$I'(\alpha) = \int_{x_1}^{x_2} \left(\frac{\partial f}{\partial Y}\frac{\partial Y}{\partial \alpha} + \frac{\partial f}{\partial Y'}\frac{\partial Y'}{\partial \alpha} \right) dx \qquad ...(1.8)$$

Differentiating (1.4) with respect to x, we get

$$Y' = y'(x) + \alpha g'(x) \qquad ...(1.9)$$

Again differentiating (1.9) with respect to α, we get

$$\frac{\partial Y'}{\partial \alpha} = g'(x) \qquad ...(1.10)$$

Now differentiating (1.4) with respect to α, we have

$$\frac{\partial Y}{\partial \alpha} = g(x) \qquad ...(1.11)$$

Using equation (1.8), we have

$$I'(\alpha) = \int_{x_1}^{x_2} \left[\frac{\partial f}{\partial Y} g(x) + \frac{\partial f}{\partial Y'} g'(x) \right] dx$$

Integrating the second term on the right by parts, we have

$$= \int_{x_1}^{x_2} \frac{\partial f}{\partial Y} g(x)\, dx + \left[\left\{ \frac{\partial f}{\partial Y'} g(x) \right\}_{x_1}^{x_2} - \int_{x_1}^{x_2} \frac{d}{dx}\left(\frac{\partial f}{\partial Y'} \right) g(x)\, dx \right]$$

$$= \int_{x_1}^{x_2} \frac{\partial f}{\partial Y} g(x)\, dx + \left[\frac{\partial f}{\partial Y'} g(x_2) - \frac{\partial f}{\partial Y'} g(x_1) \right] - \int_{x_1}^{x_2} \frac{d}{dx}\left(\frac{\partial f}{\partial Y'} \right) g(x)\, dx$$

$$= \int_{x_1}^{x_2} \frac{\partial f}{\partial Y} g(x)\, dx + 0 - \int_{x_1}^{x_2} \frac{d}{dx}\left(\frac{\partial f}{\partial Y'} \right) g(x)\, dx \qquad \text{(Using 1.5)}$$

$$= \int_{x_1}^{x_2} \left\{ \frac{\partial f}{\partial Y} - \frac{d}{dx}\left(\frac{\partial f}{\partial Y'} \right) \right\} g(x)\, dx$$

For extremum value, we have

$$I'(\alpha) = 0$$

$$\Rightarrow \qquad 0 = \int_{x_1}^{x_2} \left\{ \frac{\partial f}{\partial Y} - \frac{d}{dx}\left(\frac{\partial f}{\partial Y'} \right) \right\} g(x)\, dx$$

Since, $g(x)$ is an arbitrary differentiable function then, we have

$$\boxed{\frac{\partial f}{\partial Y} - \frac{d}{dx}\left(\frac{\partial f}{\partial Y'} \right) = 0} \qquad ...(1.12)$$

which is the required Euler's equation.

1.3. EQUIVALENT FORMS OF EULER'S EQUATION

1. Let $$I = \int_{x_1}^{x_2} f(x, y, y')\, dx$$

Here f is a function of x, y and y'

Differentiating f with respect to x, we get

$$\frac{df}{dx} = \frac{\partial f}{\partial x} + \frac{\partial f}{\partial y}\frac{dy}{dx} + \frac{\partial f}{\partial y'}\frac{dy'}{dx}$$

or $$\frac{df}{dx} = \frac{\partial f}{\partial x} + \frac{\partial f}{\partial y} y' + \frac{\partial f}{\partial y'} y'' \qquad ...(1.13)$$

Now consider $$\frac{d}{dx}\left(y' \frac{\partial f}{\partial y'} \right) = y' \frac{d}{dx}\left(\frac{\partial f}{\partial y'} \right) + y'' \frac{\partial f}{\partial y'} \qquad ...(1.14)$$

Subtracting (1.14) from (1.13), we get

$$\frac{df}{dx} - \frac{d}{dx}\left(y' \frac{\partial f}{\partial y'} \right) = \frac{\partial f}{\partial x} + \frac{\partial f}{\partial y} y' - y' \frac{d}{dx}\left(\frac{\partial f}{\partial y'} \right)$$

$$\frac{d}{dx}\left(f - y'\frac{\partial f}{\partial y'}\right) - \frac{\partial f}{\partial x} = y'\left\{\frac{\partial f}{\partial y} - \frac{d}{dx}\left(\frac{\partial f}{\partial y'}\right)\right\}$$

$$\frac{d}{dx}\left(f - y'\frac{\partial f}{\partial y'}\right) - \frac{\partial f}{\partial x} = 0 \qquad \text{(Using Euler's equation)}$$

Hence
$$\boxed{\frac{d}{dx}\left[f - y'\frac{\partial f}{\partial y'}\right] - \frac{\partial f}{\partial x} = 0} \qquad ...(1.15)$$

2. Since $\frac{\partial f}{\partial y'}$ is also a function of x, y, y'.

Let
$$\frac{\partial f}{\partial y'} = h\,(x, y, y') \qquad ...(1.16)$$

Then
$$\frac{d}{dx}\left(\frac{\partial f}{\partial y'}\right) = \frac{dh}{dx} = \frac{\partial h}{\partial x} + \frac{\partial h}{\partial y}\frac{dy}{dx} + \frac{dh}{dy'}\frac{dy'}{dx}$$

$$= \frac{\partial}{\partial x}\left(\frac{\partial f}{\partial y'}\right) + \frac{\partial}{\partial y}\left(\frac{\partial f}{\partial y'}\right)y' + \frac{\partial}{\partial y'}\left(\frac{\partial f}{\partial y'}\right)y''$$

$$= \frac{\partial^2 f}{\partial x \partial y'} + \frac{\partial^2 f}{\partial y \partial y'}y' + \frac{\partial^2 f}{\partial y'^2}y'' \qquad ...(1.17)$$

Using (1.12) and (1.17), we get

$$\frac{\partial f}{\partial y} - \left\{\frac{\partial^2 f}{\partial x \partial y'} + \frac{\partial^2 f}{\partial y \partial y'}y' + \frac{\partial^2 f}{\partial y'^2}y''\right\} = 0$$

or
$$\boxed{y''\frac{\partial^2 f}{\partial y'^2} + y'\frac{\partial^2 f}{\partial y \partial y'} + \frac{\partial^2 f}{\partial x \partial y'} - \frac{\partial f}{\partial y} = 0} \qquad ...(1.18)$$

1.4. SOLUTION OF EULER'S EQUATIONS

There are several situations in which we can easily be obtain the solution of Euler's equation in following cases:

Case I: When f is independent of x then $\frac{\partial f}{\partial x} = 0.$

By equation (1.15), we have

$$\frac{d}{dx}\left(f - y'\frac{\partial f}{\partial y'}\right) = 0$$

Integrating, we get $$f - y'\frac{\partial f}{\partial y'} = \text{Constant} \quad ...(1.19)$$

Case II: When f is independent of y then $\frac{\partial f}{\partial y} = 0$

By equation (1.12), we have $\frac{d}{dx}\left(\frac{\partial f}{\partial y'}\right) = 0$

Integrating, we get $$\frac{\partial f}{\partial y'} = \text{Constant} \quad ...(1.20)$$

Case III: When f is independent of y' then $\frac{\partial f}{\partial y'} = 0$. By equation (1.21), we have

$$\frac{\partial f}{\partial y} = 0 \quad ...(1.21)$$

Case IV: When f is independent of x and y then $\frac{\partial f}{\partial x} = 0$ and $\frac{\partial f}{\partial y} = 0$

$$\frac{\partial^2 f}{\partial x \partial y'} = 0 \quad \text{and} \quad \frac{\partial^2 f}{\partial y \partial y'} = 0$$

By equation (18), we have

$$y'' \frac{\partial^2 f}{\partial y'^2} = 0 \quad ...(1.22)$$

If $\frac{\partial^2 f}{\partial y'^2} \neq 0$ then $y'' = 0$

which gives a solution of the form

$$y = ax + b. \quad ...(1.23)$$

Note: Any function which satisfies Euler's equation is called a Extremal. Extremal is obtained by solving the Euler's equation.

1.5. STRONG AND WEAK VARIATIONS

Suppose $y = f(x)$ is a curve passing through $P(x_1, y_1)$ and $Q(x_2, y_2)$. Let another curve $y = f_1(x)$ through P and Q be obtained by displacing points on $y = f(x)$ parallel to y-axis, the displacement of P and Q, being zero, then variation δy of y is defined as

$$\delta y = f(x) - f_1(x)$$

and variation is defined as $\delta y' = f'(x) - f_1'(x)$

where $$\delta y = \frac{d}{dx}(\delta y')$$

Here, if δy is small then the variation is called strong variation and if $\delta y'$ is small, then the variation is called weak variation.

1.6. ISOPERIMETRIC PROBLEMS

The problem of maximum or minimum with constraints, it is required to determine the maximum or minimum of a function of several variable $g(x_1, x_2, \ldots\ldots, x_n)$ where the variables $x_1, x_2, \ldots\ldots, x_n$ are connected by some given condition or relation called a constraints. In calculus of variations, in some problems, it is necessary to determine extremum of an integral while one or more integrals involving the same variables and the same limits are to be kept constant. Such type problems are called isoperimetric problems and mostly solved by the method of Lagrange's multipliers. For example, to determine the shape of the closed curve of the given perimeter enclosing maximum area.

Let the isoperimetric problem consists of determine a function $f(x)$ which extremizes the functional

$$I = \int_{x_1}^{x_2} f(x, y, y')\, dx \qquad \ldots(1.24)$$

Subject to the condition that the another integral

$$J = \int_{x_1}^{x_2} g(x, y, y')\, dx = C \qquad \ldots(1.25)$$

To solve this type of problem, we use the method of Lagrange's multipliers and consider an integral

$$H = \int_{x_1}^{x_2} h(x, y, y')\, dx \qquad \ldots(1.26)$$

where $h = f + \lambda g$ and λ is the Lagrange's multiplier.

For H is extremum, if I is extremum because J is constant. Then, the modified Euler's equation is given by

$$\boxed{\frac{\partial h}{\partial y} - \frac{d}{dx}\left(\frac{\partial h}{\partial y'}\right) = 0}$$

1.7. VARIATIONAL PROBLEMS INVOLVING SEVERAL DEPENDENT VARIABLES

Let us consider the functional

$$I = \int_{x_1}^{x_2} f(x, y_1, y_2, \ldots\ldots, y_n, y_1', y_2', \ldots\ldots, y_n')\, dx \qquad \ldots(1.27)$$

involving n functions $y_1, y_2, \ldots\ldots, y_n$ of x.

The necessary condition for this integral (1.27) to be extremum is

$$\frac{\partial f}{\partial y_i} - \frac{d}{dx}\left(\frac{\partial f}{\partial y_i'}\right) = 0, \quad i = 1, 2, 3, \ldots\ldots, n$$

These are Euler's equations for the n functions.

1.8. FUNCTIONALS INVOLVING SECOND ORDER DERIVATIVES

Let the extremum of a functional

$$I = \int_{x_1}^{x_2} f(x, y, y', y'')\, dx$$

A necessary condition for above functional to be extremum is

$$\frac{\partial f}{\partial y} - \frac{d}{dx}\left(\frac{\partial f}{\partial y'}\right) + \frac{d^2}{dx^2}\left(\frac{\partial f}{\partial y''}\right) = 0$$

Proof: Given that

$$I = \int_{x_1}^{x_2} f(x, y, y', y'')\, dx \qquad \ldots(1.28)$$

Let $y(x)$ be the function which makes (1.28) extremum and satisfies the boundary conditions

$$y(x_1) = y_1, \quad y(x_2) = y_2, \quad y'(x_1) = y'_1 \text{ and } y'(x_2) = y'_2$$

Now consider a differentiable function $g(x)$ such that

$$\left.\begin{aligned} g(x_1) &= 0 = g(x_2) \\ \text{and} \quad g'(x_1) &= 0 = g'(x_2) \end{aligned}\right\} \qquad \ldots(1.29)$$

Putting $y + \alpha g$ in place of y in (1.28), we get

$$I(\alpha) = \int_{x_1}^{x_2} f(x, y + \alpha g, y' + \alpha g', y'' + \alpha g'')\, dx = \int_{x_1}^{x_2} F\, dx \qquad \ldots(1.30)$$

Differentiating (1.30) under the integral sign, we get

$$\frac{dI}{d\alpha} = \int_{x_1}^{x_2} \frac{dF}{d\alpha}\, dx = \int_{x_1}^{x_2} \left(\frac{\partial F}{\partial y}\frac{\partial y}{\partial \alpha} + \frac{\partial F}{\partial y'}\frac{\partial y'}{\partial \alpha} + \frac{\partial F}{\partial y''}\frac{\partial y''}{\partial \alpha}\right) dx$$

$$= \int_{x_1}^{x_2} \left(\frac{\partial F}{\partial y} g + \frac{\partial F}{\partial y'} g' + \frac{\partial F}{\partial y''} g''\right) dx$$

For extremum value of (1.28), we have

$$\frac{dI}{d\alpha} = 0 \quad \text{when} \quad \alpha = 0$$

$$\Rightarrow \qquad \int_{x_1}^{x_2} \left(\frac{\partial F}{\partial y} g + \frac{\partial F}{\partial y'} g' + \frac{\partial F}{\partial y''} g''\right) dx = 0$$

$$\Rightarrow \qquad \int_{x_1}^{x_2} \frac{\partial F}{\partial y} g\,dx + \int_{x_1}^{x_2} \frac{\partial F}{\partial y'} g'\,dx + \int_{x_1}^{x_2} \frac{\partial F}{\partial y''} g''\,dx = 0$$

Integrating by parts with respect to x, we have

$$\int_{x_1}^{x_2} \frac{\partial f}{\partial y} g\,dx + \left[\frac{\partial f}{\partial y'} g\right]_{x_1}^{x_2} - \int_{x_1}^{x_2} \frac{d}{dx}\left(\frac{\partial f}{\partial y'}\right) g\,dx + \left[\frac{\partial f}{\partial y''} g'\right]_{x_1}^{x_2} - \int_{x_1}^{x_2} \frac{d}{dx}\left(\frac{\partial f}{\partial y''}\right) g'\,dx = 0$$

or
$$\int_{x_1}^{x_2} \frac{\partial f}{\partial y} g\,dx + \left[\frac{\partial f}{\partial y'} g\right]_{x_1}^{x_2} - \int_{x_1}^{x_2} \frac{d}{dx}\left(\frac{\partial f}{\partial y'}\right) g\,dx + \left[\frac{\partial f}{\partial y''} g'\right]_{x_1}^{x_2}$$
$$- \left[\frac{d}{dx}\left(\frac{\partial f}{\partial y''}\right) g\right]_{x_1}^{x_2} + \int_{x_1}^{x_2} \frac{d^2}{dx^2}\left(\frac{\partial f}{\partial y''}\right) g\,dx = 0$$

Using equation (1.29), we get

or
$$\int_{x_1}^{x_2} \frac{\partial f}{\partial y} g\,dx - \int_{x_1}^{x_2} \frac{d}{dx}\left(\frac{\partial f}{\partial y'}\right) g\,dx + \int_{x_1}^{x_2} \frac{d^2}{dx^2}\left(\frac{\partial f}{\partial y''}\right) g\,dx = 0$$

or
$$\int_{x_1}^{x_2} \left[\frac{\partial f}{\partial y} - \frac{d}{dx}\left(\frac{\partial f}{\partial y'}\right) + \frac{d^2}{dx^2}\left(\frac{\partial f}{\partial y''}\right)\right] g\,dx = 0$$

This equation must hold good for all values of $g(x)$, we get

$$\boxed{\frac{\partial f}{\partial y} - \frac{d}{dx}\left(\frac{\partial f}{\partial y'}\right) + \frac{d^2}{dx^2}\left(\frac{\partial f}{\partial y''}\right) = 0} \qquad ...(1.31)$$

Hence in general, a necessary condition for the functional $I = \int_{x_1}^{x_2} f(x, y, y', y'' \ldots\ldots, y^{(n)})\,dx$ to be extremum is

$$\boxed{\frac{\partial f}{\partial y} - \frac{d}{dx}\left(\frac{\partial f}{\partial y'}\right) + \frac{d^2}{dx^2}\left(\frac{\partial f}{\partial y''}\right) - \ldots\ldots + (-1)^n \frac{d^n}{dx^n}\left(\frac{\partial f}{\partial y^{(n)}}\right) = 0} \qquad ...(1.32)$$

This equation is called the Euler-Poisson equation and its solutions are called extremals.

SOLVED EXAMPLES

Example 1: *Find the curves on which the functional* $\int_0^1 [(y')^2 + 12xy]\,dx$ *with* $y(0) = 0$ *and* $y(1) = 1$ *can be extremized.*

Solution: The Euler's equation is

$$\frac{\partial f}{\partial y} - \frac{d}{dx}\left(\frac{\partial f}{\partial y'}\right) = 0 \qquad ...(i)$$

Given that $$I = \int_0^1 (y'^2 + 12xy)\, dx \qquad ...(ii)$$

with $$y(0) = 0,\ y(1) = 1$$

Here $$f = y'^2 + 12xy$$

Then $$\frac{\partial f}{\partial y} = 12x \quad \text{and} \quad \frac{\partial f}{\partial y'} = 2y'$$

Putting these values of $\frac{\partial f}{\partial y}$ and $\frac{\partial f}{\partial y'}$ in (*i*), we get

$$12x - \frac{d}{dx}(2y') = 0 \qquad \text{or} \qquad 12x - 2y'' = 0$$

or $$y'' - 6x = 0 \qquad \text{or} \qquad y'' = 6x$$

or $$y' = 3x^2 + c_1 \qquad \text{or} \qquad y = x^3 + c_1 x + c_2 \qquad ...(iii)$$

Applying the given boundary condition in (*iii*), we get

$$y(0) = 0 = 0 + c_1 . 0 + c_2 \quad \Rightarrow \quad c_2 = 0$$

$$y(1) = 1 = 1^3 + c_1 . 1 + c_2 \quad \Rightarrow \quad c_1 + c_2 = 0$$

Solving these equations, we get $c_1 = c_2 = 0$

Hence, the required equation from (*iii*) is

$$y = x^3.$$

Example 2: *Test for an extremum the functional*

$$I[y(x)] = \int_0^1 (xy + y^2 - 2y^2 y')\, dx,\ y(0) = 1,\ y(1) = 2.$$

Solution: The Euler's equation is

$$\frac{\partial f}{\partial y} - \frac{d}{dx}\left(\frac{\partial f}{\partial y'}\right) = 0 \qquad ...(i)$$

Given that $$I = \int_0^1 (xy + y^2 - 2y^2 y')\, dx \qquad ...(ii)$$

with $$y(0) = 1, \quad y(1) = 2$$

Here $$f = xy + y^2 - 2y^2 y'$$

Then $$\frac{\partial f}{\partial y} = x + 2y - 4yy'$$

and $$\frac{\partial f}{\partial y'} = -2y^2$$

Putting these values of $\frac{\partial f}{\partial y}$ and $\frac{\partial f}{\partial y'}$ in (*i*), we get

$$x + 2y - 4yy' - \frac{d}{dx}(-2y^2) = 0$$

or $$x + 2y - 4yy' + 4yy' = 0$$

or $$x + 2y = 0 \quad \text{or} \quad y = -\frac{x}{2} \qquad ...(iii)$$

Applying the given boundary condition in (*iii*), we get

$$y(0) = 1 = 0, \quad \text{which is not possible}$$

$$y(1) = 2 = -\frac{1}{2}, \quad \text{again contradiction.}$$

Hence, these are no extremal because it is not satisfy the boundary condition.

Example 3: *Prove that the shortest distance between two points in a plane is a straight line.*

Solution: Let $P(x_1, y_1)$ and $Q(x_2, y_2)$ be the two given points and s is the length of the arc joining these points.

Then $$S = \int_{x_1}^{x_2} ds = \int_{x_1}^{x_2} \sqrt{1 + \left(\frac{dy}{dx}\right)^2}\, dx = \int_{x_1}^{x_2} \sqrt{1 + y'^2}\, dx \qquad ...(i)$$

with $$y(x_1) = y_1, \quad y(x_2) = y_2.$$

If S satisfies the Euler's equation then it will be minimum.

The Euler's equation is $$\frac{\partial f}{\partial y} - \frac{d}{dx}\left(\frac{\partial f}{\partial y'}\right) = 0 \qquad ...(ii)$$

In equation (*i*), $$f = \sqrt{1 + y'^2}$$

Then $$\frac{\partial f}{\partial y} = 0, \quad \frac{\partial f}{\partial y'} = \frac{1}{2}(1 + y'^2)^{-1/2}(2y')$$

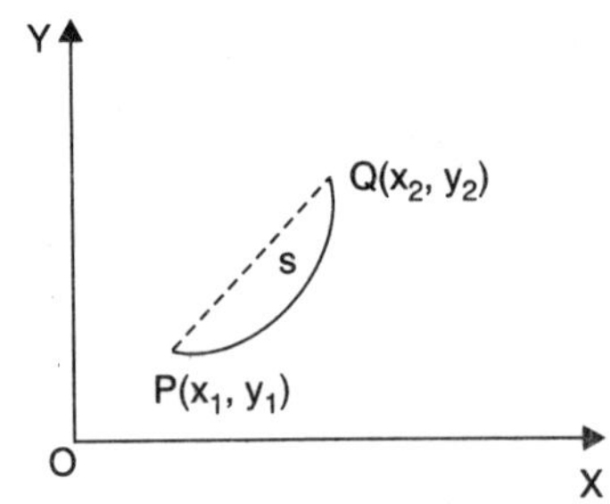

Putting these values of $\frac{\partial f}{\partial y}$ and $\frac{\partial f}{\partial y'}$ in (*ii*), we get

$$0 - \frac{d}{dx}\left(\frac{y'}{\sqrt{1 + y'^2}}\right) = 0 \quad \text{or} \quad \frac{d}{dx}\left(\frac{y'}{\sqrt{1 + y'^2}}\right) = 0$$

Integrating both sides, we get

$$\frac{y'}{\sqrt{1 + y'^2}} = a \quad \text{(Constant)}$$

Squaring both sides, we get

$$y'^2 = a^2(1 + y'^2) \quad \text{or} \quad y'^2(1 - a^2) = a^2$$

or $$y'^2 = \frac{a^2}{1 - a^2} = m^2 \text{ (Let)} \quad \text{or} \quad y'^2 = m^2$$

or $$y' = m \quad \text{on integrating, we get}$$

or $$y = mx + c \qquad ...(iii)$$

which is a straight line.

Now $\quad y(x_1) = y_1$ and $y(x_2) = y_2$

By (*iii*), we have

$$\left.\begin{aligned} mx_1 + c &= y_1 \\ \text{and} \quad mx_2 + c &= y_2 \end{aligned}\right\} \quad \text{...(iv)}$$

On subtracting, we get

$$y_2 - y_1 = m(x_2 - x_1) \quad \text{or} \quad m = \frac{y_2 - y_1}{x_2 - x_1}$$

Subtracting (*iv*) from (*iii*), we get

$$y - y_1 = m(x - x_1) \quad \text{or} \quad y - y_1 = \frac{y_2 - y_1}{x_2 - x_1}(x - x_1).$$

Example 4: *Find the path on which a particle in a absence of friction, will slide from one point to another in the shortest time under the action of gravity.*

Solution: Suppose the particle start sliding on the curve *OA* from *O* with zero velocity. Let *OP* = *S* and *t* be the time taken from *O* to *P*. Using the principle of work and energy, we have

K.E. at *P* – K.E. at *O* = Work done in moving the particle from *O* to *P*.

$$\Rightarrow \quad \frac{1}{2}mv^2 - 0 = mgy$$

$$\text{or} \quad v^2 = 2gy$$

$$\text{or} \quad \left(\frac{ds}{dt}\right)^2 = 2gy$$

$$\text{or} \quad \frac{ds}{dt} = \sqrt{2gy}$$

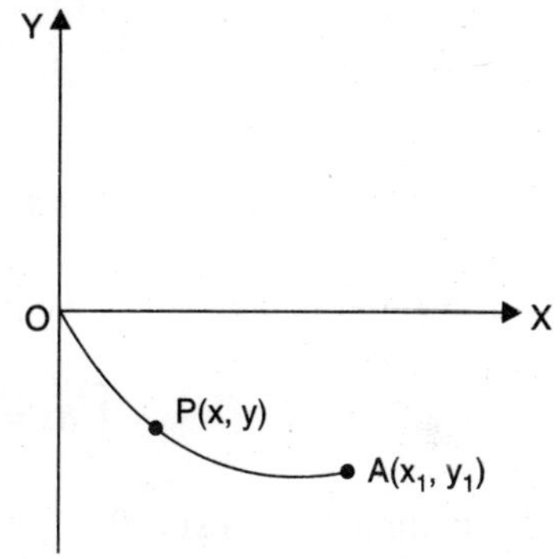

Let *T* be the time taken by the particle from *O* to *A*. Then

$$T = \int_0^T dt = \int_0^{x_1} \frac{dt}{ds}\,ds$$

$$= \int_0^{x_1} \frac{ds}{\sqrt{2gy}} = \frac{1}{\sqrt{2g}}\int_0^{x_1} \frac{ds}{\sqrt{y}}$$

$$= \frac{1}{\sqrt{2g}}\int_0^{x_1} \frac{\sqrt{1+y'^2}}{\sqrt{y}}\,dx \quad \text{...(i)}$$

Here $f = \dfrac{\sqrt{1+y'^2}}{\sqrt{y}}$ {which independent of x}

Then
$$\frac{\partial f}{\partial x} = 0 \quad \text{and} \quad \frac{\partial f}{\partial y'} = \frac{1}{2\sqrt{y}}\frac{2y'}{\sqrt{1+y'^2}} = \frac{y'}{\sqrt{y}\sqrt{1+y'^2}}$$

Since, f is independent of x then Euler's equation reduces to

$$f - y'\left(\frac{\partial f}{\partial y'}\right) = c \text{ (Constant)} \qquad ...(ii)$$

Putting the value of f and $\frac{\partial f}{\partial y'}$ in (ii), we get

$$\frac{\sqrt{1+y'^2}}{\sqrt{y}} - y'\frac{y'}{\sqrt{y}\sqrt{1+y'^2}} = c$$

$$\Rightarrow \qquad \frac{1+y'^2-y'^2}{\sqrt{y}\sqrt{1+y'^2}} = c$$

$$\Rightarrow \qquad 1 = c\sqrt{y}\sqrt{1+y'^2}$$

Squaring both sides, we get

$$1 = c^2 y\,(1+y'^2) \qquad \text{or} \qquad 1+\left(\frac{dy}{dx}\right)^2 = \frac{1}{c^2 y}$$

or $$\frac{dy}{dx} = \sqrt{\frac{1-yc^2}{c^2 y}} = \sqrt{\frac{\frac{1}{c^2}-y}{y}} \qquad \text{or} \qquad \frac{dy}{dx} = \sqrt{\frac{a-y}{y}} \qquad \left\{Let\ \frac{1}{c^2} = a\right\}$$

or $$dx = \sqrt{\frac{y}{a-y}}\,dy$$

Integrating both sides, we get

$$\int_0^x dx = \int_0^y \sqrt{\frac{y}{a-y}}\,dy$$

Putting $y = a\sin^2\theta$ ...(iii)

$$\Rightarrow \qquad dy = 2a\sin\theta\cos\theta\,d\theta$$

$$x = \int_0^\theta \sqrt{\frac{a\sin^2\theta}{a-a\sin^2\theta}}\,.\,2a\sin\theta\cos\theta\,d\theta$$

$$= \int_0^\theta 2a\sin^2\theta\,d\theta = a\int_0^\theta (1-\cos 2\theta)\,d\theta$$

$$= a\left(\theta - \frac{\sin 2\theta}{2}\right)_0^\theta$$

$$x = \frac{a}{2}(2\theta - \sin 2\theta) \qquad ...(iv)$$

Putting $\frac{a}{2} = b$ and $2\theta = \phi$. Then by equations (*iii*) and (*iv*), we get

$$x = b\,(\phi - \sin\phi)$$

and $$y = b\,(1 - \cos\phi)$$

$$\left[\begin{aligned} \therefore\ \ y &= a\sin^2\theta \\ y &= \frac{a}{2}(1-\cos 2\theta) \\ &= b\,(1-\cos\phi) \end{aligned}\right]$$

which is a cycloid.

Example 5: *Show that the geodesics on a sphere of radius a are its great circles.*

Solution: We know that in spherical coordinates (r, θ, ϕ),

$$x = r\sin\theta,\ \cos\phi, \quad y = r\sin\theta\sin\phi, \quad z = r\cos\theta \qquad ...(i)$$

The arc element on a sphere of radius a is given by

$$ds^2 = dr^2 + (rd\theta)^2 + (r\sin\theta\ d\phi)^2$$

$$= a^2\,d\theta^2 + (a\sin\theta)^2\,d\phi^2 \qquad \left[\begin{aligned} \because\ r &= a \\ \text{then } dr &= 0 \end{aligned}\right]$$

or $$ds = a\sqrt{1+\sin^2\theta\,.\left(\frac{d\phi}{d\theta}\right)^2}\,d\theta$$

$$s = \int_{\theta_1}^{\theta_2} a\sqrt{1+\sin^2\theta\,.\,\phi'^2}\,d\theta \qquad ...(ii)$$

Now the geodesic on the sphere $r = a$ is the curve for which s is minimum. From (*ii*), we have

$$f = a\sqrt{1+\sin^2\theta\,.\,\phi'^2}$$

Which is a function of θ and ϕ' while ϕ is absent. Then the Euler's equation reduces to

$$\frac{\partial f}{\partial \phi'} = c_1 \text{ (Constant)}$$

$$\frac{a\sin^2\theta\,.\,\phi'}{a\sqrt{(1+\sin^2\theta\,.\,\phi'^2)}} = c_1$$

Squaring both sides, we get

or $$\sin^4\theta\,.\,\phi'^2 = c_1^2\,(1 + \sin^2\theta\,.\,\phi'^2)$$

or $$\sin^2\theta\,.\,(\sin^2\theta - c_1^2)\,\phi'^2 = c_1^2$$

or $$\frac{d\phi}{d\theta} = \frac{c_1}{\sin\theta\sqrt{\sin^2\theta - c^2}}$$

or $$\frac{d\phi}{d\theta} = \frac{c_1\,\text{cosec}^2\,\theta}{\sqrt{(1-c^2\,\text{cosec}^2\,\theta)}}$$

Integrating both sides, we get

$$\int d\phi = \int \frac{c_1\,\text{cosec}^2\,\theta}{\sqrt{(1-c_1^2\,\text{cosec}^2\,\theta)}}\,d\theta + c_2$$

$$\phi = \int \frac{c_1 \operatorname{cosec}^2 \theta \, d\theta}{\sqrt{(1 - c_1^2) - (c \cot \theta)^2]}} + c_2$$

$$= -\sin^{-1}\left\{\frac{c_1 \cot \theta}{\sqrt{[(1 - c_1^2)]}}\right\} + c_2$$

or $$\sin(\phi + c_2) = \frac{c_1 \cot \theta}{\sqrt{[(1 - c_1^2)]}}$$

$$\sin \phi \cos c_2 + \cos \phi \sin c_2 = \frac{c_1 \cot \theta}{\sqrt{(1 - c_1^2)}}$$

$$\cot \theta = \frac{\cos c_2 \sqrt{(1 - c_1^2)}}{c_1} \sin \phi + \frac{\sin c_2 \sqrt{(1 - c_1^2)}}{c_1} \cos \phi$$

$$\cot \theta = A \sin \phi + B \cos \phi$$

or $$a \cos \theta = Aa \sin \theta \sin \phi + B\, a \cos \phi \sin \theta$$

or $$Z = A\, y + Bx \qquad [\because \; r = a]$$

which is the required geodesics arcs of the great circles.

Example 6: *Find the plane curve of fixed perimeter and maximum area.*

Solution: Let l be the perimeter of the closed curve, then

$$l = \int_{x_1}^{x_2} \sqrt{1 + y'^2}\, dx \qquad ...(i)$$

The area enclosed by the curve, x-axis and two perpendicular lines is

$$A = \int_{x_1}^{x_2} y\, dx \qquad ...(ii)$$

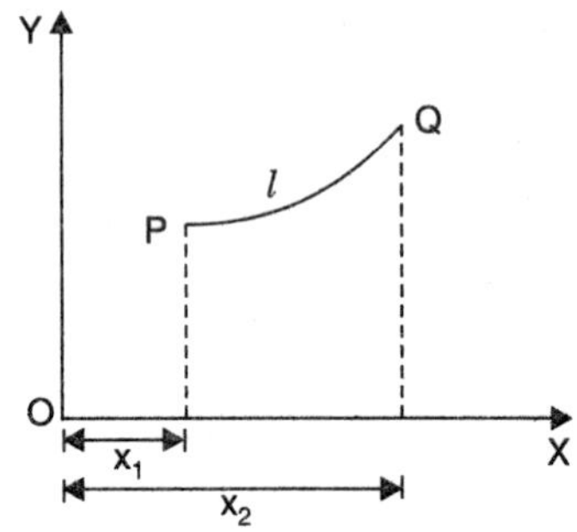

Now we have to find maximum value of (*ii*) subject to constraint (*i*). Using Lagrange's multipliers, we have

$$f = y + \lambda\sqrt{1 + y'^2}$$

For maximum or minimum value of A, f must satisfy the Euler's equation

$$\frac{\partial f}{\partial y} - \frac{d}{dx}\left(\frac{\partial f}{\partial y'}\right) = 0$$

$$\Rightarrow \qquad 1 - \lambda \frac{d}{dx}\left\{\frac{1}{2}(1 + y'^2)^{-1/2}(2y')\right\} = 0$$

$$\Rightarrow \qquad 1 - \lambda \frac{d}{dx}\left(\frac{y'}{\sqrt{1 + y'^2}}\right) = 0$$

Integrating with respect to x, we get

$$x - \frac{\lambda y'}{\sqrt{1+y'^2}} = c_1$$

$$\Rightarrow \qquad \frac{\lambda y'}{\sqrt{1+y'^2}} = x - c_1$$

Squaring both sides, we get

$$\Rightarrow \qquad \lambda^2 y'^2 = (x-c_1)^2 (1+y'^2)$$

$$\Rightarrow \qquad \{\lambda^2 - (x-c_1)^2\} y'^2 = (x-c_1)^2$$

$$\Rightarrow \qquad y'^2 = \frac{(x-c_1)^2}{\{\lambda^2 - (x-c_1)^2\}}$$

$$y' = \frac{(x-c_1)}{\{\lambda^2 - (x-c_1)^2\}^{1/2}}$$

Integrating with respect to x both sides, we get

$$y = -\{\lambda^2 - (x-c_1)^2\}^{1/2} + c_2$$

or

$$y - c_2 = -\{\lambda^2 - (x-c_1)^2\}^{1/2}$$

Squaring both sides, we get

$$(y-c_2)^2 = +\{\lambda^2 - (x-c_1)^2\} \quad \text{or} \quad (x-c_1)^2 + (y-c_2)^2 = \lambda^2$$

This is the equation of a circle whose centre is (c_1, c_2) and radius is λ.

Example 7: *Prove that the sphere is the solid figure of revolution which, for a given surface area, has maximum volume.*

Solution: Let us consider the arc *OPA* of the curve which rotates about the x-axis.

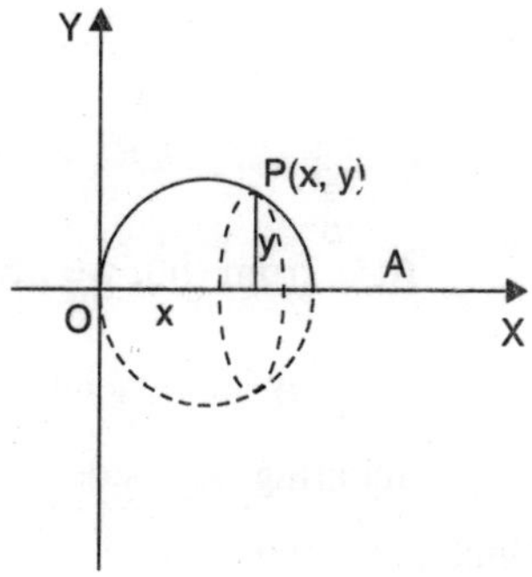

Then the surface

$$S = \int_0^a 2\pi y \, ds = \int_0^a 2\pi y \sqrt{1+y'^2} \, dx \qquad ...(i)$$

and volume $V = \int_0^a \pi y^2 \, dx$ ...(ii)

Now we have to maximize V with the given surface S.

Let

$$f = \pi y^2, \quad g = 2\pi y \sqrt{1+y'^2}$$

and

$$h = f + \lambda g = \pi y^2 + \lambda 2\pi y \sqrt{1+y'^2}$$

Since, h is independent of x *i.e.*, h does not contain x. Then the Euler equation reduce to

$$h - y'\left(\frac{\partial h}{\partial y'}\right) = c_1 \text{ (Constant)} \qquad ...(iii)$$

or $$\pi y^2 + \lambda 2\pi\, y\sqrt{1+y'^2} - y'\,\frac{1}{2}\,\frac{2\pi y\lambda(2y')}{\sqrt{1+y'^2}} = c_1$$

or $$\pi y^2 + 2\pi y\,\lambda\,\sqrt{1+y'^2} - \frac{2\pi\lambda\, y\, y'^2}{\sqrt{1+y'^2}} = c_1$$

or $$\pi y^2 + \frac{2\pi y\lambda\,(1+y'^2) - 2\pi y\lambda y'^2}{\sqrt{1+y'^2}} = c_1$$

or $$\pi y^2 + \frac{2\pi y\lambda}{\sqrt{1+y'^2}} = c_1$$

Since, the curve passes through the origin O and A for which $y = 0$, so $c_1 = 0$

$$\pi y^2 + \frac{2\pi y\lambda}{\sqrt{1+y'^2}} = 0 \quad \text{or} \quad y + \frac{2\lambda}{\sqrt{1+y'^2}} = 0$$

or $$y\sqrt{1+y'^2} = -\,2\lambda$$

squaring both sides, we get

$$1 + \left(\frac{dy}{dx}\right)^2 = \frac{4\lambda^2}{y^2} \quad \text{or} \quad \frac{dy}{dx} = \sqrt{\frac{4\lambda^2 - y^2}{y^2}}$$

or $$\frac{y\,dy}{\sqrt{4\lambda^2 - y^2}} = dx$$

Integrating both sides, we get

$$-\sqrt{4\lambda^2 - y^2} = x + c_2$$

or $$\sqrt{4\lambda^2 - y^2} = -\,(x + c_2) \qquad \ldots(iv)$$

At 0, from (*iv*), we get $\qquad c_2 = -\,2\,\lambda$

Then by (*iv*), we have $\qquad \sqrt{4\lambda^2 - y^2} = -\,(x - 2\lambda)$

Squaring both sides, we get $\;4\lambda^2 - y^2 = (x - 2\lambda)^2\;$ or $\qquad (x - 2\lambda)^2 + y^2 = 4\lambda^2$

which is a circle

Hence, on revolving the circle about x-axis, the solid formed is a sphere.

Example 8: *Show that the functional* $\displaystyle\int_0^{\pi/2}\left\{2xy + \left(\frac{dx}{dt}\right)^2 + \left(\frac{dy}{dt}\right)^2\right\}dt$ *such that* $x(0) = 0$, $x(\pi/2) = -1$, $y(0) = 0$, $y(\pi/2) = 1$ *is extremum for* $x = -\sin t$, $y = \sin t$.

Solution: The Euler's equations are

$$\frac{\partial f}{\partial x} - \frac{d}{dt}\left(\frac{\partial f}{\partial x'}\right) = 0 \qquad ...(i)$$

$$\frac{\partial f}{\partial y} - \frac{d}{dt}\left(\frac{\partial f}{\partial y'}\right) = 0 \qquad ...(ii)$$

Given that $\int_0^{\pi/2} \left\{2xy + \left(\frac{dx}{dt}\right)^2 + \left(\frac{dy}{dt}\right)^2\right\} dt$

Here $f = 2xy + x'^2 + y'^2$

Then $\frac{\partial f}{\partial x} = 2y,\ \frac{\partial f}{\partial x'} = 2x',\ \frac{\partial f}{\partial y} = 2x$ and $\frac{\partial f}{\partial y'} = 2y'$

Putting these values of $\frac{\partial f}{\partial x}$ and $\frac{\partial f}{\partial x'}$ in (*i*), we get

$$2y - \frac{d}{dt}(2x') = 0 \quad \Rightarrow \quad \frac{d^2x}{dt^2} - y = 0$$

$$\Rightarrow \qquad \frac{d^2x}{dt^2} = y \qquad ...(iii)$$

Also putting the values of $\frac{\partial f}{\partial y}$ and $\frac{\partial f}{\partial y'}$ in (*ii*), we get

$$2x - \frac{d}{dt}(2y') = 0 \quad \Rightarrow \quad \frac{d^2y}{dt^2} - x = 0$$

$$\Rightarrow \qquad \frac{d^2y}{dt^2} = x \qquad ...(iv)$$

From (*iii*) and (*iv*), we have

$$\frac{d^4x}{dt^2} = \frac{d^2y}{dt^2} = x \qquad \text{or} \quad (D^4 - 1)x = 0$$

Its solution is

$$x(t) = c_1e^t + c_2e^{-t} + c_3\cos t + c_4\sin t \qquad ...(v)$$

Now from (*iii*), we have

$$y(t) = x''$$
$$= c_1 e^t + c_2 e^{-t} - c_3\cos t - c_4\sin t \qquad \text{[using (v)]} \qquad ...(vi)$$

Applying the given boundary conditions in (*v*) and (*vi*), we have

$$x(0) = 0 = c_1 + c_2 + c_3$$
$$x(\pi/2) = -1 = c_1 e^{\pi/2} + c_2 e^{-\pi/2} + c_4$$
$$y(0) = 0 = c_1 + c_2 - c_3$$

$$y(\pi/2) = 1 = c_1 e^{\pi/2} + c_2 e^{-\pi/2} - c_4$$

Solving these equations, we get

$$c_1 = c_2 = c_3 = 0 \quad \text{and} \quad c_4 = -1$$

Hence, the required solution from (v) and (vi) is

$$x = -\sin t \quad \text{and} \quad y = \sin t.$$

Example 9: *Show that the curve which extremizes the functional* $I = \int_0^{\pi/4} (y''^2 - y^2 + x^2)\, dx$ *under the conditions* $y(0) = 0,\ y'(0) = 1,\ y(\pi/4) = y'(\pi/4) = \frac{1}{\sqrt{2}}$ *is* $y = \sin x$.

Solution: The Euler's equation for second order derivatives is

$$\frac{\partial f}{\partial y} - \frac{d}{dx}\left(\frac{\partial f}{\partial y'}\right) + \frac{d^2}{dx^2}\left(\frac{\partial f}{\partial y''}\right) = 0 \qquad ...(i)$$

Given that $\qquad I = \int_0^{\pi/4} (y''^2 - y^2 + x^2)\, dx \qquad ...(ii)$

Here $\qquad f = y''^2 - y^2 + x^2$

Then $\qquad \frac{\partial f}{\partial y} = -2y, \quad \frac{\partial f}{\partial y'} = 0 \quad \text{and} \quad \frac{\partial f}{\partial y''} = 2y''$

Putting these values in (i), we get

$$-2y - \frac{d}{dx}(0) + \frac{d^2}{dx^2}(2y'') = 0 \quad \text{or} \quad -2y + 2y^{(4)} = 0$$

or $\qquad y^{(4)} - y = 0 \quad \text{or} \quad (D^4 - 1)\, y = 0$

Its solution is

$$y(x) = c_1 e^x + c_2 e^{-x} + c_3 \cos x + c_4 \sin x \qquad ...(iii)$$

$$\therefore \qquad y'(x) = c_1 e^x - c_2 e^{-x} - c_3 \sin x + c_4 \cos x \qquad ...(iv)$$

Applying the given boundary condition in (iii) and (iv), we get

$$y(0) = 0 = c_1 + c_2 + c_3$$

$$y'(0) = 1 = c_1 - c_2 + c_4$$

$$y(\pi/4) = \frac{1}{\sqrt{2}} = c_1 e^{\pi/4} + c_2 e^{-\pi/4} + \frac{1}{\sqrt{2}} c_3 + \frac{1}{\sqrt{2}} c_4$$

$$y'(\pi/4) = \frac{1}{\sqrt{2}} = c_1 e^{\pi/4} - c_2 e^{-\pi/4} - \frac{1}{\sqrt{2}} c_3 + \frac{1}{\sqrt{2}} c_4$$

Solving these equations, we get

$$c_1 = c_2 = c_3 = 0 \quad \text{and} \quad c_4 = 1$$

Hence, the required solution from (iii) is $y = \sin x$.

EXERCISE 1.1

1. On which curve the functional $\int_0^{\pi/2} (y'^2 - y^2 + 2xy)\,dy$ with $y(0) = 0$ and $y(\pi/2) = 0$, be extremized?

2. Find a complete solution of the Euler's equation for $\int_{x_1}^{x_2} (y^2 - y'^2 - 2y \cosh x)\,dx$.

3. Find the extremal of the functional $I = \int_{x_1}^{x_2} \frac{1+y^2}{y'^2}\,dx$.

4. Find the extremals of the functional $\int_{x_1}^{x_2} \frac{y'^2}{x^3}\,dx$.

5. Solve the Euler's equation for $\int_{x_1}^{x_2} (x + y')y'\,dx$.

6. Solve the Euler's equation for $\int_{x_1}^{x_2} (1 + x^2 y')y'\,dx$.

7. Find the extremal of the function and extremum value of the

$$\int_0^{\pi/2} (y'^2 - y^2)\,dx$$

subject to $y(0) = 0$, $y(\pi/2) = 1$.

8. Solve the variational problem

$$\int_1^2 [x^2 y'^2 + 2y(x + y)]\,dx = 0, \text{ given } y(1) = y(2) = 0.$$

9. Find the geodesics on a right circular cylinder of radius a.

10. Find the extremal of the function $I = \int_0^{\pi} (y'^2 - y^2)\,dx$ with boundary conditions $y(0) = 0$, $y(\pi) = 1$ and subject to the constraint $\int_0^{\pi} y\,dx = 1$.

11. Find a function $y(x)$ for which $\int_0^1 (x^2 + y'^2)\,dx$ is extremum, given that $\int_0^1 y^2\,dx = 2$, $y(0) = 0$, $y(1) = 0$.

12. Show that the functional $\int_0^1 \left\{2x + \left(\frac{dx}{dt}\right)^2 + \left(\frac{dy}{dt}\right)^2\right\} dt$, such that $x(0) = 1$, $y(0) = 1$, $x(1) = 1.5$, $y(1) = 1$ is extremum for $x = \frac{1+t^2}{2}$, $y = t$.

13. Find the extremals of the functionals $\int_0^{\pi/2} (y''^2 - y^2 + x^2)\,dx$ that satisfies the conditions $y(0) = 1$, $y'(0) = 0$, $y(\pi/2) = 0$, $y'(\pi/2) = -1$.

14. Find the extremal of $\int_{x_1}^{x_2} (16y^2 - y''^2 + x^2)\,dx$.

15. Find the extremal of $\int_{x_1}^{x_2} (2xy - y'''^2)\,dx$.

16. Find the curve passing through the points (x_1, y_1) and (x_2, y_2) which when rotated about the x-axis gives a minimum surface area.

ANSWERS

1. $y = x - \dfrac{\pi}{2} \sin x.$

2. $y = c_1 \cos x + c_2 \sin x + \dfrac{1}{2} \cosh x.$

3. $y = \sinh (c_1 x + c_2).$

4. $y = c_1 x^4 + c_2.$

5. $y = \dfrac{-x^4}{4} + c_1 x + c_2.$

6. $y = c_1 x^{-1} + c_2.$

7. $y = \sin x$, value = 0.

8. $y = \dfrac{1}{21} \{8 \log 2 (x^{-2} - x) + 7x \log x\}.$

9. $z = c_1 \phi + c_2.$

10. $y = -\dfrac{1}{2} \cos x + \left(\dfrac{1}{2} - \dfrac{\pi}{4}\right) \sin x + \dfrac{1}{2}.$

11. $y = \pm 2 \sin m\pi x$, where m is an integer.

12. $y = (c_1 + c_2 x) \cos x + (c_3 + c_4 x) \sin x.$

13. $y = \cos x.$

14. $y = c_1 e^{2x} + c_2 e^{-2x} + c_3 \cos 2x + c_4 \sin 2x.$

15. $y = \dfrac{x^7}{7!} + c_1 x^5 + c_2 x^4 + c_3 x^3 + c_4 x^2 + c_5 x + c_6.$

16. $y = c \cosh \left(\dfrac{x + a}{c}\right).$

CHAPTER 2

Z-Transform

INTRODUCTION

In this chapter, we shall discuss a new type of transform, Z-transform. The progress of communication engineering is based on discrete analysis. Z-transform plays an important role in solving difference equation which represent a discrete system. Thus, the study of Z-transform is necessary part for engineers and scientists. Z-transform has many properties like Laplace transform but Z-transform operates on a sequences $\{u_n\}$ of discrete integer-(½) valued arguments ($n = 0, \pm 1, \pm 2, ...$) and Laplace transform operates on a continuous function that is the main difference between these transforms.

2.1. Z-TRANSFORM

Let $\{u_n\}$ be a sequence defined for discrete values $n = 0, 1, 2, 3, ...$ then, the Z-transform of sequence $\{u_n\}$ is denoted by $Z(u_n)$ and defined as

$$Z(u_n) = \sum_{n=-\infty}^{\infty} u_n \, z^{-n} = U(z) \qquad ...(2.1)$$

Provided the infinite series (2.1) is convergent and U is a function of complex number z.

For $n < 0, \quad u_n = 0$

The Z-transform is defined as

$$Z(u_n) = \sum_{n=0}^{\infty} u_n \, z^{-n}$$

2.2. LINEARITY PROPERTIES

Let $\{u_n\}$ and $\{v_n\}$ be discrete sequences then

$$Z\,(au_n + bv_n) = a\,Z(u_n) + bZ(v_n)$$

where a and b are constants.

Proof: By the definition, we have

$$Z(au_n + bv_n) = \sum_{n=0}^{\infty} (au_n + bv_n)z^{-n} = \sum_{n=0}^{\infty} a\, u_n z^{-n} + \sum_{n=0}^{\infty} b\, v_n z^{-n}$$

$$= a\sum_{n=0}^{\infty} u_n z^{-n} + b\sum_{n=0}^{\infty} v_n z^{-n} = a\, Z(u_n) + b\, Z(v_n).$$ **Hence proved.**

2.3. CHANGE OF SCALE PROPERTY OR DAMPING RULE

If $\{u_n\}$ be any discrete sequence and $Z(u_n) = U(z)$ then

(i) $Z(a^{-n} u_n) = U(az)$ (ii) $Z(a^n u_n) = U\left(\frac{z}{a}\right)$

Proof: (i) By the definition, we have

$$Z(a^{-n} u_n) = \sum_{n=0}^{\infty} (a^{-n} u_n)z^{-n} = \sum_{n=0}^{\infty} u_n (az)^{-n} = U(az).$$ **Hence proved.**

(ii) By the definition, we have

$$Z(a^n u_n) = \sum_{n=0}^{\infty} (a^n u_n)z^{-n} = \sum_{n=0}^{\infty} u_n \left(\frac{z}{a}\right)^{-n} = U\left(\frac{z}{a}\right).$$ **Hence proved.**

2.4. SOME STANDARD Z-TRANSFORMS

(i) $Z(a^n) = \frac{z}{z-a}, \quad n \geq 0$

Proof: By the definition, we have

$$Z(a^n) = \sum_{n=0}^{\infty} a^n z^{-n} = \sum_{n=0}^{\infty} \left(\frac{a}{z}\right)^n = 1 + \frac{a}{z} + \left(\frac{a}{z}\right)^2 + \left(\frac{a}{z}\right)^3 \ldots\ldots$$

$$= \frac{1}{1-\left(\frac{a}{z}\right)}$$ [The sum of infinite term of G.P.]

$$= \frac{z}{z-a}.$$ **Hence proved.**

Some Particular Cases

(1) $$Z(1) = \frac{z}{z-1}$$

Proof: We have $Z(1) = \sum_{n=0}^{\infty} 1\, z^{-n}$

$$= \sum_{n=0}^{\infty} \frac{1}{z^n} = 1 + \frac{1}{z} + \frac{1}{z^2} + \frac{1}{z^3} \ldots\ldots$$

$$= \frac{1}{1 - \frac{1}{z}} = \frac{z}{z-1}.$$ **Hence proved.**

(2) $Z(k) = \frac{kz}{z-1}$, where k is constant

Proof: We have $Z(k) = \sum_{n=0}^{\infty} k\, z^{-n} = k \sum_{n=0}^{\infty} z^{-n}$

$$= \frac{kz}{z-1}$$ (By above case). **Hence proved.**

(3) $Z[(-1)^n] = \frac{z}{z+1}$

Proof: Taking $a = -1$ in $Z(a^n) = \frac{z}{z-a}$

We get $Z(-1)^n = \frac{z}{z+1}$.

(*ii*) $Z(n^p) = -z \frac{d}{dz} [Z(n^{p-1})],\ p \in N$

Proof: By the definition, we have

$$Z(n^p) = \sum_{n=0}^{\infty} n^p z^{-n} \qquad \ldots(2.2)$$

and $$Z(n^{p-1}) = \sum_{n=0}^{\infty} n^{p-1} z^{-n} \quad \text{(on replacing } p \text{ by } p-1) \qquad \ldots(2.3)$$

Differentiating (2.3) w.r.t. z, we get

$$\frac{d}{dz}[Z(n^{p-1})] = \frac{d}{dz}\left[\sum_{n=0}^{\infty} n^{p-1} z^{-n}\right] = \sum_{n=0}^{\infty} n^{p-1}(-n)\, z^{-n-1}$$

$$= -z^{-1} \sum_{n=0}^{\infty} n^p z^{-n} = -z^{-1}[Z(n^p)] \qquad \text{[By (2.2)]}$$

Thus $$Z[n^p] = -z\frac{d}{dz}[Z(n^{p-1})].$$ **Hence proved.** ...(2.4)

Some Particular Cases

(1) $$Z(n) = \frac{z}{(z-1)^2}$$

Proof: Taking $p = 1$ in (2.4), we get

$$Z(n) = -z\frac{d}{dz}[Z(1)] = -z\frac{d}{dz}\left[\frac{z}{z-1}\right]$$

$$= -z\frac{(z-1-z)}{(z-1)^2} = \frac{z}{(z-1)^2}.$$ **Hence proved.**

(2) taking $p = 2$ in (2.4), we get

$$Z(n^2) = -z\frac{d}{dz}[Z(n)] = -z\frac{d}{dz}\left[\frac{z}{(z-1)^2}\right]$$

$$= -z\frac{1(z-1)^2 - z2(z-1)}{(z-1)^4} = \frac{z^2+z}{(z-1)^3}.$$ **Hence proved.**

Similarly,

(3) $Z(n^3) = \dfrac{z^3+4z^2+z}{(z-1)^4}.$

(4) $Z(n^4) = \dfrac{z^4+71z^3+11z^2+z}{(z-1)^5}.$

2.5. SHIFTING U_N TO THE RIGHT

If $Z(u_n) = U(z)$ then $Z(u_{n-r}) = z^{-r}U(z)$, $r > 0$

Proof: By the definition, we have

$$Z(u_{n-r}) = \sum_{n=0}^{\infty} u_{n-r}z^{-n} = \sum_{n=0}^{\infty} u_{n-r}z^{-(n-r)-r} = z^{-r}\sum_{n=0}^{\infty} u_{n-r}z^{-(n-r)}$$

$$= z^{-r}\sum_{k=-r}^{\infty} u_k z^{-k}$$ [take $n - r = k$]

$$= z^{-r}\sum_{k=0}^{\infty} u_k z^{-k}$$ [$\because$ $u_n = 0$ if $n < 0$]

$$Z(u_{n-r}) = z^{-r}U(z).$$ **Hence proved.**

2.6. SHIFTING U_n TO THE LEFT

If $Z(u_n) = U(z)$ then $Z(u_{n+r}) = z^r[U(z) - u_0 - u_1 z^{-1} - u_2 z^{-2} \ldots\ldots u_{r-1} z^{-(r-1)}]$

Proof: By the definition, we have

$$Z(u_{n+r}) = \sum_{n=0}^{\infty} u_{n+r}\, z^{-n} = \sum_{n=0}^{\infty} u_{n+r}\, z^{-(n+r)+r}$$

$$= z^r \sum_{n=0}^{\infty} u_{n+r}\, z^{-(n+r)} = z^r \sum_{k=r}^{\infty} u_k\, z^{-k} \qquad \text{(take } n + r = k\text{)}$$

$$= z^r \left[\sum_{k=0}^{\infty} u_k\, z^{-k} - \sum_{k=0}^{r-1} u_k\, z^{-k} \right]$$

$$Z(u_{n+r}) = z^r[U(z) - u_0 - u_1 z^{-1} - u_2 z^{-2} \ldots\ldots u_{r-1} z^{-(r-1)}].$$ **Hence proved.**

Particular Case

(1) $Z(u_{n+1}) = z\,[U(z) - u_0]$.

(2) $Z(u_{n+2}) = z^2\,[U(z) - u_0 - z^{-1}u_1]$.

2.7. MULTIPLICATION BY *n*

If $Z(u_n) = U(z)$

Then $$Z(nu_n) = -z \frac{d}{dz} U(z)$$

Proof: By the definition, we have

$$Z(nu_n) = \sum_{n=0}^{\infty} n u_n\, z^{-n} = -z \sum_{n=0}^{\infty} u_n\, (-n)\, z^{-n-1}$$

$$= -z \sum_{n=0}^{\infty} u_n \frac{d}{dz}(z^{-n}) = -z \frac{d}{dz} \sum_{n=0}^{\infty} u_n\, z^{-n}$$

$$Z(nu_n) = -z \frac{d}{dz} U(z).$$ **Hence proved.**

Particular case

(1) $Z(n^2 u_n) = (-z)^2 \dfrac{d^2}{dz^2}\, U(z)$

(2) $Z(n^k u_n) = (-z)^k \dfrac{d^k}{dz^k}\, U(z)$

2.8. DIVISION BY *n*

If $Z(u_n) = U(z)$. Then $Z\left(\frac{u_n}{n}\right) = -\int^{n} \frac{U(z)}{z}\, dz$.

Proof: By the definition, we have

$$Z\left(\frac{u_n}{n}\right) = \sum_{n=0}^{\infty} \frac{u_n}{n} z^{-n} = \sum_{n=0}^{\infty} u_n \frac{1}{n} z^{-n} = -\sum_{n=0}^{\infty} u_n \int_0^{n} z^{-n-1}\, dz$$

$$= -\int^{n} \sum_{n=0}^{\infty} u_n z^{-n-1}\, dz = -\int^{n} z^{-i} \sum_{n=0}^{\infty} u_n z^{-n}\, dz = -\int^{n} \frac{U(z)}{z}\, dz$$

$$Z\left(\frac{u_n}{n}\right) = -\int^{n} \frac{U(z)}{z}\, dz.$$ **Hence Proved.**

2.9. INITIAL VALUE THEOREM

If $Z(u_n) = U(z)$ then $u_0 = \lim_{z \to \infty} U(z)$.

Proof: By the definition, we have

$$Z(u_n) = U(z) = \sum_{n=0}^{\infty} u_n z^{-n}$$

or

$$U(z) = u_0 + \frac{u_1}{z} + \frac{u_2}{z^2} + \ldots\ldots$$

taking limit on both sides as $z \to \infty$, we get

$$\lim_{z \to \infty} U(z) = \lim_{z \to 0} \left[u_0 + \frac{u_1}{z} + \frac{u_2}{z^2} + \ldots\ldots \right] = u_0$$

Hence $u_0 = \lim_{z \to \infty} U(z)$. **Hence proved.**

Some Particular Cases

(1) $u_1 = \lim_{z \to \infty} z\,[U(z) - u_0]$

(2) $u_2 = \lim_{z \to \infty} z^2\,[U(z) - u_0 - u_1(z)]$

2.10. FINAL VALUE THEOREM

If $Z(u_n) = U(z)$, then $\lim_{n \to \infty} u_n = \lim_{z \to 1} (z-1)\, U(z)$

Proof: By the definition, we have

$$Z(u_{n+1} - u_n) = \sum_{n=0}^{\infty} (u_{n+1} - u_n) z^{-n}$$

or $$Z(u_{n+1}) - Z(u_n) = \sum_{n=0}^{\infty} (u_{n+1} - u_n) z^{-n}$$

$$z[U(z) - u_0] - U(z) = \sum_{n=0}^{\infty} (u_{n+1} - u_n) z^{-n}$$

or $$(z-1)\, U(z) - zu_0 = \sum_{n=0}^{\infty} (u_{n+1} - u_n) z^{-n}$$

Taking limit on both sides as $z \to 1$, we get

$$\lim_{z \to 1} [(z-1)\, U(z) - zu_0] = \lim_{z \to 1} \sum_{n=0}^{\infty} (u_{n+1} - u_n) z^{-n}$$

$$\lim_{z \to 1} (z-1)\, U(z) - u_0 = \sum_{n=0}^{\infty} (u_{n+1} - u_n)$$

$$= \lim_{n \to \infty} [(u_1 - u_0) + (u_2 - u_1) + \ldots\ldots + (u_{n+1} - u_n)] = u_\infty - u_0$$

or $$u_\infty = \lim_{z \to 1} (z-1)\, U(z)$$

$$\lim_{n \to \infty} u_n = \lim_{z \to 1} (z-1)\, U(z).$$ **Hence proved.**

SOLVED EXAMPLES

Example 1: *Find the Z-transform of the following sequence:*

(i) $\{15, 10, 7, 4, \underset{\uparrow}{1}, -1, 0, 3, 6\}$ *(ii)* $\left\{\frac{1}{2^n}\right\}\ -2 \le n < 4$

(iii) $\left\{\frac{1}{3^n}\right\}$.

Solution: (*i*) We have

$$Z\{u_n\} = \sum_{n=-\infty}^{\infty} u_n z^{-n}$$

$$= 15z^4 + 10z^3 + 7z^2 + 4z + 1 - \frac{1}{z} + \frac{0}{z^2} + \frac{3}{z^3} + \frac{6}{z^4}. \quad \textbf{Ans.}$$

(*ii*) We have $$Z\{u_n\} = \sum_{n=-2}^{4} \frac{1}{2^n} z^{-n} = 4z^2 + 2z + 1 + \frac{1}{2z} + \frac{1}{4z^2} + \frac{1}{8z^3} + \frac{1}{16z^4}. \quad \textbf{Ans.}$$

(*iii*) We have $$Z\{u_n\} = \sum_{-\infty}^{\infty} \frac{1}{3^n} = \ldots\ldots + 9z^2 + 3z + 1 + \frac{1}{3z} + \frac{1}{9z^2} + \frac{1}{27z^3} + \ldots\ldots \quad \textbf{Ans.}$$

Example 2: *Find the Z-transform of* $\left\{\frac{a^n}{n!}\right\}$, $n \geq 0$.

Solution: We have

$$Z\left\{\frac{a^n}{n!}\right\} = \sum_{n=0}^{\infty} \frac{a^n}{n!} z^{-n} = \sum_{n=0}^{\infty} \frac{1^n}{n!}\left(\frac{a}{z}\right)^n$$

$$= 1 + \frac{a}{z} + \frac{1}{2!}\left(\frac{a}{z}\right)^2 + \frac{1}{3!}\left(\frac{a}{z}\right)^3 \ldots\ldots = e^{\frac{a}{z}}. \quad \textbf{Ans.}$$

Example 3: *Find the Z-transform of* $\left\{\frac{a^n}{n}\right\}$, $n \geq 1$.

Solution: We have

$$Z\left\{\frac{a^n}{n}\right\} = \sum_{n=1}^{\infty} \frac{a^n}{n} z^{-n} = \sum_{n=1}^{\infty} \frac{1}{n}\left(\frac{a}{2}\right)^n = \frac{a}{z} + \frac{1}{2}\left(\frac{a}{z}\right)^2 + \frac{1}{3}\left(\frac{a}{z}\right)^3 \ldots\ldots$$

$$= -\log\left(1 - \frac{a}{z}\right). \quad \textbf{Ans.}$$

Example 4: *Find the Z-transform of* $2n + 5\sin\frac{n\pi}{4} - 3a^4$.

Solution: We know that

$$Z\left(2n + 5\sin\frac{n\pi}{4} - 3a^4\right) = 2Z(n) + 5Z\left(\sin\frac{n\pi}{4}\right) - 3a^4\, Z(1)$$

$$= 2\frac{z}{(z-1)^2} + 5\frac{z\,.\sin\frac{\pi}{4}}{z^2 - 2z\cos\frac{\pi}{4} + 1} - 3a^4\,\frac{z}{z-1}$$

$$= \frac{2z}{(z-1)^2} + \frac{5z \cdot \frac{1}{\sqrt{2}}}{z^2 - 2z\frac{1}{\sqrt{2}} + 1} - 3a^4 \frac{z}{z-1}$$

$$= \frac{2z}{(z-1)^2} + \frac{5z}{\sqrt{2}(z^2 - \sqrt{2}z + 1)} - \frac{3a^4 z}{z-1}. \quad \textbf{Ans.}$$

Example 5: *Find the Z-transform of $(n - 1)^2$.*

Solution: We know that

$$Z[(n-1)^2] = Z(n^2 - 2n + 1) = Z(n^2) - 2Z(n) + Z(1)$$

$$= \frac{z^2 + z}{(z-1)^3} - 2\frac{z}{(z-1)^2} + \frac{z}{(z-1)}$$

$$= \frac{z^2 + z - 2z(z-1) + z(z-1)^2}{(z-1)^3} = \frac{z^3 - 3z^2 + 4z}{(z-1)^3}. \quad \textbf{Ans.}$$

Example 6: *Find the Z-transform of sin $(an + b)$, where a and b are constants.*

Solution: We know that

$$Z[\sin(an + b)] = Z(\sin an \cos b + \cos an \sin b)$$

$$= \cos b\, Z(\sin an) + \sin b\, Z(\cos an)$$

$$= \cos b \cdot \frac{z \sin a}{z^2 - 2z\cos a + 1} + \sin b \frac{z^2 - z\cos a}{z^2 - 2z\cos a + 1}$$

$$= \frac{z(\sin a \cos b + z \sin b - \cos a \sin b)}{z^2 - 2z\cos a + 1}$$

Thus $$Z[\sin(an + b)] = \frac{z[\sin(a-b) + z\sin b]}{z^2 - 2z\cos a + 1}. \quad \textbf{Ans.}$$

Example 7: *Find the Z-transform of the following:*

(i) 2^n *(ii)* $\sin 2n$ *(iii)* a^{n+3}

Solution: (*i*) We have $Z(a^n) = \dfrac{z}{z-a}$

Then $$Z(2^n) = \frac{z}{z-2}$$

Aliter: We have $$Z(a^n) = \sum_{n=0}^{\infty} a^n z^{-n} = \sum_{n=0}^{\infty} \frac{a^n}{z^n}$$

$\therefore$ $$Z(2^n) = \sum_{n=0}^{\infty} \frac{2^n}{z^n} = 1 + \frac{2}{z} + \left(\frac{2}{z}\right)^2 + \left(\frac{2}{z}\right)^3 \ldots$$

$$= \frac{1}{1-\frac{2}{z}} = \frac{z}{z-2}. \quad \textbf{Ans.}$$

(*ii*) We have $$Z(\sin n\theta) = \frac{z \sin\theta}{z^2 - 2z\cos\theta + 1}$$

Then $$Z(\sin 2n) = \frac{z \sin 2}{z^2 - 2z\cos 2 + 1}. \quad \textbf{Ans.}$$

(*iii*) We have $$Z(a^{n+3}) = z(a^n a^3) = a^3\, z(a^n)$$

$$= \frac{a^3 z}{z-a}$$

Thus, $$z(a^{n+3}) = \frac{a^3 z}{z-a}. \quad \textbf{Ans.}$$

Example 8: *Find the Z-transform of the following:*

(*i*) na^n (*ii*) $n^2 a^n$.

Solution: (*i*) We have $Z(n) = U(z) = \dfrac{z}{(z-1)^2}$

Using damping rule, we get

$$Z(na^n) = U\left(\frac{z}{a}\right) = \frac{z/a}{\left(\frac{z}{a}-1\right)^2} = \frac{az}{(z-a)^2}. \quad \textbf{Ans.}$$

(*ii*) We have $$Z(n^2) = U(z) = \frac{z^2+z}{(z-1)^3}$$

Using damping rule, we get

$$Z(n^2 a^n) = U\left(\frac{z}{a}\right) = \frac{\left(\frac{z}{a}\right)^2 + \left(\frac{z}{a}\right)}{\left(\frac{z}{a}-1\right)^3} = \frac{az^2 + a^2 z}{(z-a)^3}. \quad \textbf{Ans.}$$

Example 9: *Find the Z-transform of the following:*

(*i*) $\cos n\theta$ (*ii*) $a^n \cos n\theta$

Solution: (*i*) We have $Z(1) = U(z) = \dfrac{z}{z-1}$

Then $$Z(e^{-in\theta}) = Z\left[\left(e^{i\theta}\right)^{-n}.1\right] = U(e^{i\theta} z)$$ (by damping rule)

$$= \frac{e^{i\theta} z}{e^{i\theta} z - 1} = \frac{z}{z-e^{-i\theta}} = \frac{z(z-e^{i\theta})}{(z-e^{-i\theta})(z-e^{i\theta})}$$

$$= \frac{z^2 - ze^{i\theta}}{z^2 - z(e^{i\theta} + e^{-i\theta}) + 1}$$

or $$Z(\cos n\theta - i \sin n\theta) = \frac{z^2 - z(\cos\theta + i\sin\theta)}{z^2 - 2z\cos\theta + 1}$$

Compair on both sides real parts, we get

$$Z(\cos n\theta) = \frac{z^2 - z\cos\theta}{z^2 - 2z\cos\theta + 1}. \quad \textbf{Ans.}$$

(*ii*) Let $Z(\cos n\theta) = U(z)$

Then $Z(a^n \cos n\theta) = U\left(\frac{z}{a}\right)$ [by damping rule]

$$= \frac{\left(\frac{z}{a}\right)^2 - \left(\frac{z}{a}\right)\cos\theta}{\left(\frac{z}{a}\right)^2 - 2\left(\frac{z}{a}\right)\cos\theta + 1}$$

$$Z(a^n \cos n\theta) = \frac{z^2 - az\cos\theta}{z^2 - 2az\cos\theta + a^2}. \quad \textbf{Ans.}$$

Example 10: *Find the Z-transform of*

(*i*) *sinh* $n\theta$ (*ii*) a^n *sinh* $n\theta$

Solution: (*i*) We have

$$Z(\sinh n\theta) = Z\left(\frac{e^{n\theta} - e^{-n\theta}}{2}\right) = \frac{1}{2}[Z(e^{\theta})^n - Z(e^{-\theta})^n]$$

$$= \frac{1}{2}\left[\frac{z}{z - e^{\theta}} - \frac{z}{z - e^{-\theta}}\right] = \frac{1}{2}\,\frac{z(z - e^{-\theta}) - z(z - e^{\theta})}{(z - e^{\theta})(z - e^{-\theta})}$$

$$= \frac{1}{2}\,\frac{z(e^{\theta} - e^{-\theta})}{z^2 - z(e^{\theta} + e^{-\theta}) + 1} = \frac{1}{2}\,\frac{2z\sinh\theta}{z^2 - 2z\cosh\theta + 1}$$

$$= \frac{z\sinh\theta}{z^2 - 2z\cosh\theta + 1}. \quad \textbf{Ans.}$$

(*ii*) We have $Z(a^n \sinh n\theta) = Z[(a^{-1})^{-n} \sinh n\theta] = U(a^{-1}z)$ (By damping rule)

where $U(z) = Z(\sinh n\theta)$

$\therefore$ $$Z(a^n \sinh n\theta) = \frac{a^{-1}z\sinh\theta}{(a^{-1}z)^2 - 2(a^{-1}z)\cosh\theta + 1}$$

$$(a^n \sinh n\theta) = \frac{za \sin h\,\theta}{z^2 - 2az\cos\theta + a^2}.\quad \textbf{Ans.}$$

Example 11: *Show that* $Z\left(\frac{1}{n!}\right) = e^{1/z}$.

Hence, evaluate (i) $Z\left(\frac{1}{n+1!}\right)$ *(ii)* $Z\left(\frac{1}{n+2!}\right)$.

Solution: We have

$$Z\left(\frac{1}{n!}\right) = \sum_{n=0}^{\infty} \frac{1}{n!} z^{-n} = 1 + \frac{1}{1!}\frac{1}{z} + \frac{1}{2!}\frac{1}{z^2} + \frac{1}{3!}\frac{1}{z^3} \ldots.$$

$$= e^{1/z}.\ \textbf{Hence proved.}$$

(*i*) Shifting $\frac{1}{n!}$ one unit to the left, we get

$$Z\left(\frac{1}{(n+1)!}\right) = z\left[Z\left(\frac{1}{n!}\right) - 1\right] = z(e^{1/z} - 1)$$

(*ii*) Shifting $\frac{1}{(n+1)!}$ one unit to the left, we get

$$Z\left(\frac{1}{(n+2)!}\right) = z\left[Z\left(\frac{1}{(n+1)!}\right) - 1\right] = z\,[z(e^{1/z} - 1) - 1]$$

$$= z^2\,[e^{1/z} - 1 - z^{-1}].\quad \textbf{Ans.}$$

Example 12: *If* $U(z) = \frac{2z^2 + 3z + 4}{(z-3)^3}$ *then find* u_1, u_2 *and* u_3.

Solution: We have $U(z) = \frac{2z^2 + 3z + 4}{(z-3)^3} = \frac{1}{z}\frac{(2 + 3z^{-1} + 4z^{-2})}{(1 - 3z^{-1})^3}$

Then by initial value theorem

$$u_0 = \lim_{z\to\infty} U(z) = \lim_{z\to 0} \frac{1}{z}\frac{2 + 3z^{-1} + 4z^{-2}}{(1 - 3z^{-1})^3} = 0$$

$$u_1 = \lim_{z\to\infty} z[U(z) - u_0] = \lim_{z\to 0} z\frac{1}{z}\frac{2 + 3z^{-1} + 4z^{-2}}{(1 - 3z^{-1})} = 2$$

$$u_2 = \lim_{z\to\infty} z^2[U(z) - u_0 - u_1 z^{-1}]$$

$$= \lim_{z\to\infty} z^2\left[\frac{1}{z} \times \frac{2 + 3z^{-1} + 4z^{-2}}{(1 - 3z^{-1})^3} - 0 - \frac{2}{z}\right] = 21$$

$$u_3 = \lim_{z \to \infty} z^3 [U(z) - u_0 - u_1 z^{-1} - u_2 z^{-2}]$$

$$= \lim_{z \to \infty} z^3 \left[\frac{1}{z} \times \frac{2 + 3z^{-1} + 4z^{-2}}{(1 - 3z^{-1})^3} - 0 - \frac{2}{z} - \frac{21}{z^2} \right] = 139.$$

EXERCISE 2.1

1. Find the Z-transform of the following:

(*i*) 5^n (*ii*) $\sin 3n$ (*iii*) $(n + 1)^2$

(*iv*) $(\cos \theta + i \sin \theta)^n$ (*v*) $3n - 4 \sin \frac{n\pi}{4} + 5a$ (*vi*) $\frac{a^n e^{-a}}{n!}$

(*vii*) $\sin (n + 1) \theta$ (*viii*) $\sinh \left(\frac{n\pi}{2}\right)$ (*ix*) $\cos \left(\frac{k\pi}{2} + \frac{\pi}{4}\right)$.

2. Find the Z-transform of the following:

(*i*) e^{an} (*ii*) ne^{an} (*iii*) $n^2 e^{an}$.

3. Find the Z-transform of (*i*) $\sin n\theta$ (*ii*) $a^n \sin n\theta$

4. Find the Z-transform of (*i*) $\cosh n\theta$ (*ii*) $a^n \cosh n\theta$.

5. Find the Z-transform of (*i*) $\cos \left(\frac{n\pi}{2} + \frac{\pi}{4}\right)$ (*ii*) $\cosh \left(\frac{n\pi}{2} + \theta\right)$.

6. Find the Z-transform of $n \sin n\theta$.

7. If $U(z) = \frac{2z^2 + 5z + 14}{(z-1)^4}$, find u_0, u_1, u_2 and u_3.

8. Find the Z-transform of $e^t \sin 2t$.

9. Using $Z(n) = \frac{z}{(z-1)^2}$, show that $Z(n \cos n\theta) = \frac{(z^3 + z) \cos \theta - 2z^2}{(z^2 - 2z \cos \theta + 1)^2}$.

10. Using $Z(n^2) = \frac{(z^2 + z)}{(z^2 - 1)^3}$, show that $Z[(n + 1)^2] = \frac{(z^3 + z^2)}{(z-1)^3}$.

11. Find $Z(-e^{-an} \sin n\theta)$.

12. If $Z\{u_n\} = \frac{z}{z-1} + \frac{z}{z^2+1}$, then find $Z\{u_{n+2}\}$.

13. Find the Z-transform of unit impulse function which is given by

$$\delta_n = \begin{cases} 1 & \text{if } n = 0 \\ 0 & \text{if } n \neq 0 \end{cases}.$$

14. Find the Z-transform of $\{u_n\}$ where

$$\{u_n\} = \begin{cases} 4^n & \text{if } n < 0 \\ 3^n & \text{if } n \geq 0 \end{cases}.$$

15. Find the Z-transform of the discrete unit step function given by

$$u_n = \begin{cases} 1 & \text{if } n \geq 0 \\ 0 & \text{if } n < 0 \end{cases}.$$

ANSWERS

1. (*i*) $\dfrac{z}{z-5}$ (*ii*) $\dfrac{z \sin 3}{z^2 - 2z \cos 3 + 1}$ (*iii*) $\dfrac{z^2(2z+1)}{(z-1)^3}$

(*iv*) $\dfrac{z}{z - e^{i\theta}}$ (*v*) $\dfrac{(3-5a)z + 5az^2}{(z-1)^2} - \dfrac{2\sqrt{2}z}{z^2 - \sqrt{2}z + 1}$ (*vi*) $e^{a(z^{-1}-1)}$

(*vii*) $\dfrac{z^2 \sin\theta}{z^2 - 2z\cos\theta + 1}$ (*viii*) $\dfrac{z \sinh \pi/2}{z^2 - 2z \cosh \pi/2 + 1}$ (*ix*) $\dfrac{z^2 - z}{\sqrt{2}\,(z^2+1)}$.

2. (*i*) $\dfrac{z}{z - e^a}$ (*ii*) $\dfrac{e^a z}{(z - e^a)^2}$ (*iii*) $\dfrac{ze^a(z + e^a)}{(z - e^a)^3}$

3. (*i*) $\dfrac{z \sin\theta}{z^2 - 2z\cos\theta + 1}$ (*ii*) $\dfrac{az\sin\theta}{z^2 - 2az\cos\theta + 1}$.

4. (*i*) $\dfrac{z^2 - z\cosh\theta}{z^2 - 2z\cosh\theta + 1}$ (*ii*) $\dfrac{z(z - a\cosh\theta)}{z^2 - 2az\cosh\theta + a^2}$.

5. (*i*) $\dfrac{z(z-1)}{\sqrt{2(z^2+1)}}$ (*ii*) $\dfrac{z^2\cosh\theta - z\cosh\left(\dfrac{\pi}{2} - \theta\right)}{z^2 - 2z\cosh\left(\dfrac{\pi}{2}\right) + 1}$.

6. $\dfrac{z(z^2-1)\sin\theta}{(z^2 - 2z\cos\theta + 1)^2}$ **7.** $u_0 = 0$, $u_1 = 0$, $u_2 = 2$ and $u_3 = 13$.

8. $\dfrac{ez \sin 2}{z^2 - 2ez + e^2}$ **11.** $\dfrac{e^a z \sin\theta}{e^{2a}z^2 - 2e^a z\cos\theta + 1}$.

12. $\dfrac{z(z^2 - z + 2)}{(z-1)(z^2+1)}$ **13.** 1

14. $\dfrac{-z}{z^2 - 7z + 12}$.

2.11. INVERSE Z-TRANSFORM

It is denoted by $z^{-1}U(z) = u_n$ and defined as $u_n = Z^{-1}[U(z)] = Z^{-1}\left[\sum_{n=0}^{\infty} u_n z^{-n}\right]$.

2.12. CONVOLUTION THEOREM

If $Z^{-1}[U(z)] = u_n$ and $Z^{-1}[V(z)] = v_n$ then $Z^{-1}[U(z).\ V(z)] = \sum_{m=0}^{\infty} u_m v_{n-m} = u_n * v_n$

Where, the symbol $*$ denotes the convolution operation.

Proof: We have $Z(u_n) = U(z) = \sum_{n=0}^{\infty} u_n z^{-n}$ and $Z(v_n) = V(z) = \sum_{m=0}^{\infty} v_n z^{-n}$

$$\therefore \quad U(z)\,V(z) = \sum_{n=0}^{\infty} u_n z^{-n} \cdot \sum_{n=0}^{\infty} v_n z^{-n}$$

$$= (u_0 + u_1 z^{-1} + u_2 z^{-2} + \ldots\ldots + u_n z^{-n} \ldots\ldots)$$
$$\times (v_0 + v_1 z^{-1} + v_2 z^{-2} + \ldots\ldots + v_n z^{-n} \ldots\ldots)$$

$$= \sum_{n=0}^{\infty} (u_0 v_n + u_1 v_{n-1} + \ldots\ldots. + u_{n-1} v_1 + u_n v_0) z^{-n}$$

$$= Z(u_0 v_n + u_1 v_{n-1} + \ldots\ldots + u_{n-1} v_1 + u_n v_0) = Z\left(\sum_{m=0}^{n} u_m v_{n-m}\right)$$

Hence $Z^{-1}[U(z)\,V(z)] = \sum_{m=0}^{n} u_m v_{n-m}$. **Hence proved.**

SOLVED EXAMPLES

Example 1: *Find* $Z^{-1}\left[\frac{z^2}{(z-2)^2}\right]$ *by convolution theorem.*

Solution: We have $Z^{-1}\left[\frac{z^2}{z-2}\right] = 2^n$

$$\therefore \quad Z^{-1}\left[\frac{z^2}{(z-2)^2}\right] = Z^{-1}\left[\left(\frac{z}{z-2}\right)\left(\frac{z}{z-2}\right)\right]$$

$$= \sum_{m=0}^{n} 2^m \, 2^{n-m} \qquad \text{(By Convolution theorem)}$$

$$= 2^n \sum_{n=0}^{n} 2^m \, 2^{-m} = 2^n \sum_{n=0}^{n} 1 = 2^n (n+1). \quad \textbf{Ans.}$$

Example 2: *Using convolution theorem evaluate* $Z^{-1}\left[\dfrac{z^2}{(z-a)(z-b)}\right]$.

Solution: We have $Z^{-1}\left[\dfrac{z}{z-a}\right] = a^n$ and $Z^{-1}\left[\dfrac{z}{z-b}\right] = b^n$

$$\therefore \quad Z^{-1}\left[\frac{z^2}{(z-a)(z-b)}\right] = Z^{-1}\left[\frac{z}{z-a} \times \frac{z}{z-b}\right]$$

$$= \sum_{m=0}^{n} a^m b^{n-m} \qquad \text{(By Convolution theorem)}$$

$$= b^n \sum_{m=0}^{n} \left(\frac{a}{b}\right)^m = b^n \left[1 + \frac{a}{b} + \left(\frac{a}{b}\right)^2 + \dots + \left(\frac{a}{b}\right)^n\right]$$

$$= b^n \left\{\frac{\left(\frac{a}{b}\right)^{n+1} - 1}{\frac{a}{b} - 1}\right\} = b^n \left\{\frac{a^{n+1} - b^{n+1}}{b^n (a-b)}\right\}$$

$$Z^{-1}\left[\frac{z^2}{(z-a)(z-b)}\right] = \frac{a^{n+1} - b^{n+1}}{a-b}. \quad \textbf{Ans.}$$

Example 3: *Find* $Z^{-1}\left[\dfrac{2z(2z-1)}{z^3 - 5z^2 + 8z - 4}\right]$.

Solution: We have

$$\frac{2z(2z-1)}{z^3 - 5z^2 + 8z - 4} = \frac{2z(2z-1)}{(z-1)(z-2)^2}$$

$$= \frac{2z}{(z-1)} - \frac{2z}{(z-2)} + \frac{6z}{(z-2)^2} \qquad \text{(By partial fraction)}$$

Then, $Z^{-1}\left[\dfrac{2z(2z-1)}{z^3 - 5z^2 + 8z - 4}\right] = Z^{-1}\left[\dfrac{2z}{z-1} - \dfrac{2z}{z-2} + \dfrac{6z}{(z-2)^2}\right]$

$$= 2Z^{-1}\left[\frac{z}{z-1}\right] - 2Z^{-1}\left[\frac{z}{z-2}\right] + 3Z^{-1}\left[\frac{2z}{(z-2)^2}\right]$$

$$= 2 \,.\, 1^n - 2.2^n + 3n \,.\, 2^n. \quad \textbf{Ans.}$$

Examples 4: *Find the inverse Z-transform of* $\dfrac{2z^2+3z}{(z+2)(z-4)}$.

Solution: We have $$\frac{2z^2+3z}{(z+2)(z-4)} = \frac{1}{6}\frac{z}{(z+2)} + \frac{11}{6}\left(\frac{z}{z-4}\right)$$

Then $$Z^{-1}\left[\frac{2z^2+3z}{(z+2)(z-4)}\right] = Z^{-1}\left[\frac{1}{6}\frac{z}{z+2} + \frac{11}{6}\frac{z}{z-4}\right]$$

$$= \frac{1}{6}(-2)^n + \frac{11}{6}(4)^n. \quad \textbf{Ans.}$$

Example 5: *Find the inverse Z-transform of log* $\left(\dfrac{z}{z+1}\right)$ *by power series method.*

Solution: We have $$U(z) = \log\left(\frac{z}{z+1}\right) = \log\frac{1}{\left(1+\frac{1}{z}\right)} = -\log\left(1+\frac{1}{z}\right)$$

$$= -\left[\frac{1}{z} - \frac{1}{2z^2} + \frac{1}{3z^3} \ldots\right]$$

$$= -\frac{1}{z} + \frac{1}{2z^2} - \frac{1}{3z^3} \ldots = \sum_{n=0}^{\infty} \frac{(-1)^n}{n} z^{-n}$$

$\therefore$ $$Z^{-1}[U(z)] = Z^{-1}\left[\sum_{n=0}^{\infty} \frac{(-1)^n}{n} z^{-n}\right]$$

$\Rightarrow$ $$u_n = \frac{(-1)^n}{n} \text{ if } n \geq 1. \quad \textbf{Ans.}$$

Example 6: *Find* $Z^{-1}\left[\dfrac{z}{(z-1)(z-2)}\right]$ *for* $|z| > 2$.

Solution: We have $$U(z) = \frac{z}{(z-1)(z-2)} = \frac{2}{z-2} - \frac{1}{z-1}$$

Since, $$|z| > 2 \Rightarrow |z| > 1 \text{ and } \left|\frac{2}{z}\right| < 1 \Rightarrow \left|\frac{1}{z}\right| < 1$$

Now $$U(z) = \frac{2}{z\left(1-\frac{2}{z}\right)} - \frac{1}{z\left(1-\frac{1}{z}\right)} = \frac{2}{z}\left(1-\frac{2}{z}\right)^{-1} - \frac{1}{z}\left(1-\frac{1}{z}\right)^{-1}$$

$$= \frac{2}{z}\left[1+\frac{2}{z}+\frac{2^2}{z^2}+\ldots\ldots+\frac{2^{n-1}}{z^{n-1}}\right] - \frac{1}{z}\left[1+\frac{1}{z}+\frac{1}{z^2}+\ldots\ldots+\frac{1}{z^{n-1}}\ldots\ldots\right]$$

Now coefficient $\quad Z^{-n} = 2^{n-1} - 1 \quad n \geq 1$

$\therefore \quad Z^{-1}[U(z)] = \{2^{n-1} - 1)\}$. **Ans.**

EXERCISE 2.2

1. Use convolution theorem, to evaluate inverse Z-transform of the following:

   (i) $\left(\dfrac{z}{z-a}\right)^2$ $\qquad$ (ii) $\dfrac{z^2}{(z-\alpha)(z-\beta)}$.

2. Prove that $\dfrac{1}{n!} * \dfrac{1}{n!} = \dfrac{2^n}{n!}$.

3. Find the inverse Z-transform of the following:

   (i) $\dfrac{z^3 - 20z}{(z-2)^3 (z-4)}$ $\qquad$ (ii) $\dfrac{z}{(z-1)^2}$ $\qquad$ (iii) $\dfrac{z}{(z+3)^2 (z-2)}$.

4. Find the inverse Z-transform of $\dfrac{z}{z^2 + 11z + 24}$.

5. Find $Z^{-1}\left[\dfrac{1}{(z-2)(z-3)}\right]$ for (i) $|z| < 2$ (ii) $2 < |z| 3$ (iii) $|z| > 3$.

6. Find the $Z^{-1}\left[\dfrac{z}{(z+2)^2}\right]$.

7. Evaluate $Z^{-1}\left[\dfrac{z}{(2-z)(2z-1)}\right]$.

8. Find the inverse Z-transform of $U(z) = \dfrac{z^2}{\left(z - \frac{1}{4}\right)\left(z - \frac{1}{5}\right)}$

   (i) $\dfrac{1}{5} < |z| < \dfrac{1}{4}$ $\qquad$ (ii) $|z| < \dfrac{1}{5}$.

9. Obtain $Z^{-1}\left[\dfrac{2z}{(z-2)^2}\right]$, $|z| > 2$.

ANSWERS

1. (i) $a^n (n+1)$ $\qquad$ (ii) $\dfrac{\alpha^{n+1} - \beta^{n+1}}{\alpha - \beta}$

3. (i) $\dfrac{1}{2}(2^n + 2 \,.\, n^2\, 2^n) - 4^n$ $\qquad$ (ii) $\{n\}$ $\qquad$ $\{$(iii) $-\dfrac{1}{25}(-3)^n - \dfrac{1}{5} n (-3)^n + \dfrac{1}{25} 2^n$

4. $\dfrac{1}{5}[(-3)^n - (-8)^n]$

5. (*i*) $-\left(\frac{1}{3}+\frac{z}{3^2}+\frac{z^2}{3^3}+\frac{z^3}{3^4}.....\right)+\left(\frac{1}{2}+\frac{z}{2^2}+\frac{z^2}{2^3}+\frac{z^3}{2^4}......\right)$

(*ii*) $(-2^{n-1})\quad n>0$ (*iii*) $\begin{cases}(3^{n-1}-2^{n-1}) & n\geq 1\\ 0 & n<0\end{cases}$.

6. $\begin{cases}n(-2)^{n-1} & \text{if } n\neq 0\\ 0 & \text{if } n=0\end{cases}$ **7.** $-\frac{1}{3}(2)^n+\frac{1}{3}\left(\frac{1}{2}\right)^n$

8. (*i*) $-5\left(\frac{1}{4}\right)^n-4\left(\frac{1}{5}\right)^n$ (*ii*) $4\left(\frac{1}{5}\right)^n-5\left(\frac{1}{4}\right)^n$

9. $\{n.2^n\}$.

2.13. SOLUTION OF DIFFERENCE EQUATION BY Z-TRANSFORM

Example 1: *Using the Z-transform, Solve* $y_{n+2}-5y_{n+1}+6y_n=u_n$ *with* $y_0=0$, $y_1=1$ *and* $u_n=1$ *for* $n=0, 1, 2,.....$.

Solution: We have $y_{n+2}-5y_{n+1}+6y_n=u_n$ ...(2.5)

with $y_0=0$, $y_1=1$ and $u_n=1$ for $n=0, 1, 2, ...$

Taking the Z-transform on both sides of (2.5), we have

$$Z(y_{n+2}-5y_{n+1}+6y_n)=Z(u_n)$$

or
$$Z(y_{n+2})-5Z(y_{n+1})-6Z(y_n)=Z(1)$$

or
$$z^2[Y(z)-y_0-y_1z^{-1}]-5z[Y(z)-y_0]-6Y(z)=\frac{z}{z-1}\qquad [\because\ Z(y_n)=Y(z)]$$

or
$$(z^2-5z+6)\,Y(z)=\frac{z}{z-1}+z\qquad [\because\ y_0=0, y_1=1]$$

or
$$Y(z)=\frac{z^2}{(z-1)(z^2-5z+6)}$$

$$Z(y_n)=\frac{z^2}{(z-1)(z-2)(z-3)}$$

Taking inverse Z-transform, on both sides, we get

$$y_n=Z^{-1}\left[\frac{z^2}{(z-1)(z-2)(z-3)}\right]$$

$$=Z^{-1}\left[\frac{1}{2}\frac{z}{(z-1)}-2\frac{z}{(z-2)}+\frac{3}{2}\frac{z}{(z-3)}\right]$$

$$= \frac{1}{2} z^{-1}\left[\frac{z}{z-1}\right] - 2z^{-1}\left[\frac{z}{z-2}\right] + \frac{3}{2} z^{-1}\left[\frac{z}{z-3}\right]$$

$$y_n = \frac{1}{2} \cdot 1 - 2(2)^n + \frac{3}{2}(3)^n. \quad \textbf{Ans.}$$

Example 2: *Solve* $y_{n+2} - 3y_{n+1} - 4y_n = 3^n$ *by Z-transform.*

Solution: We have $y_{n+2} - 3y_{n+1} - 4y_n = 3^n$ ...(2.6)

Taking the Z-transform on both sides of (2.6), we get

$$Z[y_{n+2} - 3y_{n+1} - 4y_n] = Z[3^n]$$

or $$Z[y_{n+2}] - 3Z[y_{n+1}] - 4Z[y_n] = \frac{z}{z-3} \qquad [\because \; Z(y_n) = y(Z)]$$

or $$Z^2[Y(z) - y_0 - y_1 z^{-1}] - 3z[Y(z) - y_0] - 4Y(z) = \frac{z}{z-3} \qquad [\because \; Z(y_n) = Y(z)]$$

or $$(z^2 - 3z - 4)\, Y(z) = \frac{z}{z-3} + (z^2 - 3z)\, y_0 + z y_1$$

or $$Y(z) = \frac{z}{(z-3)(z^2-3z-4)} + \frac{z(z-3)}{z^2-3z-4} y_0 + \frac{z}{z^2-3z-4} y_1$$

$$= \frac{z}{(z-3)(z+1)(z-4)} + \frac{z(z-3)}{(z+1)(z-4)} y_0 + \frac{z}{(z+1)(z-4)} y_1$$

$$= -\frac{1}{4}\left(\frac{z}{z-3}\right) + \frac{1}{5}\left(\frac{z}{z-4}\right) + \frac{1}{20}\left(\frac{z}{z+1}\right) + \left[\frac{1}{5}\left(\frac{z}{z-4}\right) + \frac{4}{5}\left(\frac{z}{z+1}\right)\right] y_0 + \left[\frac{1}{5}\left(\frac{z}{z-4}\right) - \frac{1}{5}\left(\frac{z}{z+1}\right)\right] y_1$$

Taking inverse Z-transform on both sides, we get

$$y_n = Z^{-1}\left\{\left[-\frac{1}{4}\left(\frac{z}{z-3}\right) + \frac{1}{5}\left(\frac{z}{z-4}\right) + \frac{1}{20}\left(\frac{z}{z+1}\right)\right] + \left[\frac{1}{5}\left(\frac{z}{z-4}\right) + \frac{4}{5}\left(\frac{z}{z+1}\right)\right] y_0 + \left[\frac{1}{5}\left(\frac{z}{z-4}\right) - \frac{1}{5}\left(\frac{z}{z+1}\right)\right] y_1\right\}$$

$$= -\frac{1}{4}3^n + \frac{1}{5}4^n + \frac{1}{20}(-1)^n + \left[\frac{1}{5}4^n + \frac{4}{5}(-1)^n\right] y_0 + \left[\frac{1}{5}4^n - \frac{1}{5}(-1)^n\right] y_1$$

$$= -\frac{1}{4}3^n + \left(\frac{1}{5} + \frac{y_0}{5} + \frac{y_1}{5}\right)4^n + \left(\frac{1}{20} + \frac{4y_0}{5} - \frac{y_1}{5}\right)(-1)^n$$

$$y_n = -\frac{1}{4}3^n + c_1 4^n + c_2(-1)^n. \quad \textbf{Ans.}$$

Example 3: *Solve* $y_{n+2} + 2y_{n+1} + y_n = n$ *with* $y_0 = 0 = y_1$.

Solution: We have $y_{n+2} + 2y_{n+1} + y_n = n$ ...(2.7)

with $y_0 = 0, y_1 = 1$

Taking the Z-transform on both sides of (2.7), we get

$$Z[y_{n+2} + 2y_{n+1} + y_n] = Z[n]$$

or $$Z[y_{n+2}] + 2Z[y_{n+1}] + Z[y_n] = \frac{z}{(z-1)^2}$$

or $$z^2 [Y(z) - y_0 - y_1 z^{-1}] + 2z[Y(z) - y_0] + Y(z) = \frac{z}{(z-1)^2} \qquad [\because Z[y_n] = Y(z)]$$

or $$(z^2 + 2z + 1)\, y(z) = \frac{z}{(z-1)^2}$$

or $$Y(z) = \frac{z}{(z-1)^2 (z+1)^2}$$

$$Y(z) = \frac{1}{4}\left[-\frac{z}{z-1} + \frac{z}{(z-1)^2} + \frac{z}{z+1} + \frac{z}{(z+1)^2}\right]$$

Taking inverse Z-transform on both sides, we get

$$y_n = \frac{1}{4}[-(1)^n + n + (-1)^n - n(-1)^n] = \frac{1}{4}(n-1)[1-(-1)^n]. \quad \textbf{Ans.}$$

EXERCISE 2.3

Solving the following difference equation:

1. $y_{n+1} + \frac{1}{4} y_n = \left(\frac{1}{4}\right)^n$ $n \ge 0$ with $y = 0$.
2. $6y_{n+2} - y_{n+1} - y_n = 0$ given $y_0 = 0, y_1 = 1$.
3. $y_{n+2} - 4y_{n+1} + 3y_n = 5^n$.
4. $y_{n+2} + 6y_{n+1} + 9y_n = 2^n$ with $y_0 = 0 = y_1$.
5. $y_{n+2} - 4y_n = n - 1$.
6. $y_{n+2} + 4y_{n+1} + 3y_n = 3^n$ with $y_0 = 0, y_1 = 1$.
7. $y_{n+2} - 2y_{n+1} + y_n = n^2 2^n$.

ANSWERS

1. $y_n = 2\left(\frac{1}{4}\right)^n - 2\left(-\frac{1}{4}\right)^n$
2. $y_n = \frac{5}{6}\left[\left(\frac{1}{2}\right)^n - \left(-\frac{1}{3}\right)^n\right]$
3. $y_n = c_1 + c_2 3^n + \frac{1}{8} 5^n$.
4. $y_n = \frac{1}{25}\left[2^n - (-3)^n + \frac{5}{3} n(-3)^n\right]$
5. $y_n = c_1 2^n + \left(c_2 + \frac{1}{9}\right)(-2)^n + \frac{1}{9}(1)^n \frac{1}{3}$
6. $y_n = \frac{3}{8}(-1)^n + \frac{1}{24} 3^n - \frac{5}{12}(-3)^n$
7. $y_n = c_1 + nc_2 + (n^2 - 8n + 20)2^n$.

Table 1

(Area under standard normal curve from 0 to Z)

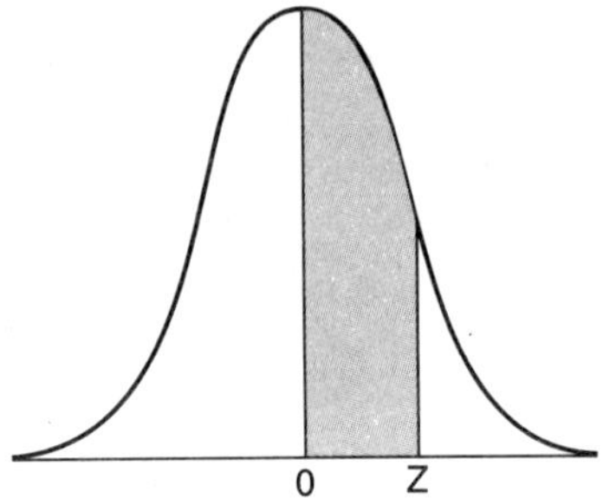

Z	*0.00*	*0.01*	*0.02*	*0.03*	*0.04*	*0.05*	*0.06*	*0.07*	*0.08*	*0.09*
0.0	0.0000	0.0040	0.0080	0.0120	0.0160	0.0199	0.0239	0.0279	0.0319	0.0359
0.1	.0398	.0438	.0478	.0517	.0557	.0596	.0636	.0675	.0714	.0753
0.2	.0793	.0832	.0871	.0910	.0948	.0987	.1026	.1064	.1103	.1141
0.3	.1179	.1217	.1255	.1293	.1331	.1368	.1406	.1443	.1480	.1517
0.4	.1554	.1591	.1628	.1664	.1700	.1736	.1772	.1808	.1844	.1879
0.5	.1915	.1950	.1985	.2019	.2054	.2088	.2123	.2157	.2190	.2224
0.6	.2257	.2291	.2324	.2357	.2389	.2422	.2454	.2486	.2517	.2549
0.7	.2580	.2611	.2642	.2673	.2704	.2734	.2764	.2794	.2823	.2852
0.8	.2881	.2910	.2939	.2967	.2995	.3023	.3051	.3078	.3106	.3133
0.9	.3159	.3186	.3212	.3238	.3264	.3289	.3315	.3340	.3365	.3389
1.0	.3413	.3438	.3461	.3485	.3508	.3531	.3554	.3577	.3599	.3621
1.1	.3643	.3665	.3686	.3708	.3729	.3749	.3770	.3790	.3810	.3830
1.2	.3849	.3869	.3888	.3907	.3925	.3944	.3962	.3980	.3997	.4015
1.3	.4032	.4049	.4066	.4082	.4099	.4115	.4131	.4147	.4162	.4177
1.4	.4192	.4207	.4222	.4236	.4251	.4265	.4279	.4292	.4306	.4319
1.5	.4332	.4345	.4357	.4370	.4382	.4394	.4406	.4418	.4429	.4441
1.6	.4452	.4463	.4474	.4484	.4495	.4505	.4515	.4525	.4535	.4545
1.7	.4554	.4564	.4573	.4582	.4591	.4599	.4608	.4616	.4625	.4633
1.8	.4641	.4649	.4656	.4664	.4671	.4678	.4686	.4693	.4699	.4706
1.9	.4713	.4719	.4726	.4732	.4738	.4744	.4750	.4756	.4761	.4767
2.0	.4772	.4778	.4783	.4788	.4793	.4798	.4803	.4808	.4812	.4817
2.1	.4821	.4826	.4830	.4834	.4838	.4842	.4846	.4850	.4854	.4857
2.2	.4861	.4864	.4868	.4871	.4875	.4878	.4881	.4884	.4887	.4890
2.3	.4893	.4896	.4898	.4901	.4904	.4906	.4909	.4911	.4913	.4916
2.4	.4918	.4920	.4922	.4925	.4927	.4929	.4931	.4932	.4934	.4936
2.5	.4938	.4940	.4941	.4943	.4945	.4946	.4948	.4949	.4951	.4952
2.6	.4953	.4955	.4956	.4957	.4959	.4960	.4961	.4962	.4963	.4964
2.7	.4965	.4966	.4967	.4968	.4969	.4970	.4971	.4972	.4973	.4974
2.8	.4974	.4975	.4976	.4977	.4977	.4978	.4979	.4979	.4980	.4981
2.9	.4981	.4982	.4982	.4983	.4984	.4984	.4985	.4985	.4986	.4986
3.0	.4987	.4987	.4987	.4988	.4988	.4989	.4989	.4989	.4990	.4990
3.1	.4990	.4991	.4991	.4991	.4992	.4992	.4992	.4992	.4993	.4993
3.2	.4993	.4993	.4994	.4994	.4994	.4994	.4994	.4995	.4995	.4995
3.3	.4995	.4995	.4995	.4996	.4996	.4996	.4996	.4996	.4996	.4997
3.4	.4997	.4997	.4997	.4997	.4997	.4997	.4997	.4997	.4997	.4998
3.6	.4998	.4998	.4999	.4999	.4999	.4999	.4999	.4999	.4999	.4999
3.9	.5000									